Flash Ironmaking

This book addresses the two major issues faced by the modern steel industry: CO_2 emissions and energy consumption. The steel industry accounts for 6.7% of the anthropogenic CO_2 emissions and consumes 6% of the total energy consumed in manufacturing. In response to these critical issues, a new technology called flash ironmaking has been developed, aimed at producing iron directly from iron ore concentrate using gaseous reductants/fuels such as natural gas or hydrogen. This ironmaking technology takes advantage of the rapid reaction rate of fine particles and bypasses the palletization process. This book discusses the principles of flash ironmaking, laboratory experiments, and design and operation of a prototype flash reactor.

- Provides theories and principles of ironmaking and a novel ironmaking technology.
- Includes laboratory experiments to establish the kinetic feasibility of flash ironmaking.
- Covers the design and operation of a prototype flash reactor as well as the design of industrial-size flash ironmaking reactors.
- Describes various cases of flow sheet development, which forms the basis for process analysis and simulation.
- Presents economic analysis case studies.

Presenting a novel technology that addresses contemporary issues facing one of the largest manufacturing industries, this book is aimed at professionals and researchers in metallurgy, materials engineering, manufacturing engineering, and related disciplines.

Flash Ironmaking

H. Y. Sohn

CRC Press
Taylor & Francis Group
Boca Raton London New York

CRC Press is an imprint of the
Taylor & Francis Group, an **informa** business

First edition published 2023
by CRC Press
6000 Broken Sound Parkway NW, Suite 300, Boca Raton, FL 33487-2742

and by CRC Press
4 Park Square, Milton Park, Abingdon, Oxon, OX14 4RN

CRC Press is an imprint of Taylor & Francis Group, LLC

ISBN: 978-1-032-37775-9 (hbk)
ISBN: 978-1-032-37838-1 (pbk)
ISBN: 978-1-003-34219-9 (ebk)

DOI: 10.1201/9781003342199

Typeset in Times
by codeMantra

To Victoria Bee-Tuan, Berkeley Jihoon, and Edward Jihyun
for their Sacrifices, Triumphs, and Love

Contents

Abbreviations

BF	Blast Furnace
BOF	Basic Oxygen Furnace
CFD	Computational Fluid Dynamics
COG	Coke Oven Gas
DR	Direct Reduction or Direct Reduced
DRI	Direct Reduced Iron
DSC	Differential Scanning Calorimetry
DTA	Differential Thermal Analysis
DTR	Drop-Tube Reactor (same as LFR)
EAF	Electric Arc Furnace
EDC	Eddy Dissipation Concept
EDF	Excess Driving Force, defined by Eq. (7.10)
EPA	Environmental Protection Agency
FIT	Flash Ironmaking Technology
F-O-F	Fuel–Oxygen–Fuel Gas Injection Mode (See Section 8.2.2.)
HHV	High Heating Value
HMI	Human-Machine Interface
H-O-H	Hydrogen–Oxygen–Hydrogen Gas Injection Mode (See Section 8.2.2.)
HyFIT	Hydrogen-Based Flash Ironmaking Technology
IC	Integrated Circuits
ICP	Inductively Coupled Plasma
LFR	Laminar-Flow Reactor (same as DTR)
LHS	Left-Hand Side
LHV	Low Heating Value
MMT	Million Tonnes
M-O-M	Methane–Oxygen–Methane Gas Injection Mode (See Section 8.2.2.)
MPFR	Mini-Pilot Flash Reactor (See Section 9.1.)
NG	Natural Gas
NPV	Net Present Value
OES	Optical Emission Spectroscopy
O-F-O	Oxygen–Fuel–Oxygen Gas Injection Mode (See Section 8.2.2.)
O-H-O	Oxygen–Hydrogen–Oxygen Gas Injection Mode (See Section 8.2.2.)
O-M-O	Oxygen–Methane–Oxygen Gas Injection Mode (See Section 8.2.2.)
PIV	Particle Image Velocimetry
PLC	Programmable Logic Controller
PSA	Pressure Swing Adsorption
RD	Reduction Degree
RHS	Right-Hand Side
RTE	Radiative Transfer Equation
SCR	Semiconductor (or Silicon)-Controlled Rectifier
SEM	Scanning Electron Microscope or Micrscopy
SLPM	Standard Liters per Minute

SMR	Steam-Methane Reforming
SR	Smelting Reduction
UDF	User-Defined Function
USGS	United States Geological Survey
WGS	Water-Gas Shift
WSGGM	Weighted-Sum-of-Gray-Gases Model
XRD	X-ray Diffraction

Nomenclature

a	Heating Rate
A	Reactant Gas
A_p	Original Surface Area of the Solid Reactant
b	Stoichiometric Coefficient in Reaction (6.99)
B	$\equiv \dfrac{\beta E}{RT_o}$ Defined by Eq. (6.129) or Solid Reactant B
c	Stoichiometric Coefficient in Reaction (6.99)
$c_{p,i}$	Specific Heat of Species i in the Mixture (J/kg·K)
C	Product Gas
C_D	Drag Coefficient
C_i	Concentration of Gas Species i
d_p	Geometric Mean Particle Diameter (m)
D_A	Solid-State Diffusivity of the Diffusing Species
D_e	Effective Diffusivity
e	Base of the Natural Logarithm
E	Activation Energy
$f(X)$	Differential Conversion Function, variously defined depending on the reaction system
F_D	Drag Force [kg/m s^2]
F_p	Shape Factor that has the values of 1, 2, and 3, respectively, for Flat Plates, Infinite Cylinders, and Equidimensional Solids (Spheres, Right Cylinders, Cubes, other Regular Polyhedral, etc.)
$F_{p,i}$	External Body Force Arising from Gas/Particle Interactions due to Drag Forces [J/m]
g	Gravitational Acceleration
$g(X)$	Integral Conversion Function, variously defined depending on the reaction system
h	Heat Transfer Coefficient
h_g	Specific Enthalpy of Gas Mixture
ΔH	Enthalpy Change
I	Radiative Intensity (W/m^2)
k	Reaction-Rate Constant or Thermal Conductivity
k_{app}	Apparent Rate Constant
k_{eff}	Effective Thermal Conductivity (W/m·K): $=$kg$+$kt
k_g	Gas Thermal Conductivity (W/m·K)
k_i	Rate Constant for Reduction of Concentrate Particles by H_2 or CO (1/atm·s)
k_m	External Mass Transfer Coefficient
k_o	Pre-Exponential Factor
K	Equilibrium Constant
m	Reaction Order with Respect to Reactant Gas Partial Pressure

m_i	Initial Mass of the Sample
m_p	Particle Mass (kg)
m_t	Mass of the Sample after Reduction
$M_{w,i}$	Molecular Weight of Species *i*
n	Avrami Parameter in Eq. (6.103)
$\dot{n}_{H_2,min}$	Minimum Moles of H_2 Required for Complete Reduction
$\dot{n}_{H_2O}$	Moles of Water Vapor Produced (from reduction and flame)
$\dot{n}_O$	Atoms of Removable O in the Feed
Nu	Nusselt Number
p_i	Partial Pressure of Species i
P	Total Pressure (pa)
Pr	Prandtl Number
Q	Heat Flux
r, r_p	Length Coordinate Perpendicular to the Solid Surface, and Radius of Solid Particle, respectively
R	Gas Constant
R_e	Reynolds Number
S_p	Net Rate of Mass Addition to Gas Phase per Unit Volume due to Gas–Solid Reaction
Sh	Sherwood Number
t	Time
T	Temperature
T_p	Particle Temperature (K)
u_i	Gas Phase Velocity Components (m/s)
u_p	Particle Velocity (m/s)
u_t	Terminal Velocity of Particle (m/s)
V_p	Original Volume of the Solid Reactant
X	Fractional Conversion of the Solid Reactant
Y_i	Mass Fraction of Species *i*
Z	Volume of Product Solid Produced from Unit Volume of Reactant Solid

GREEK SYMBOLS

α, β	Constants in Eq. (6.124)
α_B	Volume fraction of the solid occupied by reactant *B*
δ_{ij}	Kronecker delta [-]
ε	Turbulence dissipation rate (J·kg/s)
ε_p	Particle emissivity
ρ	Mass density
ρ_B	True molar density of the solid reactant *B*
μ	Viscosity
σ	Stefan-Boltzmann constant ($W/m^2 \cdot K^4$)
σ_p	Equivalent particle scattering factor (1/m)
τ	Particle residence time (s)

SUBSCRIPTS

B	Pertains to Solid Reactant B
e	Equilibrium
o	Bulk property

SUPERSCRIPTS

$^\circ$	Denotes Standard State
e	Refers to Equilibrium

Preface

The production of iron and steel makes up a huge and indispensable industry. It has a long chapter in human history and contributes enormously to the elevation of the standard of living. More importantly, it will continue to play this role and indeed expand its contributions to modern living. Because of this expectation, it is imperative to continue to develop and adopt innovative new ideas and technologies. The two most critical issues that the steel industry faces today are energy consumption and environmental effects, which are inter-related and largely concentrated in the ironmaking part of the steel industry.

In response to such needs, the Flash Ironmaking Technology (FIT) has been developed at the author's laboratory at the University of Utah under the auspices of the U.S. Department of Energy and American Iron and Steel Institute. The project went through several phases for 13 years and has attracted wide national and international interest, including a U.S. Congressional Briefing in 2008. (See the photograph below.)

2008 U.S. Congressional Briefing on Environmental Issues in the Steel Industry. (Andrew G. Sharkey, III, President and CEO, American Iron & Steel Institute on left; Author in middle; Lawrence Kavanagh, VP Environment and Technology, AIST on right)

This book chronicles the process of the development of the FIT together with the related theories and scientific principles. The development of the FIT was based on the author's research experience in the field of copper flash smelting. The two technologies share rather similar physical configurations, but the similarities end right there. Copper flash smelting is an oxidation, and thus highly exothermic and rapid, reaction process. However, the flash reduction of iron ore concentrate is an endothermic reaction, which tends to be slower and requires a supply of heat. In addition, the oxidation/combustion reactions are irreversible and use up essentially all the gaseous reactants as long as there is a continuous supply of solid or liquid reactant. On the other hand, the gaseous reduction of iron oxide is significantly limited by equilibrium. Thus, the recycling of the unused gaseous reactant is necessary. This adds to the increased complexity of the process and makes it essential to clearly understand the principles of reaction thermochemistry together with the generation and transfer of heat. Thus, this book contains a substantive discussion of these related basic principles.

Acknowledgments

The development of the FIT was based on the author's research experience in the field of copper flash smelting, which benefited greatly from the author's professional association with David B. George, formerly of Kennecott Copper Co. (U.S.); Juho Makinen, Markku Kyto, and Pekka Taskinen, formerly of Outotec (Finland), and Frank Jorgensen, formerly of CSIRO (Australia). It has been my great fortune and personal pleasure having known and worked with these giants of the field.

At the outset of the conception of the idea regarding flash ironmaking, the author wished to seek confirmation from an industrial person. For this, I reached out to Dr. Pinakin Chaubal of ArcelorMittal, whose affirmative response and encouragements were instrumental in finalizing the overall concept of flash ironmaking.

I wish to add my deepest gratitude to my own graduate students and postdoctoral research associates, past and present, who have contributed immeasurably to my lifelong work on the subject of flash reaction processes in general and flash ironmaking in particular: Y. B. Hahn, K. W. Seo, Y. Yasuda, M. Pérez-Tello, G. Han, M. E. Choi, H. Wang, M. Olivas-Martinez, S. E. Perez-Fontes, Z. Yuan, M. Elzohiery, D.-Q. Fan, A. Abdelghany, R. Sarkar, Y. Mohassab, and S. Roy. I would also like to express a special thanks John Otero and Syamantak Roy for their considerable help with the preparation and improvement of many of the drawings in the book.

The most direct support for the development of the FIT came from the American Iron and Steel Institute and the U.S. Department of Energy. Over the years, my work relevant to the subject of this book has been generously supported by various federal agencies, the State of Utah, and industrial firms; some of the most relevant to the work described in this book include National Science Foundation, U.S. Department of the Interior, U.S. Bureau of Mines, Camille and Henry Dreyfus Foundation (through a Teacher-Scholar Award Grant), ArcelorMittal Co., LS-Nikko, and the University of Utah. I am deeply grateful to these agencies and firms.

Author

H. Y. Sohn, Ph.D. Chem. Eng. from the University of California at Berkeley, is a Distinguished Professor of Materials Science & Engineering and an Adjunct Professor of Chemical Engineering at the University of Utah, USA, He is also an Honorary Professor of Metallurgy at Kunming University of Science and Technology in China. He has more than 40 years of experience in research and teaching and has received more than a dozen prestigious awards for his work, including being named four-time winner of the Extraction and Processing Science Award from the Minerals, Metals and Materials Society (TMS). In 2006, TMS honored him with "Sohn International Symposium on Advanced Processing of Metals and Materials: Principles, Technologies and Industrial Practice". He has authored six monographs and more than 600 journal articles throughout his career, as well as 17 edited books and 24 book chapters. His primary research interests encompass chemical synthesis of inorganic nanomaterials, metallurgical process engineering including CFD modeling, nonferrous metal production, fluid–solid reaction engineering, synthesis and processing of ceramic and intermetallic compounds, hydrogen storage materials development, solvent extraction, and combustion of solids and liquids.

1 Introduction

Iron has been used by mankind since the beginning of recorded history. Iron was processed in the 1700s in small scales via smelting processes. The most common forms of iron ores on earth are hematite (Fe_2O_3) and magnetite (Fe_3O_4). To produce iron from its ore, smelting processes are used where the iron oxide contained in the ore is heated with a reducing agent to remove oxygen.

The most commonly used form of iron is steel, which is an alloy of iron with carbon and/or other elements. Steel has great properties and high tensile strength and is used in all different types of applications such as infrastructure, tools, vehicles, machines, and appliances. In 2017, the iron production from iron ore was 1,250 million tonnes [World Steel Association, 2018a, b].

Steel is produced by two major routes:

1. The first is integrated steel production in which steel is produced from iron ore using:
   - Blast furnace (BF) combined with basic oxygen furnace (BOF) or
   - Direct reduction (DR) process combined with electric arc furnace (EAF) treatment.
2. The second route is based on steel scraps through the EAF steelmaking.

The first route of steelmaking is a multi-step process and requires a large amount of energy compared with the second route. However, the first route will continue to be the major steel producer. This is attributed to the fact that most steel products remain in use for decades, as in buildings and bridges, before they can be recycled and there is not enough steel scrap to meet the rapidly growing demand for steel [World Steel Association, 2018b].

The BF-BOF steelmaking route consumes 19.8–31.2 GJ of energy per tonne of crude steel, while the EAF route using 100% steel scrap consumes 9.1–12.5 GJ per tonne of steel [Yang et al., 2014].

Table 1.1 shows a comparison between the three major steelmaking routes in terms of the primary energy intensity, defined as the energy used for the production facility as well as to generate the electricity consumed at the facility to produce a metric tonne of mild steel [Worrell et al., 2008]. These numbers were based on the best practice of modern plants. Although the BF consumes around 78% of the total energy consumed in the BF-BOF route, this route still has lower energy consumption compared with the DR-EAF route.

Raw material preparation consumes much energy and emits a large amount of CO_2 gas. The United States emitted 55 MMT CO_2 from iron and steel production in 2014 [Bains et al., 2017]. Pelletizing, sintering, and cokemaking emit ~20% of the total CO_2 in the BF-BOF route and the BF contributes ~70%, whereas the steelmaking step adds only ~11% of the total emissions [Orth et al., 2007].

DOI: 10.1201/9781003342199-1

TABLE 1.1
World Best Practice Primary Energy Intensity Values for Iron and Steel in GJ per Metric Ton of Steel [Worrell et al., 2008]

Production Step	Process	BF-BOF	DR-EAF	Scrap-EAF
Material preparation	Sintering	2.2	2.2	
	Pelletizing		0.8	
	Coking	1.1		
Ironmaking	Blast furnace (BF)	12.4[a]		
	Direct reduction (DR)		9.2	
Steelmaking	Basic oxygen furnace (BOF)	−0.3		
	Electric arc furnace (EAF)		5.9	5.5
	Refining	0.4		
Total (GJ/t)		**15.8**	**18.1**	**5.5**

[a] See comments regarding energy requirements in Section 3.3.

Iron is produced from iron ore via BF, DR, and recently smelting reduction (SR), with the BF as the main means to make iron, producing more than 90% of the world production, after which comes DR and SR. In 2016, a total of 1,174 million tonnes of pig iron were produced from blast furnace route in the world compared with 73 million tonnes of direct reduced iron [World Steel Association, 2018a]. In the next chapter, these major ironmaking processes will be discussed.

In the subsequent chapters, the development of the novel flash ironmaking technology is described starting from the major issues facing the current ironmaking processes, the need for a new technology that overcomes these issues resulting in the concept of the flash ironmaking process, the kinetic feasibility determination to the laboratory flash furnace work, and finally the tests in a pilot plant. Process simulation and economic analysis of the new process [Pinegar et al., 2011, 2012, 2013a, 2013b] will also be summarized.

2 Current Technologies for Ironmaking

2.1 BLAST FURNACE PROCESS

Steel is one of the most useful metals, which cannot readily be substituted by other materials. The annual worldwide output of steel is forecast to grow to 2,200 million tonnes in 2050 [Hasanbeigi et al., 2012].

Currently, the blast furnace (BF) process produces more than 90% of iron worldwide, while the rest is produced by the direct reduction (DR) (about 7%) and other processes [Worldsteel Association, 2019]. A schematic diagram of a BF is shown in Figure 2.1. The BF produces crude iron (hot metal or pig iron) of ~1,500 ± 20°C to be refined to steel, which is the final product used in many modern applications. Steelmaking is done by the basic oxygen furnace (BOF) or electric arc furnace (EAF) process, which uses hot metal and steel scrap in varying proportions. The hot metal fed to these steelmaking processes typically contains 4%–4.5% C, 0.3%–0.7%

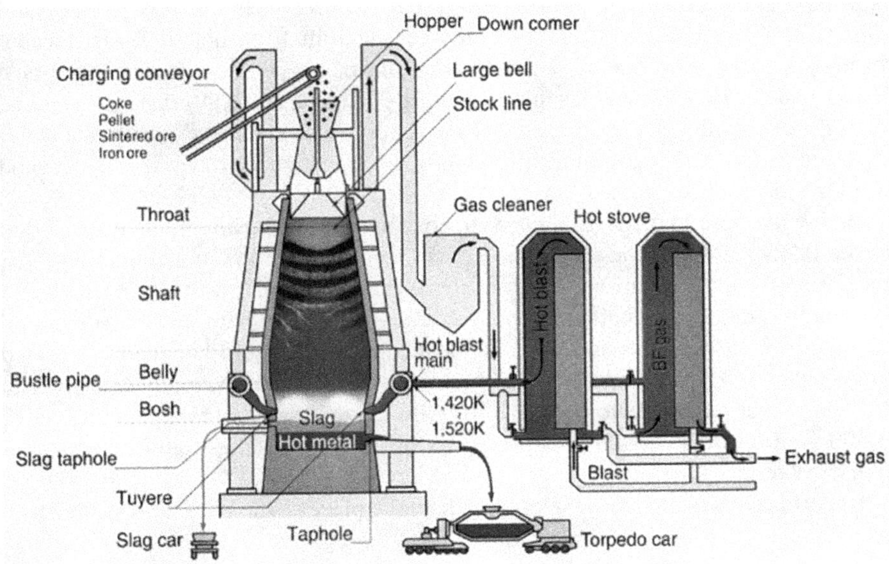

FIGURE 2.1 Schematic diagram of an iron blast furnace. (With permission from JFE 21st Century Foundation. https://www.google.com/search?biw=1282&bih=609&tbm=isch&sa=1&ei=tAUeXp-LGbrL0PEPjseaqAg&q=JFE+steel+blast+furnace&oq=JFE+steel+blast+furnace&gs_l=img.12...65027.66330..69331...0.0..0.111.325.3j1......0....1..gws-wiz-img.......0i8i7i30.Dz46jibTUSg&ved=0ahUKEwiftbDU2YPnAhW6JTQIHY6jBoUQ4dUDCAc#imgrc=, accessed January 14, 2020).

DOI: 10.1201/9781003342199-2

Si, 0.2%–0.4% Mn, and 0.06%–0.13% P. The BF is fed with high-grade lump iron ores, iron-ore pellets or sinter, which may range in size from 1 to 10 cm, as the source of iron and coke as the main fuel as well as reducing agent. With high-grade lump ores being depleted, pellets have become the main raw materials. Furthermore, pulverized coal and other hydrocarbon materials are injected as supplementary fuels with the oxygen-containing process gas to reduce the consumption of coke, which is expensive. Limestone is added as the flux to make slag.

The BF capacity ranges from 0.50 to 5.6 Mt pig iron per year in the largest-scale, state of-the-art furnaces. The currently largest BF in the world is the No. 1 BF at POSCO's Gwangyang Steelworks in Korea, with a volume of around 6,000 m^3. It has an annual production rate of about 6 million tonnes of iron [e-metallicus; Matejak, 2018], enough to produce 6 million passenger cars yearly.

The BF is continuously charged with coke and burden (sinter, pellets, lump ore, and flux) at the top in alternating cycles. The iron ore in the burden is mainly hematite (Fe_2O_3) or magnetite (Fe_3O_4). The alternating coke layers contribute to maintaining permeability of the process gas through the packed shaft. One of the reasons for using coke in BF is that it keeps its strength at elevated temperatures in contrast to coal, which softens at high temperatures. Fine ores cannot directly be used in this process. Therefore, sintering or pelletizing of fine ore is required to render the fine charge into strong and porous pellets with a size of >2 cm to enhance the burden permeability and facilitate the reduction. The flux typically contains limestone, BOF slag, and sand (silica), which form a slag with the right chemistry. Preheated blast (1,200°C–1,300°C) of air or oxygen-enriched air from hot stoves is injected through tuyeres. The hot blast reacts with the coke, forming carbon monoxide (2,100°C–2,300°C) in a hot gas mixture (mostly N_2, CO, and CO_2) that flows up and heats the descending raw materials. The descending charges also experience a series of physiochemical changes, most importantly the reduction of iron oxides by CO and H_2 in the gas (indirect reduction).

Below the bosh, the charge other than coke melts. The carbon in the coke or dissolved in molten hot metal (pig iron) reduces iron oxide in the molten burden (DR) and forms molten slag in the hearth of the furnace. Carbon monoxide present also reduces iron oxide, producing carbon dioxide which reacts with carbon to regenerate carbon monoxide through the Boudouard reaction. Liquid hot metal saturated with carbon (~1,500°C) and slag (1,550°C) are tapped intermittently (typically at 2–5 hour intervals). The top gas (100°C–300°C) that still contains heating value (3,000–4,000 kJ/Nm^3) in addition to the sensible heat is used as fuel in the process or elsewhere.

Important reactions involving carbon that take place in the BF are as follows:

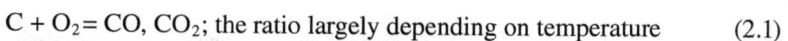

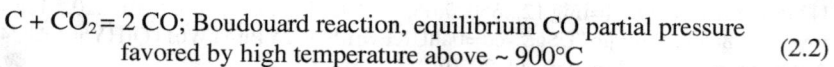

$$C + O_2 = CO, CO_2\text{; the ratio largely depending on temperature} \quad (2.1)$$

$$C + CO_2 = 2\,CO\text{; Boudouard reaction, equilibrium CO partial pressure favored by high temperature above} \sim 900°C \quad (2.2)$$

$$C + H_2O = CO + H_2O; \text{ favored by high temperature} \quad (2.3)$$

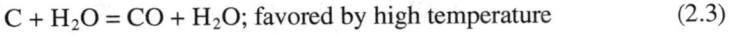

Near the belly, the charge other than coke becomes soft (cohesive zone) and starts melting, forming molten iron and slag. The slag contains iron oxide, FeO (5%–25%), which is reduced by solid carbon as follows:

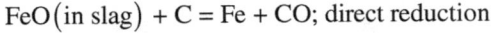

$$FeO(\text{in slag}) + C = Fe + CO; \text{ direct reduction}$$

$$\Delta H° = +36{,}130 \text{ cal } @ 1{,}200°C, \Delta G° = +36{,}610 - 36.57 \text{ T cal} \quad (2.4)$$

As the hot gas rises above the belly, CO and H_2 reduce iron oxide in the burden, which by this position has been reduced mostly to FeO in solid state, as follows:

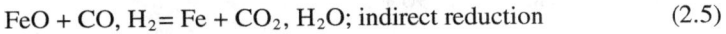

$$FeO + CO, H_2 = Fe + CO_2, H_2O; \text{ indirect reduction} \quad (2.5)$$

(Here, a note regarding the terminology may be in order. The term "DR" used here for the reduction reactions taking place in the lower part of a BF should be distinguished from the term "DR" that denotes the conversion of iron ore to metallic iron in the solid state, in one step, as an alternate, non-BF process. Chemically, this latter "DR" is similar to the "indirect reduction" taking place in the shaft region of a BF.)

Further up, hematite is first reduced to magnetite, which in turn is reduced to wustite and eventually to iron in the solid state. Wustite is thermodynamically not stable below 570°C [Roy, 2022: Spreitzer and Schenk, 2019], and magnetite is reduced directly to iron below this temperature. This can easily be seen in the phase diagram shown in Figure 2.2.

Thus, the following reactions take place in various temperature ranges (298 < T in K < 1,600):

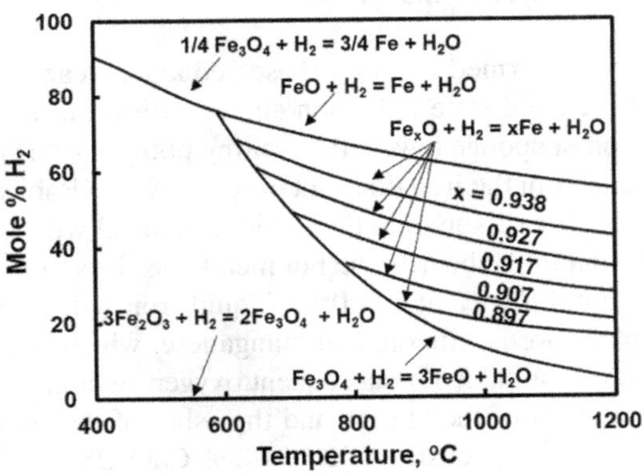

FIGURE 2.2 Equilibrium of iron–oxygen–hydrogen system adopted and redrawn from Evans and Koo [1979]. (x is the index in Fe_xO).

$$Fe_2O_3 + 1/3CO = 2/3Fe_3O_4 + 1/3CO_2;\ \Delta H° = -2{,}394 \text{ cal @570°C},$$
$$\Delta G° = -2{,}626 - 4.29\ T \text{ cal} \tag{2.6}$$

$$Fe_3O_4 + 4CO = 3Fe + 4CO_2; \quad \Delta H° = -10{,}630 \text{ cal @570°C},$$
$$\Delta G° = -28{,}520 + 20.61\ T \text{ cal} \tag{2.7}$$

$$Fe_3O_4 + CO = 3FeO + CO_2; \quad \Delta H° = +2{,}320 \text{ cal @800°C},$$
$$\Delta G° = -15{,}950 + 5.22\ T \text{ cal} \tag{2.8}$$

$$FeO + CO = Fe + CO_2; \quad \Delta H° = -3{,}741 \text{ cal @1,000°C},$$
$$\Delta G° = -4{,}190 + 5.13\ T \text{ cal} \tag{2.9}$$

$$Fe_2O_3 + 1/3H_2 = 2/3Fe_3O_4 + 1/3H_2O;\ \Delta H° = +513 \text{ cal @570°C},$$
$$\Delta G° = +207 - 6.85\ T \text{ cal} \tag{2.10}$$

$$Fe_3O_4 + 4H_2 = 3Fe + 4H_2O; \quad \Delta H° = +24{,}255 \text{ cal @570°C},$$
$$\Delta G° = +5880 - 9.99\ T\text{cal} \tag{2.11}$$

$$Fe_3O_4 + H_2 = 3FeO + H_2O; \quad \Delta H° = +10{,}474 \text{ cal @800°C},$$
$$\Delta G° = -7{,}350 - 2.65\ T \text{ cal} \tag{2.12}$$

$$FeO + H_2 = Fe + H_2O; \quad \Delta H° = +4{,}600 \text{ cal @1,127°C},$$
$$\Delta G° = +4{,}410 - 2.52\ T \text{ cal} \tag{2.13}$$

Metallic iron is largely formed through these reduction reactions, approximately two-thirds through the solid-state reduction and one-third through DR. Dissolution of carbon in solid iron or sponge lowers the melting point of iron significantly. In the shaft, the carbon content in the iron does not exceed 1 wt% [Babich et al., 2009], but at high temperatures, coke dissolves in the liquid iron to 4.3 wt% at the eutectic temperature of 1,153°C and in the hearth, the hot metal dissolves more carbon, reaching the saturation of about 4.5 wt% at 1,500°C. Liquid iron will also absorb a certain amount of sulfur, phosphorus, silicon, and manganese, which need to be controlled in sufficiently low levels suitable for subsequent oxygen steelmaking. Slag is formed from the gangue materials of the burden and the ashes of the coke and other metal oxides. The typical final slag consists of 34%–42% CaO, 28%–38% SiO_2, 6%–12% MgO, and 8%–20% Al_2O_3, making up about 96% of the slag, and contains varying amounts of minor components and impurities MnO, TiO_2, K_2O, Na_2O, S, and P [Geerdes et al., 2009]. The melting point of the final slag is about 1,300°C, while the primary slag containing 5%–25% of FeO has a lower melting point of about 1,200°C. The slag amounts to 150–300 kg/tonne of hot metal, depending on the burden composition [Geerdes et al., 2009].

Over the long history of its existence, the BF has gone through numerous significant improvements and innovations. These include the following: (1) a tremendous increase in the furnace size, (2) the increase of hot blast temperature, (3) the oxygen enrichment of injected air, (4) the injection of auxiliary fuels (hydrocarbon gases, liquid fuel, tar, and/or pulverized coal), (5) improvements in the burden preparation, (6) burden charging, including bell-less charging, (7) improvements in refractories and cooling systems, (8) improvements in burden distribution, and (9) improvements in instrumentation and automatic control. These improvements have contributed to making the BF acquire very high levels of thermal and chemical efficiencies.

Although the BF is, thus, highly developed and robust, the BF ironmaking is the most energy-intensive step in steelmaking and requires serious additional attention to environmental problems. Two major factors that contribute to these problems are the requirements of coke and iron ore pellets, which significantly contribute to the increased energy consumption of CO_2 emissions from the BF operation. Also, the BF process is highly capital and energy-intensive requiring large-scale infrastructure and operation. Those constraints limit the flexibility of the BF process in operation and choice of materials [Chatterjee, 1993; Manning and Fruehan, 2001].

Accordingly, a number of new ironmaking technologies such as DR and smelting reduction (SR) processes are growing in adoption. These alternative processes are aimed at reducing greenhouse gas emissions and energy consumption. Some of the newer processes can use coal instead of coke. In addition, the elimination of nitrogen prevents the generation and release of nitrous oxide (NO) gas into the environment. These new, alternative ironmaking processes are discussed in the subsequent sections.

Further details about the BF and associated processes can be found in a large number of comprehensive articles on the topic in the literature (e.g., Geerdes et al., 2020; Yang et al., 2014; Biswas, 1981).

2.2 DIRECT REDUCTION PROCESSES

Because of the need for smaller, lower capital-cost operations in the beginning but more recently for additional reasons of lower greenhouse gas emissions, the DR processes for alternate ironmaking have seen rapid growth, to about 7% of worldwide iron production of about 1.6 billion tonnes in 2021. The DR process for ironmaking is a coal- or gas-based process in which iron ore is reduced to sponge iron in solid state using coal or a gas mixture of H_2 and CO produced from the reforming of natural gas. DR is an attractive ironmaking route as it requires lower capital investment and uses a fuel that is cheaper than coke and results in lower CO_2 emissions by eliminating the use of coke and the need for the cokemaking step.

Despite these advantages, the DR route suffers from drawbacks such as the small scale of operations, which acts as a barrier to energy efficiency investments, and the produced DRI is pyrophoric and tends to re-oxidize unless passivated or briquetted. Also, as the iron oxide is reduced below the iron melting point, no slag phase is formed. Therefore, all gangue elements of the iron ores remain in the DRI and need to be separated via the slag in the next steelmaking process, which is mainly the EAF, increasing the energy and electrode consumption of the EAF compared with steel scrap melting (second route steelmaking). Currently, this process is more expensive than the BF process as it requires higher-quality iron ore, limiting its flexibility. Producing iron in the solid state requires a larger amount of electricity for melting DRI in EAF, which makes the process less efficient in terms of energy use.

The major DR processes are the gas-based MIDREX [Atsushi et al., 2010; Midrex, 2016], Energiron (HYL)-ZR (Zero Reforming) processes [Duarte et al., 2010], and the rotary-kiln process. The production of DRI by the shaft reactor using natural gas reached 110 million tonnes in 2021 [Sarkar et al., 2018].

Other DR processes, such as the Rotary Hearth Furnace (RHF) [Kikuchi et al., 2010] and fluidized bed processes (e.g., FIOR [Brent et al., 1999], FINMET [Brent et al., 1999], CIRCORED [Husain et al., 1999], and SPIREX [Macauley, 1997]), have been developed. However, most fluidized bed processes, other than a few exceptions, have not been successful commercially for various reasons.

The kiln processes like SL/RN uses a rotary kiln to reduce lump ore and pellets with coal. This process suffers from relatively large heat loss and requires a large facility size and, hence, finds limited applications.

Comprehensive reviews on the subject, including these other processes, have been presented by Battle et al. [2014] and, earlier, Stephenson and Smailer [1980].

The gas-based processes both use vertical shaft furnaces charged with pellets. The difference between these gas-based shaft furnace processes is that the MIDREX process externally reforms natural gas with the CO_2 and H_2O contained in the off-gas from the reduction shaft furnace and Energiron-ZR feeds unreformed natural gas together with oxygen into the furnace where partial oxidation of natural gas produces H_2/CO mixture that heats and reduces the iron oxide pellets. Reducing gas flows upward counter-currently to the descending iron ore. The off-gas exiting the top of the furnace contains a considerable amount of residual reducing gas and thus is recycled to a reformer to be reacted with fresh natural gas. The range of operating temperature is 650°C–1,200°C, below which the reduction rate is too slow and above which iron softens, especially when it absorbs carbon which lowers the melting point. Softening, sticking, or melting of the charge causes serious operating problems. Natural gas cannot be used directly in the reduction of iron ore because it decomposes to form soot but is used to produce the reducing gas and heat, and as a coolant and carburizing agent

The reactions that take place in the MIDREX process are as follows:

Overall reduction reactions:

$$Fe_2O_3 + 3H_2 = 2Fe + 3H_2O; \qquad \text{Endothermic} \tag{2.14}$$

$$Fe_2O_3 + 3CO = 2Fe + 3CO_2; \qquad \text{Exothermic} \tag{2.15}$$

Carburization reactions:

$$3Fe + 2CO = Fe_3C + CO_2 \tag{2.16}$$

$$3Fe + CO + H_2 = Fe_3C + H_2O \tag{2.17}$$

$$3Fe + CH_4 = Fe_3C + 2H_2 \quad (2.18)$$

Reforming reactions:

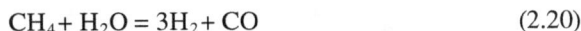

$$CH_4 + CO_2 = 2H_2 + 2CO \quad (2.19)$$

$$CH_4 + H_2O = 3H_2 + CO \quad (2.20)$$

The reactions that take place in the Energiron-ZR process are as follows:

Partial oxidation (reforming) reactions:

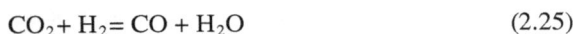

$$2CH_4 + O_2 = 2CO + 4H_2 \quad (2.21)$$

$$CH_4 + CO_2 = 2CO + 2H_2 \quad (2.22)$$

$$CH_4 + H_2O = CO + 3H_2 \quad (2.23)$$

$$2H_2 + O_2 = 2H_2O \quad (2.24)$$

$$CO_2 + H_2 = CO + H_2O \quad (2.25)$$

The overall reduction and carburization reactions are the same as in the MIDREX process given above.

Although natural gas is most widely used in the gas-based DR processes, syngases such as coke oven gas and coal gas are increasingly used in these processes.

2.3 SMELTING REDUCTION PROCESSES

The SR processes are the latest development in pig iron production, which emerged during the 1990s. SR combines iron ore pre-reduction with smelting in a bed of char generated by the devolatilization of coal. The product is molten iron to be refined into steel just as hot metal from the BF. Today, the major commercial SR processes are COREX and FINEX, although other processes such as Hismelt and HIsarna have been developed. The first two processes have been in operation in various industrial scales, including 1.5 Mtpy and 2.0 Mtpy FINEX plants in continuous commercial operation [Yi et al., 2019]. In these processes, iron ore particles are pre-reduced to about 60%–85% reduction degree in a series of fluidized bed reactors (FINEX) [Yi et al., 2019] or in a shaft furnace (COREX). The reduced iron along with coal and limestone is fed to a separate melter-gasifier (FINEX and COREX) to complete the reduction to molten iron. Further in this part, all metallurgical metal and slag reactions take place and generate a reducing gas by the gasification of coal that is used in the pre-reduction step. Thus, these processes can essentially be considered as having separate sections of the shaft and belly/hearth zones of the BF, as shown in Figure 2.3.

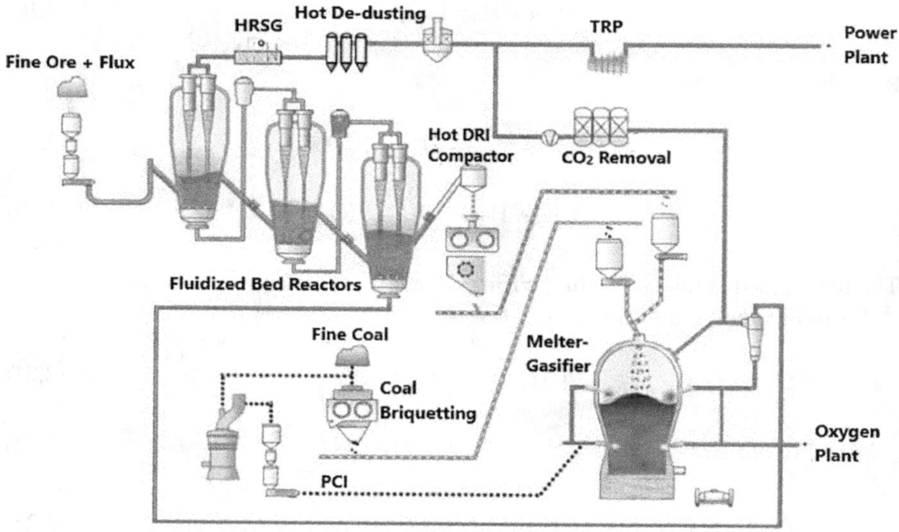

FIGURE 2.3 FINEX® process flowsheet.

This eliminates the requirements of coke and high-strength pellets to bear the high load of charges in the BF. As a result, these can be replaced, respectively, by coal (even low-quality) and less strong pellets or ore fines. Furthermore, these processes allow the use of industrial or tonnage oxygen, with many accompanying benefits such as a reduction in gas volume to be handled and an increase in energy efficiency. A detailed review of the FINEX process is available in the literature [Yi et al., 2019].

The Hismelt and HIsarna processes have been developed more recently [Hasanbeigi et al., 2012] and employ an SR vessel in which iron oxide fines, flux, and coal are injected into molten iron, producing iron and carbon monoxide.

Although individual SR processes have somewhat different features and attributes, they have the following general advantages and disadvantages compared with the BF process.

2.3.1 Advantages

- The use of non-coking coal in the form of briquettes and/or pulverized fines presents large benefits from the elimination of the cost for coking coal, energy consumption, and CO_2 and other pollutant emissions associated with cokemaking process.
- The direct utilization of iron ore fines, which eliminates the need for ore pelletization and/or sintering and the associated costs, energy consumption, and CO_2 and other pollutant emissions.
- Higher smelting intensity because of the faster reaction rates. The specific melting capacity is at least twice that of the BF.
- The process produces molten iron that has the same quality as the hot metal from a BF.

- The process permits the use of tonnage oxygen with the associated benefits of low gas volumes and thermal efficiency.
- Reduced process steps.

2.3.2 Disadvantages

- Being coal-based, the process still generates large quantities of CO_2 gas.
- The maximum size of an SR unit is smaller than most modern BFs.
- The process requires highly efficient post-combustion.
    - The operation of the pre-reduction step requires careful control. The pre-reduction shaft of COREX must overcome non-uniform flow of gas flow and pellet disintegration. The fluidized bed pre-reduction step of FINEX requires fine particle sizes.

3 Issues Facing the Steel Industry

The currently dominant blast furnace (BF) process needs the solid charge to be sinters or pellets and coke. The preparation of these two solid charges requires much energy and is pollution-prone. The direct reduction (DR) processes also require pellets. The use of coke as the main fuel and reducing agent is the root cause of the large emissions of greenhouse gas, mostly carbon dioxide, from the steel industry. In addition, the requirement of pellets in all the currently dominant ironmaking technologies, the BF and DR processes adds to the large energy consumption by the steel industry. The two major issues facing the steel industry, that is, greenhouse gas emissions and the high requirement of energy, are discussed below.

3.1 RAW MATERIALS

The two main features associated with the currently dominant ironmaking process, the BF process, that present a possible opportunity toward innovation are the uses of coke and iron ore sinter or pellets. Coke not only requires a suitable coking coal, to start with but also consumes a significant amount of energy and generates considerable amounts of CO_2 and other pollutants for its production. Successful smelting reduction (SR) processes such as COREX, FINEX, and HISMELT all use coal as a source of energy and reducing agents, which still produces a considerable amount of CO_2.

With the world supply of high-grade lump ores being greatly diminished as well as for the reasons of removing impurities from the ore, most iron ores fed to current ironmaking processes are upgraded by first grinding them to concentrates or fines and then sintered or pelletized. Especially for the BFs and dominant DR processes, the sinters and pellets must be indurated for sufficient strength and to prevent disintegration while still maintaining high reactivity. Such sintering and pelletization processes consume energy and generate much CO_2 and other pollutants.

Existing alternative processes avoid one of these two issues but not both and are limited to low production capacities or lower quality products. Therefore, the steel industry would benefit from the development of a low capital cost process, which is scalable to large capacities, can take advantage of the availability of inexpensive iron ore concentrate, and can use fuels that significantly reduce potentially harmful greenhouse gas emissions.

3.2 GREENHOUSE GAS EMISSIONS

The BF route is primarily coal-based and accompanied with ~2 tonnes of CO_2 emissions per tonne of crude steel while the CO_2 emissions per tonne of crude steel for the

DOI: 10.1201/9781003342199-3

NG-DRI route are ~1.4 tonnes [Contreras et al., 2021]. Earlier, Yilmaz et al. [2017] had calculated the CO_2 emissions in the NG-DRI route by using a suitable process model (for Midrex and Energiron processes) and found that the CO_2 emissions could be reduced by up to 26.7% relative to the conventional BF route. The injection of ~76 kg H_2 was able to avoid the emissions of one tonne of CO_2. In addition, the biomass syngas containing reducing gases, such as H_2, CO, and CH_4 with a composition similar to that of the natural gas has been used in the DRI route recently [Abd Rashid et al., 2014]. The studies conducted by Guo et al. [2016] and Weiland et al. [2014] demonstrated that the biomass syngas based on sawdust, brown seaweed, and torrefied wood residues, etc. used as the reductant during DRI process obtained a closed reduction rate to that of natural gas.

The rapid increase in steel production is expected to contribute greatly to the increases in energy consumption, CO_2 emissions and fine dust generation from the steel industry. Hence, it is critical to find solutions to energy and environmental issues to sustain the iron and steel industry in the future. The iron and steel sector is the second-largest industrial user of energy and is also the largest industrial source of CO_2 emissions [UN Climate Technology Centre & Network, 2022], generating about 7% of global greenhouse gas emissions [World Steel Association, 2018b; Tacke and Steffen, 2004]. The industry is responsible for the largest share (~27%) of CO_2 emissions from the worldwide manufacturing industry [International Energy Agency, 2007].

The energy consumption and the amount of CO_2 emissions in the BF process are essentially at their theoretical minimum levels. This is because of the significant improvements made over the long years of industrial dominance in terms of efficiencies in energy use, coke rate, cokemaking technique, and increased injection of other combustibles such as coal, natural gas, and plastics. With modern BFs operating at near-theoretical conditions, there is no much room to improve in the BF itself under its fundamental operating concept and forms of raw materials used. Thus, drastically innovative new technologies must be developed to make significant further reductions in energy usage and CO_2 emissions.

The BF is quite efficient from the viewpoint of energy and productivity but requires pelletizing or sintering of iron ore and in addition must use coke as the main fuel and reducing agent. Cokemaking and sintering/pelletization are pollution prone and generate large amounts of CO_2. Steelmaking emits 1.9 tonnes of CO_2 for every tonne of steel produced, accounting for about 7% of the total manmade CO_2 emissions [World Steel Association. 2018b]. Overall, sintering (13%), pelletization (2%), and cokemaking (5%) together produce ~20% of the overall CO_2 emissions in the BF-BOF route, and the BF contributes ~70%, with 10% generated by steelmaking [Orth et al., 2007].

As seen in Table 3.1, natural gas–based DRI production generates lower CO_2 emissions, in the range of 0.77–1.1 tonnes of CO_2 per tonne of steel (in contrast to ~1.9 tonnes of CO_2 per tonne of steel for the BF-BOF process), varying with the source of electricity used [World Steel Association, 2018c]. SR processes like COREX and FINEX, which are coal-based, produce somewhat more CO_2 than DR processes but less than BF and much less than rotary kilns (SL/RN).

TABLE 3.1
CO_2 Emissions from Various Steelmaking Technologies in Tonnes per Tonne of Iron

BF-BOF[a]	Midrex Process[b]	HYL III-Energiron[a]	SL/RN Process[a]	SR[a,c]	Circored[a,d]
~1.9	~1.1	0.77–0.92	~3.2	1.3–1.8	~1.2

[a] Institute for Industrial Productivity [Institute for Industrial Productivity, 2019].

[b] Metius et al. [2006].

[c] Hasanbeigi et al. [2014].

[d] Husain et al. [1999].

TABLE 3.2
Energy Requirements by Best Practices Worldwide for Iron and Steel Production (GJ per tonne of mild steel) [Worrell et al., 2008]

Production Step	Process	BF-BOF	DR-EAF	SR-BOF[a]	EAF-Scrap
Feed preparation	Sintering	2.2	2.2		
	Pelletizing		0.8	0.8	
	Coking	1.1			
Ironmaking		12.4	9.2	17.9	
Steelmaking	Main step	−0.3	5.9	−0.3	5.5
	Refining	0.4		0.4	
Total (GJ/t)		15.8	18.1	18.8	5.5

[a] Mainly COREX; does not include FINEX or HISMELT.

3.3 ENERGY CONSUMPTION

Energy consumption is a critical issue in the steel industry. Cokemaking and sintering/pelletization that are required in the BF process are energy-intensive. Table 3.2 shows a comparison between the four major steelmaking routes in terms of the energy consumed for iron and steel production and also to generate electricity [Worrell et al., 2008]. These numbers were based on the best practice of modern plants. In terms of just ironmaking, DR consumes the least amount of energy at 12.2 GJ/tonne of steel, and SR requires the largest amount at 18.7 GJ/tonne, with BF positioned in the middle at 15.7 GJ/tonne of steel. The use of an electric arc furnace (EAF) for steelmaking is seen to be energy-intensive, whereas the use of a basic oxygen furnace (BOF) for primary steelmaking requires a minimal amount of energy. In DR processes, iron oxide is reduced in solid state and, therefore, all gangues contained in the iron ore stay in the product iron (DRI) and must be removed to the slag in the EAF. This causes greater electrical energy consumption than melt scraps [Institute for Industrial Productivity, 2019].

Note of caution regarding "energy requirement": It is noted here that caution must be exercised when comparing the presented energy intensity values or energy requirements to clearly understand the energy items included in these values [Sohn and Olivas-Martinez, 2014]. There are currently different approaches for selecting energy items to include in the overall "energy requirements" when an input fuel also serves as a reactant.

For a process involving heat generation from fuel combustion, the choice of which endothermic reactions to pick to compute the energy requirement can cause confusion. The essential question is whether the heating value of a reactant, which is also burned to generate process heat, should be added as an item in the "energy requirement." Some investigators include this item [Remus et al., 2013; Fruehan et al., 2000; Stubbles, 2000] and others do not [Pinegar et al., 2011, 2012, 2013a; Burgo, 1999]. The choice is arbitrary, but one should indicate explicitly, which approach is used in showing the energy calculations, especially in comparing different processes. The presented "energy requirement" varies with the selected approach.

The net differences in the energy requirements between technologies are not significantly affected by the selection of calculation approaches if the same method is used for different technologies, but the consumption values for individual processes themselves depend on them. Thus, when presenting "energy requirements," it is critical to definitively state the applied approach. Whether the heat of combustion of a reactant that can also be burned is included in the energy requirement should be clearly indicated. It will make it much more definitive to indicate the amounts for "process energy" that contain only the heating value of the fuel and "reductant energy" ("feedstock energy" in petrochemicals production) that include the heating value of the substance serving as a reactant.

For a detailed description of these different calculation methods and a discussion on the subject, the reader is referred to Sohn and Olivas-Martinez [2014]. An additional caution that follows from this consideration is that the comparison of different ironmaking processes for their energy requirements should only be done with the values obtained using the same method by the same investigators. In other words, the energy requirement value obtained by one investigator for a certain ironmaking process should not be compared with a value obtained by another investigator for a different ironmaking process without carefully checking the bases of the calculations.

Thus, the energy intensities listed in Table 3.2 for various technologies for making iron and steel may be used safely to obtain differences among the technologies. However, care should be exercised for example when using the energy intensity value for the BF-BOF route given in Table 3.2 to compare with the energy intensity of say the DR-EAF combination calculated or reported by a different investigator.

4 Flash Ironmaking Technology – Concept Development

This chapter describes the reasons and justification for the development of the Flash Ironmaking Technology (FIT) [Sohn, 2007], which is based on the reduction of iron oxide concentrates by gas in a flash reactor. The details of the development steps, starting from the kinetic feasibility determination to the laboratory flash furnace work and finally the tests in a pilot plant, will be presented in subsequent chapters of this book. Work on flowsheet development, process simulation, and economic feasibility analysis will then be presented.

An ideal alternate ironmaking process would replace the blast furnace and coke oven; would use iron ore concentrate, which would greatly reduce energy requirements and CO_2 emissions. It should also be a high-intensity process requiring much less capital investment than the blast furnace/coke oven combination and must be capable of producing at least 5,000 tonnes of iron per day so that it can support existing steel mills. A typical steel mill produces over 3 million tonnes of iron per year (~ 10,000 tonnes per day).

With these limitations of current ironmaking processes in mind, a novel FIT has been conceived by Sohn [2007] for producing iron directly from fine concentrates by a flash reduction process. This process uses a reductant gas such as natural gas, hydrogen, or a mixture of the two, and does not require pellets, sinters, or coke as required by other ironmaking processes [Sohn, 2007; Sohn et al., 2009a, b; Sohn and Choi, 2010]. The new technology would significantly reduce energy consumption by 30%–60% and decrease carbon dioxide emissions by 60%–96% compared with blast furnace ironmaking, depending on whether hydrogen or hydrocarbon gas is used. The FIT will not have problems like particles sticking or pellet disintegration. High-grade lump iron ore is scarce worldwide, and new reserves must be ground to finer sizes to beneficiate them. Thus, increasing amounts of concentrates that can feed the FIT reactor are expected to be produced worldwide [USGS, 2007].

The goal of the development of the novel FIT was to develop an entirely new transformative process for ironmaking based on the direct gaseous reduction of iron oxide concentrate in a flash reduction process, with the ultimate objective of significantly lowering energy consumption and reducing environmental emissions, especially CO_2 emissions, versus the conventional ironmaking route. The FIT eliminates the iron ore pelletizing and indurating steps and, of course, eliminates the need for coke and coke ovens.

This new technology uses hydrogen or natural gas as a fuel as well as a reducing agent. It does not require pellets, sinters or coke [Sohn, 2007; Sohn et al., 2009a, b;

DOI: 10.1201/9781003342199-4

Sohn and Choi, 2010]. Furthermore, iron ore concentrate that would be used in a FIT reactor is currently produced in large quantities in the United States and increasingly elsewhere in the world [USGS, 2007].

The development started with the proof of the kinetic feasibility, considering that a typical flash reactor provides only a few seconds of residence time. This was followed by tests in a laboratory flash reactor and finally a pilot plant operation. The rate equations formulated in this work were developed considering the optimum combination of temperature and residence time and reducing gas partial pressure to achieve > 95% reduction degree. Experiments in the intermediate-scale laboratory flash reactor indicated that more than 90% reduction degree could be obtained in a few seconds of residence time at a temperature as low as 1,175°C. A pilot reactor operating at 1,200°C–1,550°C was installed and run to collect data necessary for scaling up the process. The tests in this large reactor validated the design concept in terms of heat supply and residence time and identified technical hurdles. This investigation proved the technical feasibility of the FIT. The results of this work facilitate the design of the industrial flash ironmaking reactor. The novel technology is expected to decrease the energy consumption in ironmaking by up to 44% compared with the average blast furnace process and will reduce CO_2 emissions by up to 51%. When hydrogen is used, the proposed process would use up to 60% less energy with little carbon dioxide emissions. However, it is noted that the energy requirements and CO_2 emissions during the production of natural gas, hydrogen, or coal must be added for a comprehensive comparison.

A sketch of the Flash Ironmaking process is shown in Figure 4.1. A gaseous fuel is partially oxidized with industrial oxygen to generate a reducing gas at 1,600–1,900 K. Iron ore concentrate is fed from the top. In one version of the process, the reduced product is collected as solid particles and briquettes to be used in separate steelmaking furnaces. In another version, the reduced iron is collected as a molten bath as part of a continuous direct steelmaking process, as sketched in Figure 4.1.

Natural gas is also plentiful in the United States and could easily supply potential ironmaking based on the FIT [U.S. Energy Information Administration, 2011]. Hydrogen would be cleaner, once the hydrogen economy is developed [U.S. Department of Energy, 2006; Züttel et al., 2008].

These benefits of the FIT come largely from the elimination of cokemaking and sintering/pelletization, by using gaseous reduction of iron oxide concentrate in a flash reactor. It takes advantage of the large specific area of fine particles of the concentrate, much like the flash smelting of copper or nickel sulfide concentrate. The process would use inexpensive, abundant natural gas or syngas to produce hot reducing gas. It would also directly use magnetite concentrates that are currently widely used in the steel industry. The availability of large quantities of iron ore concentrate is also a backbone for the possible development of this technology. About 60%–70% of U.S. iron production is based on taconite concentrates (50–55 million tonnes/year) [USGS, 2007]. Newer reserves also need to be ground down to concentrate sizes to beneficiate them and make them acceptable feed materials for ironmaking.

The FIT is initially expected to be operated using natural gas as a fuel and reducing agent. Natural gas is an abundant readily available resource in the United States and elsewhere in the world [U.S. Energy Information Administration, 2011; Pittsburgh

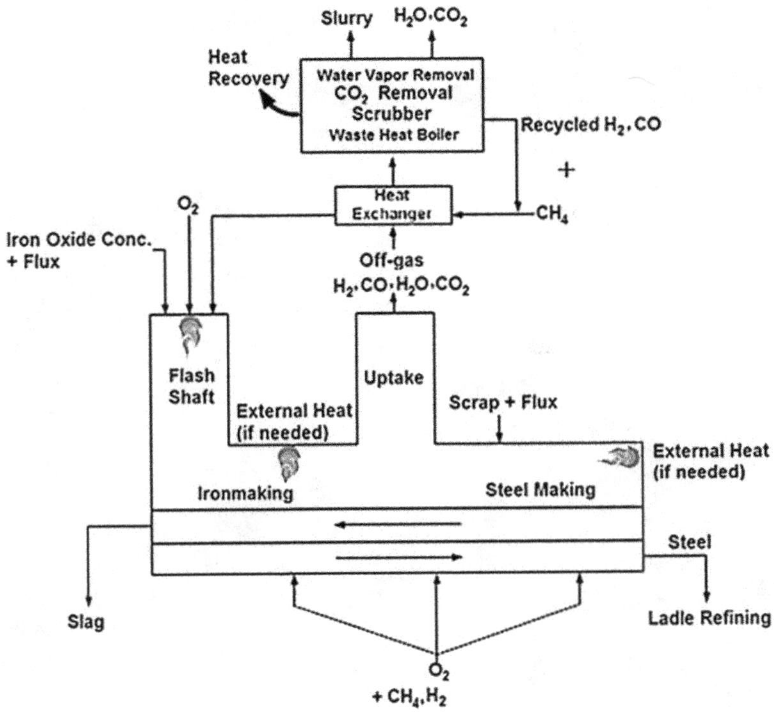

FIGURE 4.1 A sketch of direct steelmaking process based on the Flash Ironmaking Technology. (Concept first proposed by H.Y. Sohn in a research proposal submitted to AISI in November 2003. Figure adapted from Sohn and Choi [2012] and Sohn [2020b]).

Post-Gazette, 2011]. For an annual production of 20 million tonnes of iron, which would represent about 40% of the current U.S. production rate, the FIT would require 0.33 TCF/year (trillion cubic feet per year) of natural gas. This represents less than 1.5% of the total U.S. consumption rate (22.7 TCF/year) [U.S. Energy Information Administration, 2011]. As a comparison, this is similar to the 1.5% of the total U.S. energy consumption used by the U.S. steel industry. When natural gas is used as a reducing agent, the proposed process will produce iron with varying carbon contents such as the iron produced by HYL's ZR self-reforming process, which contains up to 5.5% carbon [Duarte et al., 2010].

Articles on the description of flash ironmaking and comparison with other processes together with a review of the literature have been published previously [Sohn et al., 2009a, b; Pinegar et al., 2011, 2012].

5 Basic Properties and Sources of Magnetite Concentrate

A naturally occurring magnetite mineral called taconite is currently the most widely used raw material for producing steel in the United States. Up to 70% of all the steel produced in the United States comes from this mineral. Taconite is a low-grade iron ore with 20%–30% iron content, from which upgraded magnetite powder called magnetite concentrate, that contains 67%–68% iron is obtained. To be used in the predominant current ironmaking process, the blast furnace process, this concentrate must first be made into green pellets with the aid of a binder followed by firing at a high temperature to produce porous but strong pellets. Many of the alternate ironmaking processes developed in recent years also use these pellets. The magnetite concentrate is also the raw material for the Flash Ironmaking Technology.

Magnetite is one of the two most important iron ores, the other being hematite. Although hematite is more abundant worldwide, magnetite is more widely used in North America. Some of the notable iron ranges in the United States include the Mesabi Range in Minnesota and the Marquette and Menominee Ranges in Michigan. Magnetite is a black, opaque, magnetic mineral, and its spinal crystal structure contains both the ferrous (Fe^{+2}) and ferric (Fe^{+3}) forms of iron ions and exhibits a ferromagnetic property. Magnetite ore is upgraded by magnetic separation from a low-grade taconite ore.

Some of the basic properties of magnetite are listed below:

- Chemical formula: Fe_3O_4 ($Fe^{2+}Fe^{3+}{}_2O_4$) [*IUPAC* chemical name: *iron(II, III) oxide*. Common chemical name: ferrous-ferric oxide].
- Specific gravity: 5.18
- Mohs hardness: 5.5–6.5
- Vickers hardness: 681–792 kg/mm^2
- Color: gray; brown to black
- Luster: metallic to sub-metallic
- Transparency: opaque
- Magnetism: strongly ferromagnetic
- Curie temperature: 858 K (585°C)

Magnetite commonly occurs in various igneous, metamorphic, and sedimentary rocks, typically as disseminated crystals or grains at small weight fractions of their host rock. The discussion on the natural occurrence of magnetite mineral here will be limited to the magnetite ores that are used as a raw material for iron and steelmaking.

DOI: 10.1201/9781003342199-5

TABLE 5.1
Chemical Composition of Typical Iron Ore Concentrate

Component	Wt.%
Total Fe	68
FeO	28
SiO_2	4
CaO	0.2
MgO	0.3
Al_2O_3	0.06
TiO_2	0.02
MnO	0.02
Na_2O	0.01
K_2O	0.04
P	0.01
S	0.0005

The major ore that contains commercially viable quantities of magnetite mineral in North America is taconite, which is a low-grade iron ore (25%–30% Fe). It is one of the varieties of chert, a fine-grained sedimentary rock containing magnetite, hematite, and silica, and occurs in the Lake Superior region.

Because the magnetite mineral is distributed as fine grains in large amounts of silica and other gangue minerals in taconite, a higher-grade iron ore needs to be obtained from it by the process of concentration. The first step of concentration is grinding the taconite ore mixed with water into fine particles of about the size of magnetite mineral, typically 80% < 45 μm, so that the magnetite mineral may be "liberated" (separated) from the gangue minerals. The finely ground particles are subjected to a magnetic field to separate the magnetically susceptible magnetite particles from the gangue minerals that are not affected by it. The wet magnetic portion, called the "concentrate," contains 67%–68% iron and is used for producing iron. A typical composition of magnetite concentrate is given in Table 5.1.

To be used in the blast furnace or a DR shaft furnace, the iron concentrate is turned into strong pellets. First, it is mixed with a binder, such as bentonite, and limestone as a flux. Then, it is rolled into green pellets of 8–18 mm in diameter, which contain approximately 65% iron and fired at a high temperature of 1,200°C–1,350°C. The fired pellets must remain porous enough to have high rates of reduction and yet strong enough to support the load above without being broken into small pieces. The large presence of very small pieces of solids restricts the flow of process gas through the blast furnace and also causes operating difficulties in other alternate ironmaking processes.

6 Principles Related to Iron Oxide Reduction

6.1 THERMOCHEMISTRY

The chemistry of iron oxide reduction has been discussed in Section 2.1. Here, we will discuss further details of the thermochemistry of the reduction reactions, including the enthalpy of reaction and equilibrium aspects, which is important in the design of gas-based direct reduction processes including the flash ironmaking process.

6.1.1 THE FIRST LAW OF THERMODYNAMICS – HEAT AND HEAT CAPACITY

The first law of thermodynamics can be summarized as the statement of the equivalence of heat and work. In ironmaking, mechanical work is not important, but the work of expansion of gases is still important. To explain this law, let us consider a system that receives heat and mechanical work from the surroundings. According to the first law, heat and work play the same role and, thus, both increase the internal energy of the system. With the increase in internal energy, its temperature increases and, if it contains gas, the gas tends to expand. If such expansion is allowed, a part of the energy added to the system is spent for such expansion. For a continuous process, it is convenient to consider the process as a sum of small changes. Application of the first law to such a small change gives rise to the following relationship:

$$dU + PdV = \delta q + \delta W_e \tag{6.1}$$

In this equation, dU is the amount of increase in the internal energy, PdV represents the work of expansion done by the system, δq is the amount of heat added to the system, and δW_e denotes the mechanical work received by the system from the surroundings. As a reference, d represents a small change, and δ represents a small amount added to the system. In ironmaking, no mechanical work is involved and thus $\delta W_e = 0$.

If heat is added under a condition of constant system volume, there is no work of expansion and, thus, Eq. (6.1) becomes as follows;

$$dU = \delta q \tag{6.2}$$

If there is no phase change, the increase in internal energy caused an increase in an increase in the system temperature, expressed as follows:

$$dU = \delta q = C_v\, dT \tag{6.3}$$

DOI: 10.1201/9781003342199-6

Here, the proportionality constant C_v is called the heat capacity at constant volume, which in general is a function of temperature but could be considered approximately as a constant over a range of temperature.

When heat is added under constant system pressure, expansion of the system occurs, and Eq. (6.1) becomes

$$dU + PdV = \delta q \tag{6.4}$$

Although Eq. (6.4) could be used as is in calculating the relationship involved, the presence of P and V makes it rather complicated. If we define a new thermodynamic function enthalpy, H, as follows, the relationship involving δq becomes simple:

$$H = U + PV \tag{6.5}$$

Again considering a small change, the above relationship under a constant pressure becomes

$$dH = dU + PdV \tag{6.6}$$

From Eqs. (6.4) and (6.6), we get

$$dH = \delta q \tag{6.7}$$

In the absence of a phase change and considering that dV for a gas is proportional to dT under constant pressure, the increase in H can be related to the increase in temperature, as follows:

$$dH = \delta q = C_p \cdot dT \tag{6.8}$$

This indicates that the small amount of heat added to the system under constant pressure increases the enthalpy of the system by dH. Here, C_p is the heat capacity at constant pressure. Because most continuous processes are operated under constant pressure, C_p is used more often than C_v.

6.1.2 Physical Changes and Heat Content

The amount of heat (ΔH) that must be supplied to change the temperature of a system from T_1 to T_2 under constant pressure can be calculated by integrating Eq. (6.8), as follows:

$$\Delta H = H_{T_2} - H_{T_1} = \int_{T_1}^{T_2} C_p dT = C_{p,m}(T_2 - T_1) \tag{6.9}$$

Here, $C_{p,m}$ is the mean value of C_p between the two temperatures. Equation (6.9) can be applied to the entire system or a unit amount and is applicable in the absence of any phase change between the two temperatures. Such heat ΔH accompanying a temperature change without a phase change is called sensible heat. H is an extensive property, that is, its value depends on the amount of the material. The C_p per unit mass is called the specific heat, and the C_p per unit mol, molar heat capacity, is also used. (In this book, the C_p for the entire system is used, unless otherwise specified. It is also pointed out that specific heat is defined as the ratio of the heat capacity of a system to that of water of the same mass. Thus, one needs to pay careful attention to the units accompanying the given value of C_p).

When there is a phase change under constant pressure, the temperature remains unchanged while the heat content varies with heat addition. The reason for the variation of heat content with phase change is that the content of energy is different for different phases of the same material. For example, water molecules in the vapor state move more actively and, thus, contain more energy than those in liquid water even at the same temperature. The types of phase changes include crystal phase transformation, melting or freezing, boiling or condensation, and sublimation. The ΔH that accompanies phase changes is called latent heat.

When there are phase changes between any two temperatures, the change in the heat content is calculated as follows:

$$\Delta H = H_{T_2} - H_{T_1} = \int_{T_1}^{T_{PT1}} C_{p1}dT + \Delta H_{PT1} + \int_{T_{PT1}}^{T_{PT2}} C_{p2}dT + \Delta H_{PT2} + \cdots + \int_{T_{PTf}}^{T_2} C_{p3}dT \tag{6.10}$$

In this equation, T_{PTi} is the temperature of phase transformation, C_{pi} is the heat capacity of the matter at the corresponding phase, and ΔH_{PTi} is the latent heat accompanying the corresponding phase change.

Because ΔH is the change in the heat content contained in the system, its value is positive when $T_1 < T_2$, and it is negative when $T_1 > T_2$ since the heat content of the system will decrease. Furthermore, in the latter case of decreasing temperature, the phase changes are the opposite of the first case, and the sign of the corresponding ΔH_{PTi} becomes negative while its absolute value remains the same.

The values of $(H_T - H_{298})$ for most materials calculated by Eq. (6.10) can be found in the literature, for example, Thermodynamic Data for Mineral Technology [Pankratz et al., 1984], JANAF Thermochemical Tables. Second Edition [Stull and Prophet, 1971], JANAF Thermochemical Tables, 1982 Supplement [Chase Jr. et al., 1982], Thermochemical Properties of Inorganic Substances [Knacke et al., 1991], and Metallurgical Thermochemistry [Kubaschewski and Alcock, 1979]. Furthermore, these data are contained in many types of thermodynamic software.

A convenient representation of such data is in the graphical form given in Figure 6.1.

As can be seen in the above figure, the given data are in the form of $(H_T - H_{298})$, that is, the difference in H between any temperature T and room temperature 298 K, rather than the absolute value of H. From these data, the change of H between any two temperatures T_1 and T_2 can readily be calculated by

$$\Delta H = H_{T_2} - H_{T_1} = \left(H_{T_2} - H_{298}\right) - \left(H_{T_1} - H_{298}\right) \tag{6.11}$$

Namely, one can use the $(H_T - H_{298})$ values at the two temperatures given in the reference.

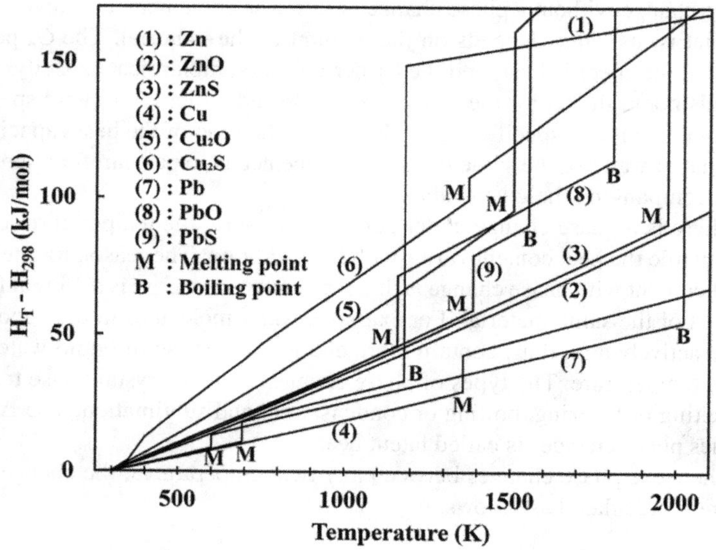

FIGURE 6.1 Graphical representation of $(H_T - H_{298})$ vs. temperature. (Prepared with data contained in HSC Chemistry software [Roine, 2002].)

6.1.3 Chemical Changes and Standard State

In addition to the physical changes described above, another type of change in the state of matter is chemical change. In other words, the heat contained in a system becomes different when there is a chemical change in the matters in the system. Such a change in the heat content is called the heat of reaction. When the reactants or products exist in solutions, rather than as pure phases, the energy contained in them is different depending on the state in which the matters exist. Thus, it is convenient to define the heat of reaction $\Delta H°$ assuming that the reactants and products all exist in certain thermodynamic standard states during the reaction. Such thermodynamic standard states are convenient in calculating changes in H as well as other thermodynamic quantities. This is true because most calculations involve the changes in such quantities, rather than their absolute values.

Thermodynamic standard states are often denoted by marking with the symbol °. It is very important to distinguish the function with and without this symbol. Different types of standard states are defined and used, but the most basic standard states are defined as pure and the most stable state of a compound under 1 atm pressure at the temperature of concern (Raoultian standard state).

Thus, we first calculate the heat of reaction $\Delta H°$ when all the species exist at their standard states. Then, if some species exist at non-standard states, the ΔH values that accompany the changes from the standard states to the actual states are taken into consideration.

6.1.4 Standard Heat of Formation or Standard Enthalpy of Formation

The standard heat of formation of a compound $\Delta H^{\circ}_{f,T}$ represents the change in enthalpy when 1 mole of the compound at standard state is formed from the corresponding elements at standard states. Many reference data books list the values of $\Delta H^{\circ}_{f,T}$ at different temperatures [Pankratz et al., 1984; Stull and Prophet, 1971; Chase Jr. et al., 1982].

For example, the standard heat of formation of CO_2 at 298 K is as follows:

$$C(\text{graphite}) + \tfrac{1}{2}\,O_2(g) = CO(g)C(\text{graphite}) + O_2(g)$$
$$= CO_2(g) : \Delta H^{\circ}_{f,298} = -94{,}051\,\text{cal/mol} \tag{6.12}$$

Here, it is noted that the standard states of oxygen and carbon dioxide are, respectively, pure O_2(g) and CO_2(g) at 1 atm and that of carbon is graphite. Carbon also exists in nature as amorphous carbon like char or soot as well as diamond, but graphite has a better defined crystalline structure and is found in nature. Since the standard heat of formation is a relative quantity, the standard heat of the elements is defined as zero by convention.

If $\Delta H^{\circ}_{f} > 0$, the product has a greater enthalpy than the reactants, and thus heat must be added from the surroundings to maintain the same temperature, making the reaction endothermic. On the contrary, the reaction is exothermic if $\Delta H^{\circ}_{f} < 0$.

6.1.5 Standard Heat of Combustion

The standard heat of reaction for completely oxidizing a compound using pure oxygen is called the standard heat of combustion of that compound (ΔH°_{c}). Thus, the standard heat of formation of CO_2 given in Eq. (6.12) is also the standard heat of combustion of graphite (usually at 298 K).

6.1.6 Hess' Law

Hess' law states that the enthalpy change of a reaction remains the same regardless of what paths the reaction takes. Thus, $\Delta H^{\circ}_{\text{reaction}}$ depends only on the initial state and the final state. This law is very convenient and allows the calculation of the enthalpy changes of reactions that cannot be measured by experiments. In addition, this law makes it possible to treat the equations for chemical reactions the same as mathematical equations.

Let us consider the formation reaction of CO gas.

$$C(\text{graphite}) + \tfrac{1}{2}O_2(g) = CO(g) \tag{6.13}$$

Experimentally, this reaction always accompanies the formation of CO_2, and the formation of pure CO at most temperatures is not possible. However, the following reactions take place quantitatively and, thus, the heat of reaction can be measured using a calorimeter.

$$C(\text{graphite}) + O_2(g) = CO_2(g); \Delta H_c^\circ = -94{,}051\,\text{cal} \quad (6.14)$$

$$CO(g) + \tfrac{1}{2}O_2(g) = CO_2(g); \Delta H_c^\circ = -67{,}634\,\text{cal} \quad (6.15)$$

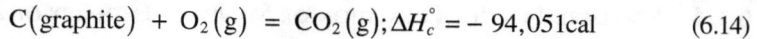

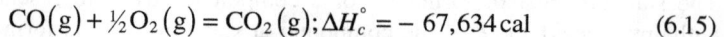

According to Lavoisier and Laplace's law of thermochemistry, the enthalpy change of a reverse reaction is the negative of that of the forward reaction, which is also directly related to the first law of thermodynamics. Thus, by considering the reverse of reaction (Eq. 6.15) with the simultaneous change of the sign of ΔH_c° and adding to reaction (Eq. 6.14) [or subtracting reaction (Eq. 6.15) from reaction (Eq. 6.14) and doing the same operation with the corresponding values of ΔH_c°], The ΔH_f° of reaction (Eq. 6.13) can be obtained as follows:

$$C(\text{graphite}) + \tfrac{1}{2}O_2(g) = CO(g); \Delta H_f^\circ = -26{,}417\ \text{cal/mol} \quad (6.16)$$

Hess' law may be explained by a graph, as shown in Figure 6.2. Like other thermodynamic functions, H is also a state function and, thus, the ΔH from C to CO_2 does not depend on the path. Therefore, the two paths shown in Figure 6.2 accompany the same ΔH, that is, the sum of ΔH changes accompanying reactions (Eqs. 6.13 and 6.15) is the same as the ΔH of reaction (Eq. 6.14). As a result, the ΔH of the desired reaction (Eq. 6.13) can be calculated from the difference of the values of ΔH for reactions (Eqs. 6.14 and 6.15), which can be measured accurately, as shown by Eq. (6.16). This method can be applied not only to the ΔH of any reactions but also to other thermodynamic functions that are state functions.

6.1.7 Heat of Chemical Reaction

Let us consider a chemical reaction expressed in the following general form:

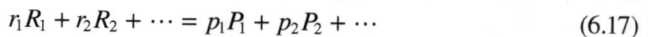

$$r_1R_1 + r_2R_2 + \cdots = p_1P_1 + p_2P_2 + \cdots \quad (6.17)$$

The collection of atoms and molecules existing in the state of R_1, R_2, … changes to the state of P_1, P_2, …. Then, as stated above, the energy, and thus the heat, contained in the system changes. In such a reaction, the reactants and/or products may exist

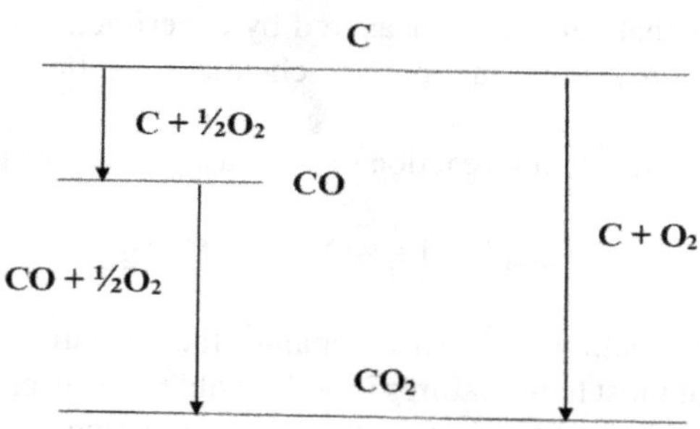

FIGURE 6.2 Graphical representation of Hess' law.

as components in a solution, rather than in pure states, and thus contain different amounts of energy than in the pure state. In such a case, it is convenient to first calculate the enthalpy change of the reaction taking place with all the species existing at thermodynamic standard states. By applying Lavoisier and Laplace's law of thermochemistry together with Hess' law to reaction (Eq. 6.17), its standard heat of reaction $\Delta H^{\circ}_{\text{reaction}}$ can be calculated as follows:

$$\Delta H^{\circ}_{\text{reaction}} = \sum_{\text{product}, i} \left(p_i.\Delta h^{\circ}_{f,P_i}\right) - \sum_{\text{reactant}, j} \left(r_j.\Delta h^{\circ}_{f,R_j}\right) \text{cal or J} \tag{6.18}$$

Here, p_i and r_j are the stoichiometric coefficients of the corresponding products and reactants, and $\Delta h^{\circ}_{f,P_i}$ and $\Delta h^{\circ}_{f,R_j}$ are the standard enthalpy of formation of the corresponding products and reactants. Considering that enthalpy has an extensive property, it is necessary to multiply the enthalpy of formation by the stoichiometric coefficient. The values of $\Delta h^{\circ}_{f,P_i}$ and $\Delta h^{\circ}_{f,R_j}$ can be found in various reference sources.

6.1.8 Heat of Reaction at Different Temperatures

Certain reference sources list the values of $\Delta h^{\circ}_{f,T}$ at various temperatures [Pankratz et al., 1984], but often, only the values at room temperature are given. When the reactants and products exist at the same temperature that is different from room temperature, the value of $\Delta H^{\circ}_{\text{reaction}}$ can be calculated as follows:

$$\Delta H^{\circ}_{\text{reaction at any T}} = \Delta H^{\circ}_{\text{reaction, 298}} + \int_{298}^{T} \Delta C_p dT \text{ cal or J} \tag{6.19}$$

Here, ΔC_p represents the difference in the total heat capacities of the reactants and products at that temperature. To apply this equation, it is convenient to use the equations for C_p expressed as a function of temperature, which typically takes the following form [Kubaschewski and Alcock, 1979]:

$$C_p = a + bT + cT^{-2} + dT^2 + \dots \tag{6.20}$$

This concept may be expanded, incorporating the fact that enthalpy is a state function, to calculate the total value $\Delta H^{\circ}_{\text{reaction}}$ of when the products exist at a different temperature from the reactants. To do this, one changes the temperature of the reactants from its value to 298 K (or any other reference temperature in general), carries out the reaction at this reference temperature, and then changes the temperature of the products to any actual temperature. The sum of the ΔH for each process yields the total value of enthalpy change.

Equation (6.19) can be derived with reference to Figure 6.3. The enthalpy change ΔH for converting the reactants at 298 K to the products at T is independent of path. Therefore, let us consider the two different paths shown in Figure 6.3. In the first path, one carries out the reaction at 298 K and then heats the products to T. The total ΔH in this case is given by

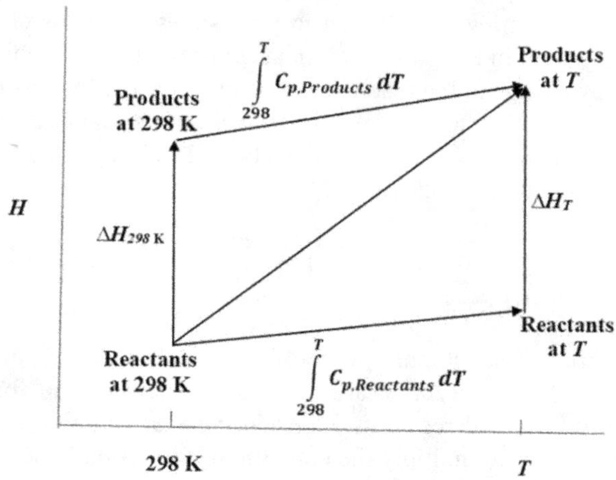

FIGURE 6.3 Calculation of the heat of reaction at any temperature T.

$$\Delta H = \Delta H^{\circ}_{\text{reaction, }298} + \int_{298}^{T} C_{p,\text{Products}}\,dT \tag{6.21}$$

In the second path, one first heats the reactants from 298 K to T and then carries out the reaction at that temperature. The total ΔH in this case is given by

$$\Delta H = \int_{298}^{T} C_{p,\text{Reactants}}\,dT + \Delta H^{\circ}_{\text{reaction, }T} \tag{6.22}$$

The values of ΔH in the two paths are equal and thus from these two equations $\Delta H^{\circ}_{\text{reaction, }T}$ is obtained, resulting in Eq. (6.19).

The same result is obtained by considering the overall path in which the temperature of the reactants is changed from T to 298 K, the reaction takes place at 298 K, and the temperature of the products is changed to T. By using this last procedure, the enthalpy changes when the reactants and the products exist at different temperatures can readily be calculated.

This figure is drawn for the case of an endothermic reaction, but the procedure may be applied equally to an exothermic reaction. The procedure is equally applicable to any reference temperature other than 298 K. It is also valid when the reactants and/or products do not exist at the standard states.

In general,

$$\Delta H_{T_2} = \Delta H_{T_1} + \int_{T_1}^{T_2} \Delta C_p\,dT \tag{6.23}$$

When there are phase changes in the reactants and/or products at T_t between T_1 and T_2, the ΔC_p at the corresponding intervals are used, as follows:

$$\Delta H_{T_2} = \Delta H_{T_1} + \int_{T_1}^{T_t} \Delta C_p dT + \Delta H_{t,\text{Product}} - \Delta H_{t,\text{Reactant}} + \int_{T_t}^{T_2} \Delta C_p dT \qquad (6.24)$$

ΔH_t in this equation is the latent heat of phase change.

6.1.9 Adiabatic Reaction Temperature

An adiabatic reaction takes place under the condition in which there is no heat exchange between the system and the surroundings. Thus, the temperature of the system changes because the heat of reaction is converted to the sensible heat of the system. In such a system, the resulting adiabatic reaction temperature is of interest. Since there is no exchange of heat in an adiabatic condition, $\Delta H = 0$, and if the initial temperature of the system is T_o, Eq. (6.21) can be rewritten as

$$-\Delta H_{T_o} = \int_{T_o}^{T_{ad}} C_{p,\text{Products}} dT \qquad (6.25)$$

Usually, T_o is 298 K. The adiabatic reaction temperature T_{ad} is the temperature that satisfies Eq. (6.25), and the adiabatic flame temperature is often of interest as the highest temperature that can be provided by the combustion of a fuel.

The adiabatic temperature is usually calculated for an exothermic reaction and, thus, by applying a figure similar to Figure 6.3, Figure 6.4 is obtained:

6.1.10 Heat of Mixing

In most real processes, mixtures of materials are involved. When pure substance i is added to a solution, the molecules of i interact with other components in addition to i and, thus, its energy level changes. The change in enthalpy when one mole of i is

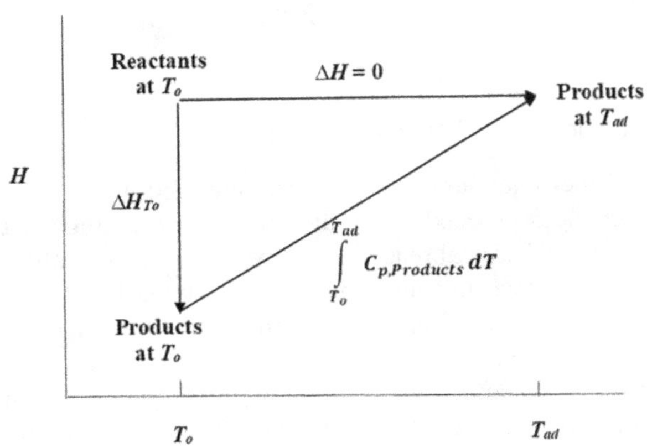

FIGURE 6.4 Calculation of the adiabatic reaction temperature.

added to a large amount of a solution so that the composition and temperature are not changed is the difference between the enthalpy of 1 mole of i in the solution at that solution composition and temperature, $\bar{H}_i$, and the enthalpy of 1 mole of pure i at that temperature, H_i°, and is called the heat of mixing $\Delta\bar{H}_i^m$ at that solution composition and temperature. Thus,

$$\Delta\bar{H}_i^m = \bar{H}_i - H_i^\circ \tag{6.26}$$

Depending on the nature of the interaction between i molecules and that between i and other components, mixing can be endothermic or exothermic. The total enthalpy of a solution of N components is the sum of the enthalpy of each component in the solution $n_i\bar{H}_i$ (n_i = number of moles of i). Thus,

$$H_{\text{solution}} = \sum_i^N n_i\bar{H}_i = \sum_i^N n_i H_i^\circ + \sum_i^N n_i\Delta\bar{H}_i^m \tag{6.27}$$

Using the quantities typically given in thermodynamic reference sources $\left(H_{i,T}^\circ - H_{i,\,298}^\circ\right)$,

$$\left(H_{\text{solution},T} - \sum_i^N n_i H_{i,\,298}^\circ\right) = \sum_i^N n_i\left(H_{i,T}^\circ - H_{i,\,298}^\circ\right) + \sum_i^N n_i\Delta\bar{H}_{i,T}^m \tag{6.28}$$

Here, the term on the left-hand side is the difference between the total enthalpy of the solution at T and the sum of the enthalpy values of all pure components at 298 K.

The heat of mixing of ideal gases is zero. Thus, the total enthalpy of a gas mixture at moderate pressures can be calculated by the sum of the enthalpy of each component gas. As a result, the heat capacity of a gas mixture can be calculated as

$$C_{p,\text{total}} = \sum_i^N n_i C_{p,i} \tag{6.29}$$

6.1.11 The Second Law of Thermodynamics

The second law of thermodynamics may be summarized by the statement that "All the spontaneous changes proceed in the direction of increasing the disorder of the universe." Applied to a chemical reaction, for example, this law allows us to predict what products we can expect starting from certain reactants by a spontaneous change consistent with the increase of disorder in the universe or to understand which impurities are expected to be distributed to which phase.

We need to develop a quantitative theory to apply this qualitative statement of the law to a real process. For this purpose, we define a thermodynamic function entropy, S. The energy level of molecules that make up matter increases and their movement becomes more active, thus increasing their degree of disorder, when heat is added to it. Considering that the effect is inversely proportional to the absolute temperature, the following relationship is established [Lewis et al., 1961]. In terms of small changes,

$$dS = \frac{\text{Added Heat}}{\text{Existing Extent of Disorder}} = \delta q / T \tag{6.30}$$

With this definition of the degree of disorder, the application of the statement of the second law to a spontaneous change yields the following relationship:

$$dS_{\text{system}} + dS_{\text{surroundings}} = dS_{\text{total}} \geq 0 \tag{6.31}$$

At equilibrium, $dS_{\text{total}} = 0$. Here, the system is the part of the universe in which the change of interest takes place, and the surroundings are the rest of the universe outside the system. Then, the entropy change in the surroundings is just due to the exchange of heat with the system, as follows:

$$dS_{\text{surroundings}} = \delta q_{\text{surr}} / T_{\text{surr}} \tag{6.32}$$

Without changes in temperature and pressure and when the temperature of the system and the surroundings are the same, this equation becomes

$$dS_{\text{surroundings}} = -\delta q_{\text{sys}} / T_{\text{sys}} = -dH_{\text{sys}} / T_{\text{sys}} \tag{6.33}$$

Substituting these two equations in Eq. (6.31), we get

$$dS_{\text{system}} - dH_{\text{sys}} / T_{\text{sys}} = dS_{\text{total}} \geq 0 \tag{6.34}$$

Thus, dS_{total} can be expressed just with the conditions of the system, which we are concerned with and can measure. By multiplying this equation by ($-T$) and considering that T is the absolute temperature, we obtain the following relationship:

$$dH - TdS = -TdS_{\text{total}} \leq 0 \tag{6.35}$$

This is the mathematical expression that defines the spontaneous change according to the second law of thermodynamics. Although this equation could be used as is in calculating the relationship involved, the presence of H, T, and S makes it rather complicated. If we define a new thermodynamic function as follows, the criterion for spontaneous changes becomes simple:

$$G = H - TS \tag{6.36}$$

The function G, thus, defined is called the Gibbs free energy. When one considers a small change taking place at a fixed temperature,

$$dG = dH - TdS: \quad T = \text{absolute temperature} \tag{6.37}$$

Considering this equation together with Eq. (6.35), the following simple statement is established for a spontaneous small change:

$$dG < 0 \tag{6.38}$$

When the overall change takes place at a fixed temperature and pressure, integration of dG results in the following:

$$\Delta G < 0 \tag{6.39}$$

Therefore,

If $\Delta G = 0$, the system is at equilibrium.

If $\Delta G < 0$, the change under consideration is spontaneous.

If $\Delta G > 0$, the change under consideration proceeds in the opposite direction.

Let us consider reaction (Eq. 6.17). It is again convenient to define ΔG of the reaction when all the components in the reaction are at their thermodynamic standard state. This term is denoted by $\Delta G°$ and is called the standard Gibbs free energy of reaction. Its value is calculated as follows, similarly to the calculation of $\Delta H^{\circ}_{\text{reaction}}$:

$$\Delta G^{\circ}_{\text{reaction}} = \sum_{\text{product, } i} \left(p_i \cdot \Delta g^{\circ}_{f,P_i}\right) - \sum_{\text{reactant, } j} \left(r_j \cdot \Delta g^{\circ}_{f,R_j}\right) \text{ cal or J} \tag{6.40}$$

In this equation, $\Delta g^{\circ}_{f,P_i}$ and $\Delta g^{\circ}_{f,R_j}$ represent the standard Gibbs free energy of formation of the corresponding species and can be found in various reference sources. G is also an extensive property and, thus, the calculation of $\Delta G^{\circ}_{\text{reaction}}$ requires the consideration of the stoichiometric coefficient.

6.1.12 Activity and Activity Coefficient

When the reactants and/or products do not exist in the standard states, such as in solutions or under pressures different from 1 atm, the ΔG values of the components are different from those at the standard states. In such a condition, a factor called activity (a) that compares the partial pressure of a gas or the concentration of a component in a condensed phase (liquid or solid) with those at standard states is introduced.

For a gas mixture, the activity a_i of component i is defined by

$$a_i = \frac{f_i}{p_{ss}} = \frac{p_i}{p_{ss}} \tag{6.41}$$

Here f_i is the fugacity (effective partial pressure), but in most processes under modest pressures, most gases behave ideally, and the fugacity may be replaced by the partial pressure, as indicated in Eq. (6.41). p_{ss} is the pressure of the gas at the standard state, that is, 1 atm or 101.3 kPa. Often, a_i is *replaced by* p_i, *but it must be remembered that this applies only if the pressures in the atm unit are used, in which case* p_{ss} *is 1 atm. If other units of pressure are used,* p_i, *must be divided by* p_{ss}. This is evident just considering the fact that activity is a ratio and thus must be unitless.

For a liquid or solid solution, the Raoultian activity a_i of component i is defined by

$$a_i = \frac{\text{effective concentration of } i \text{ in solution}}{\text{concentration of } i \text{ in standard state}} = \frac{\text{effective concentration of } i \text{ in solution}}{\text{concentration of } i \text{ in pure phase}} \tag{6.42}$$

Since a_i is a ratio, it is convenient to use mole fraction as the concentration.

A solution for which this ratio is the same as the mole fraction is called an *ideal solution,* namely

$$a_i = X_i \tag{6.43}$$

The activity, thus, defined is the Raoultian activity, and such a solution is said to follow Raoult's law. According to Raoult's law, the partial pressure of component i in an ideal solution at equilibrium with a vapor phase is given by

$$p_i = X_i \cdot p_i^{vap} \tag{6.44}$$

Here, p_i^{vap} is the vapor pressure of pure i at the same temperature. This relationship will be proved below where we discuss chemical equilibria.

In a real, non-ideal solution, the interaction between components is not directly proportional to concentration, and we need to introduce a correction factor. This factor is called the activity coefficient, as shown in the following equation:

$$a_i = \gamma_i \cdot X_i \tag{6.45}$$

Here, γ_i is the activity coefficient. The qualitative relationship a_B-X_B in a binary solution of A and B is shown in Figure 6.5. The behavior of a component with γ_i smaller than 1 is called negative deviation and that with γ_i larger than 1 is called positive deviation.

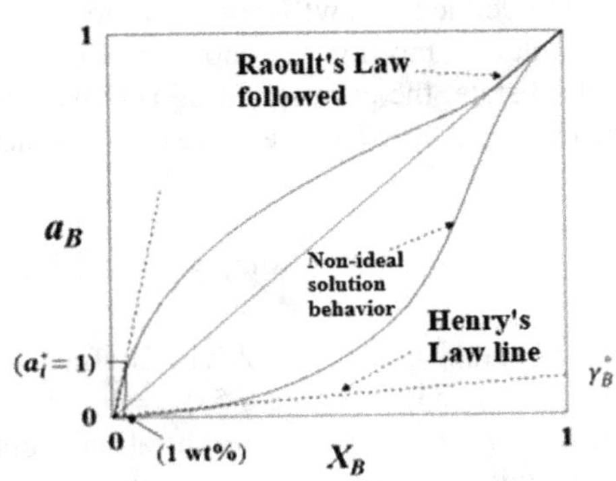

FIGURE 6.5 Relationship between Raoultian activity and mole fraction in a binary solution.

Although qualitative, Figure 6.5 shows an important characteristic of any solution, that is, as X_B approaches 1, component B behaves as in an ideal solution. In other words, $\gamma_B \to 1$ and $a_i \to X_i$ as $X_B \to 1$. When the concentration of B is very low, there is a region of linear relationship near $X_B = 0$, that is, the linearized region near the origin. This means that in this region γ_B is constant, and this relationship is called Henry's law in solution thermodynamics. The constant activity coefficient is denoted by $\overset{\circ}{\gamma}_B$ and called the Raoultian activity coefficient at infinite dilution.

It is important to understand the characteristics of dilute solutions because metals, especially steels, contain impurities or alloying components at low concentrations. In dilute solutions, the interactions among the solutes are negligibly small, even if there are many solutes, and the important interactions are between each solute and the solvent. Thus, the behavior of each solute may be described by that in a binary solution with the solvent.

Often, the concentrations in a molten metal solution are expressed by wt%, and it is convenient to define an activity different from the Raoultian activity. In general, mole fraction is not proportional to wt%, and thus, if one uses wt%, the relationship between activity and concentration is different from that in Figure 6.5. However, at very low concentration of B, X_B, and [wt%B] becomes proportional, as can be seen below:

$$X_B = \frac{\textit{Number of moles of B}}{\textit{Total number of moles}} = \frac{\left(\frac{\%B}{100}\right)\frac{W_T}{M_B}}{\left(\frac{\%A}{100}\right)\frac{W_T}{M_A} + \left(\frac{\%B}{100}\right)\frac{W_T}{M_B}}$$

$$= \frac{\frac{\%B}{M_B}}{\frac{\%A}{M_A} + \frac{\%B}{M_B}} \cong \left(\frac{\%B}{100}\right)\left(\frac{M_A}{M_B}\right) \tag{6.46}$$

The last relationship is the result of applying the condition $\%B << \%A$ to the previous term. Thus, for the solute at a low concentration, its activity is proportional to wt%. Typically, the region in which Henry's law is applicable includes 1 wt% and, thus, a new standard state is often defined as 1 wt% concentration of the solute. In the case of aqueous solutions, 1 molal = 1 mol/kg concentration is used more often.

In any type of standard state, the corresponding activity must be unity and thus when 1 wt% is used as the standard state, the corresponding new activity is defined as follows:

$$a_i^* = \gamma_i^* \cdot (wt\%\ i) \text{ or more simply } a_i^* = \gamma_i^* \cdot [\%\ i] \tag{6.47}$$

Here, a_i^* is Henrian activity and γ_i^* is Henrian activity coefficient. In the region where Henry's law is applicable, as shown in Figure 6.5, $\gamma_i^* = 1$ and at the standard state of 1 wt% $a_i^* = 1$. In a solution for which 1 wt% concentration is outside of Henry's law region, a fictitious solution with $a_i^* = 1$ at 1 wt% concentration is taken as the Henrian standard state, and the actual a_i^* at 1 wt% concentration will be γ_i^*.

6.1.13 Chemical Equilibrium

The value of ΔG for a reaction taking place with the species in any state is calculated by the use of activities (a or a^*) [e.g. Lewis et al., 1961; Moore, 1990], as follows:

$$\Delta G = \Delta G^\circ + RT \cdot \text{Ln}(Q)\text{: } T = \text{absolute temperature} \tag{6.48}$$

Here, Q is a term that represents the difference between the actual ΔG and ΔG°. Let us derive this relationship for the general chemical reaction represented by Eq. (6.17).

Consider 1 mole of component i:

$$g_i = \mu_i = \overset{\circ}{g_i} + RT \cdot \text{Ln}(a_i) \tag{6.49}$$

As usual, ° denotes the value at standard state. The free energy of a unit mole of component i is called the chemical potential μ_i of i. Derivation of the above relationship can be found in reference books on thermodynamics [e.g., Lewis et al., 1961; Smith and Van Ness, 1959].

The free energy change of reaction (Eq. 6.17) is given as follows:

$$\Delta G = \sum_{\text{product, } i} \left(p_i \cdot g_{P_i}\right) - \sum_{\text{reactant, } j} \left(r_j \cdot g_{R_i}\right) \text{ cal or J} \tag{6.50}$$

Similarly,

$$\Delta G^\circ = \sum_{\text{product, } i} \left(p_i \cdot \overset{\circ}{g}_{P_i}\right) - \sum_{\text{reactant, } j} \left(r_j \cdot \overset{\circ}{g}_{R_i}\right) \text{ cal or J} \tag{6.51}$$

Substituting Eq. (6.49) in Eq. (6.50),

$$\Delta G = \Delta G^\circ + \sum_{\text{product, } i} \left(p_i \cdot RT\text{Ln}a_{P_i}\right) - \sum_{\text{reactant, } j} \left(r_j \cdot RT\text{Ln}a_{R_i}\right) \tag{6.52}$$

Equation (6.51) can be rewritten as follows:

$$\Delta G^\circ = \sum_{\text{product, } i} \left[\left(p_i \cdot \overset{\circ}{g}_{P_i}\right) - \sum_{\substack{\text{all elements } k \text{ making} \\ \text{up products}}} \nu_k \overset{\circ}{g}_k\right] - \sum_{\text{reactant, } j} \left[\left(r_j \cdot \overset{\circ}{g}_{R_i}\right) - \sum_{\substack{\text{all elements } k \text{ making} \\ \text{up products}}} \nu_k \overset{\circ}{g}_k\right]$$

$$= \sum_{\text{product, } i} \left[\left(p_i \cdot \overset{\circ}{g}_{P_i}\right) - \sum_{\substack{\text{all elements } k \text{ making} \\ \text{up products}}} \nu_k \overset{\circ}{g}_k\right] - \sum_{\text{reactant, } j} \left[\left(r_j \cdot \overset{\circ}{g}_{R_i}\right) - \sum_{\substack{\text{all elements } k \text{ making} \\ \text{up reactants}}} \nu_k \overset{\circ}{g}_k\right] \text{cal or J} \tag{6.53}$$

In the above derivation, we have applied the fact that all the elements that make up the reactants and the products are identical. Furthermore, if we consider that $\sum_{\text{product}, i}\left[\left(p_i \cdot g^\circ_{P_i}\right) - \sum_{\text{all elements } k \text{ making up products}} \nu_k g^\circ_k\right] = \sum_{\text{product}, i}\left(p_i \cdot \Delta g^\circ_{f,P_i}\right)$ and this relationship also applies to the reactants, we can prove that ΔG° can be calculated by the following equation, which has earlier been given as Eq. (6.40):

$$\Delta G^\circ_{\text{reaction}} = \sum_{\text{product}, i}\left(p_i \cdot \Delta g^\circ_{f,P_i}\right) - \sum_{\text{reactant}, j}\left(r_j \cdot \Delta g^\circ_{f,R_j}\right) \text{ cal or J} \tag{6.40}$$

Equation (6.52) now becomes

$$\Delta G = \Delta G^\circ + RT \cdot \text{Ln}\frac{a_{P_1}^{p_1} \cdot a_{P_2}^{p_2} \cdots}{a_{R_1}^{r_1} \cdot a_{R_2}^{r_2} \cdots} \tag{6.54}$$

It appears that the third term has the units of RT, that is, cal or J per mol that is different from those of ΔG and ΔG°, which are just cal or J. However, as can be seen from Eq. (6.52), these terms have the same units, with the number of moles embedded in the Ln term due to the unique property of this function.

Therefore,

$$Q = \frac{\prod_i a_{P_i}^{p_i}}{\prod_j a_{R_j}^{r_j}} = \frac{a_{P_1}^{p_1} \cdot a_{P_2}^{p_2} \cdots}{a_{R_1}^{r_1} \cdot a_{R_2}^{r_2} \cdots} \tag{6.55}$$

Here, $a_{P_i}^{p_i}$ and $a_{R_j}^{r_j}$ are, respectively, functions of the activities of products P_i and reactants R_j, with the activities raised to the powers of the corresponding stoichiometric coefficients p_i and r_j. It is convenient to think of the activity as a measure of concentration relative to that at the standard state, being dimensionless as a ratio. The term Q thus defined is called the activity quotient.

Equation (6.54) indicates that $\Delta G = \Delta G^\circ$ when all the components are at the standard states because then all the activities are 1, consistent with the definition of ΔG°. Reaction (Eq. 6.17) is valid when the amounts of all the components are increased n times. Let us examine whether Eq. (6.54) is applicable even in such a case. In this case, the change in G becomes n times, considering that it is an extensive property. According to Eq. (6.40), the standard Gibbs free energy of reaction becomes $n\Delta G^\circ$ when the amounts of all the components increase n times, and Eq. (6.54) becomes as follows:

$$n\Delta G^\circ + RT \cdot Ln\frac{a_{P_1}^{np_1} \cdot a_{P_2}^{np_2} \cdots}{a_{R_1}^{nr_1} \cdot a_{R_2}^{nr_2} \cdots} = n\Delta G^\circ + nRT \cdot \text{Ln}\frac{a_{P_1}^{p_1} \cdot a_{P_2}^{p_2} \cdots}{a_{R_1}^{r_1} \cdot a_{R_2}^{r_2} \cdots} = n\Delta G \tag{6.56}$$

In other words, the change in G becomes n times, and Eq. (6.56) becomes identical to Eq. (6.54). Thus, as long as Eq. (6.17) is correctly balanced, that is, the numbers of the elements are identical on both sides of the equation, the equilibrium relationship remains equally applicable even if the amounts of the components change.

At equilibrium, $\Delta G = 0$, and Eq. (6.54) yields

$$\Delta G^\circ = -RT \cdot \mathrm{Ln}\left(\frac{a_{P_1}^{p_1} \cdot a_{P_2}^{p_2} \cdots}{a_{R_1}^{r_1} \cdot a_{R_2}^{r_2} \cdots}\right)_{eq} \tag{6.57}$$

The term $(Q)_{eq}$ is called the equilibrium constant (K).

$$K = \left(\frac{a_{P_1}^{p_1} \cdot a_{P_2}^{p_2} \cdots}{a_{R_1}^{r_1} \cdot a_{R_2}^{r_2} \cdots}\right)_{eq} = \exp\left(-\frac{\Delta G^\circ}{RT}\right) \tag{6.58}$$

As can be seen in this equation, K for the given chemical reaction is a function only of temperature and is independent of the actual states or concentrations (activities) of the components. This equation is sometimes called van't Hoff equation or van't Hoff isotherm.

From the above relationships, Eq. (6.54) may also be rearranged as follows:

$$\Delta G = -RT \cdot \mathrm{Ln}K + RT \cdot \mathrm{Ln}Q = RT \cdot \mathrm{Ln}\left(\frac{Q}{K}\right) \tag{6.59}$$

The values of $\Delta G^\circ_{\text{formation}}$ necessary to calculate $\Delta G^\circ_{\text{reaction}}$ can be found in various reference sources. The following references are some selected examples: Thermodynamic Data for Mineral Technology [Pankratz et al., 1984]; JANAF Thermochemical Tables. Second Edition [Stull and Prophet, 1971]; JANAF Thermochemical Tables, 1982 Supplement [Chase Jr., et al., 1982]; Thermochemical Properties of Inorganic Substances [Knacke et al., 1991]; Metallurgical Thermochemistry [Kubaschewski and Alcock 1979]. In addition, such data are included in various computer software for thermodynamic computation.

Let us prove using the above relationships that Raoult's law, that is, $p_i = X_i \cdot p_i^{\text{vap}}$, is consistent with the fact that $a_i = X_i$ for an ideal solution at equilibrium with a mixture of ideal gases.

First, consider that the following process is at equilibrium under this condition:

$$i \text{ in solution} = i \text{ in gas} \tag{6.60}$$

Applying Eq. (6.57) to this process,

$$\Delta G^\circ = -RT \cdot \mathrm{Ln}\left(\frac{p_i}{a_i}\right)_{eq} \tag{6.61}$$

When this process is applied to pure liquid,

$$\Delta G^\circ = -RT \cdot \mathrm{Ln}\left(\frac{p_i^{\text{vap}}}{1}\right)_{eq} \tag{6.62}$$

From these two equations, $p_i = a_i \cdot p_i^{\text{vap}}$ and because $a_i = X_i$ for an ideal solution, $p_i = X_i \cdot p_i^{\text{vap}}$. For a non-ideal solution, the relationship becomes

$$p_i = a_i \cdot p_i^{\text{vap}} = \gamma_i \cdot X_i \cdot p_i^{\text{vap}} \tag{6.63}$$

When certain reactants or products are treated from the viewpoint of Henry's law (standard state = 1 wt%), this must be indicated clearly. There are different symbols to indicate this, one of which is to place an underline below the component in concern, as shown below:

$$r_1 R_1 + r_2 \underline{R_2} + \cdots\cdots = p_1 \underline{P_1} + p_2 P_2 + \cdots\cdots \tag{6.64}$$

In accompanying Q, the corresponding activities are replaced by the Henrian activity, as shown below:

$$Q = \frac{a^{*p_1}_{\underline{P_1}} \cdot a^{p_2}_{P_2} \cdots}{a^{r_1}_{R_1} \cdot a^{*r_2}_{\underline{R_2}} \cdots} \tag{6.65}$$

Of course, the value of ΔG° for the reaction written as this is different from the value when all the components are treated from the Raoultian point of view (standard state = pure component under 1 arm pressure). The relation between the two types of ΔG° can be obtained as shown below. This relationship is necessary because most reference sources list the values of ΔG_f° under the Raoultian standard states, and one needs the value of ΔG° for the reaction as written in Eq. (6.64).

The chemical potential, μ_i, or the partial molar Gibbs free energy, of a component in a solution, is a given quantity, which can be related to the free energy of the component at the selected standard state and the corresponding activity, as follows:

$$\mu_i = \overset{\circ}{g_i} + RT \cdot \mathrm{Ln}(a_i) \tag{6.66a}$$

$$= \underline{\overset{\circ}{g_i}} + RT \cdot \mathrm{Ln}(a_i^*) \tag{6.66b}$$

Here, $\overset{\circ}{g_i}$ and $\underline{\overset{\circ}{g_i}}$ are, respectively, the molar Gibbs free energy of component i at Raoultian (pure component under 1 arm) and Henrian (1 wt%) standard states. The change when component i dissolves from the pure state into a solvent at 1 wt% concentration is the difference in the change of the two standard states, Thus, the difference between $\underline{\overset{\circ}{g_i}}$ and $\overset{\circ}{g_i}$ can be obtained from the above two equations, as follows:

$$\Delta \overset{\circ}{g}_{R\to H} = \underline{\overset{\circ}{g_i}} - \overset{\circ}{g_i} = RT \cdot \mathrm{Ln}\left(\frac{a_i}{a_i^*}\right) = RT \cdot \mathrm{Ln}\left(\frac{\gamma_i \cdot X_i}{\gamma_i^* \cdot [\%\, i]}\right) \tag{6.67}$$

Equations (6.66a) and (6.66b) apply at any concentration. Thus, we apply this relationship at the dilute concentration range in which Henry's law is valid, $\gamma_i^* = 1$ and $\gamma_i = \overset{\circ}{\gamma_i}$. Equation (6.67), thus, becomes

$$\Delta \overset{\circ}{g}_{R\to H} = RT \cdot \mathrm{Ln}\left[\frac{\overset{\circ}{\gamma_i}}{100}\left(\frac{M_A}{M_B}\right)\right] \tag{6.68}$$

Using this equation, the difference in the molar Gibbs free energy of component i between the Raoultian and Henrian standard states can be calculated through $\overset{\circ}{\gamma_i}$. Furthermore, this procedure can be used to calculate the difference in the molar

Gibbs free energy of component i when other standard states such as 1 molal concentration are considered.

6.1.14 Calculation of Equilibrium Composition

When more than one raw material are used in a process, it is beneficial to use the less expensive raw material in excess so that the more valuable material is used to a greater extent at equilibrium. Thus, it is necessary to calculate the expected amounts of the reactants and products at equilibrium. The equilibrium condition for reaction (Eq. 6.64) is given by Eq. (6.48) together with Eq. (6.65), as follows: (Of course, the values of ΔG° in this case must be calculated using the ΔG_f° corresponding to the Henrian standard states for the underlined components.)

$$K = \exp\left(-\frac{\Delta G^\circ}{RT}\right) = \left(\frac{a^{*p_1}_{\underline{P_1}} \cdot a_{P_2}^{p_2} \cdots}{a_{R_1}^{r_1} \cdot a^{*r_2}_{\underline{R_2}} \cdots}\right)_{\text{equil.}} \tag{6.69}$$

Consider the following reaction, as an example.

$$A(g) + b \cdot B(s) = g \cdot G(g) + h \cdot H(s) \tag{6.70}$$

Assume that the gases are ideal, and the solids are in pure states. Then,

$$p_A = \frac{n_A}{V}RT;\ p_G = \frac{n_G}{V}RT;\ P_T = \frac{n_T}{V}RT \tag{6.71}$$

Here, n_i and n_T are, respectively, the number of moles of gaseous component I and the total number of moles of the gas. Denote the moles of all the relevant components before the reaction as follows:

A: n_A^i, B: n_B^i, G: n_G^i, H: n_H^i, Inert Gas: n_I^i

Suppose that y moles of A are consumed when equilibrium is reached. Then,

$$n_A = n_A^i - y;\ n_G = n_G^i + g \cdot y;\ n_B = n_B^i - b \cdot y;\ n_H = n_H^i + h \cdot y \tag{6.72}$$

The total number of moles of the gas phase is given by

$$n_{T,\,\text{gas}} = n_A^i + n_G^i + (g-1)y + n_I^i \tag{6.73}$$

If the equilibrium is reached under a constant total pressure P_T,

$$K = \frac{p_G^g}{p_A} = \frac{\left[\left(n_G / n_{T,\,\text{gas}}\right)P_T\right]^g}{\left(n_A / n_{T,\,\text{gas}}\right)P_T} \tag{6.74}$$

If the equilibrium is reached under a constant total volume V, neglecting the volume of the solids,

$$P_T = \left(n_{T,\,\text{gas}} / V\right)RT \tag{6.75}$$

$$K = \frac{p_G^g}{p_A} = \frac{\left(n_G RT/V\right)^g}{n_A RT/V} \quad (6.76)$$

In either case, Eqs. (6.74) and (6.76) contain, under the given conditions, one unknown variable y, which when solved for yields the values of n_A, n_G, n_B, and n_H existing at equilibrium.

6.1.15 Ellingham Diagram – $\Delta G°$ – T Diagram

The Ellingham diagram is a diagram showing the values of $\Delta G°$ as functions of temperature for the formation of simple compounds from gaseous and other elements [Ellingham, 1944; Shamsuddin and Sohn, 2019]. The general equation describing such reactions for 1 mole of the reactant gas is given by

$$(2x/y)R + O_2(or\ S_2) = (2/y)R_xO_y\left[or\ (2/y)R_xS_y\right] \quad (6.77)$$

The equilibrium of such a reaction is directly and conveniently related to its partial pressure when the reaction is written based on 1 mole of the gas like O_2. In the case when R and the product R_xO_y are pure solids or liquids,

$$\Delta G° = -RT \cdot \text{Ln}\left(1/p_{O_2}\right)_{eq} \quad (6.78)$$

Thus,

$$\left(p_{O_2}\right)_{eq} = 1/K = \exp\left(\frac{\Delta G°}{RT}\right) \quad (6.79)$$

Figure 6.6 shows an example of Ellingham diagram for the formation of selected oxides.

This diagram shows $\Delta G°$ as function of temperature, and the lines represent the following relations derived from Eq. (6.37):

$$\Delta G° = \Delta H° - T \cdot \Delta S° \quad (6.80)$$

In the case where a metal is oxidized to form an oxide, the gas oxygen is incorporated into the solid and thus its disorderliness decreases resulting in $\Delta S < 0$. Furthermore, as mentioned earlier, the value of $\Delta H°$ remains almost constant relative to temperature within the range of no phase change. Thus, the lines in the Ellingham diagram are, in general, straight lines with constant positive slopes within the same condensed phase. When there is a phase change in the reactant or product, the value of ΔS and the slope change at that temperature.

Since all the lines are drawn for 1 mol of O_2, the oxide located below in Figure 6.6 is more stable than an oxide located above it, and thus the metal that forms the oxide below can reduce the oxides located above. In other words, if the oxide of metal M_1 is located below the oxide of M_2, the ΔG of the following reaction will be negative, making the reaction thermodynamically spontaneous:

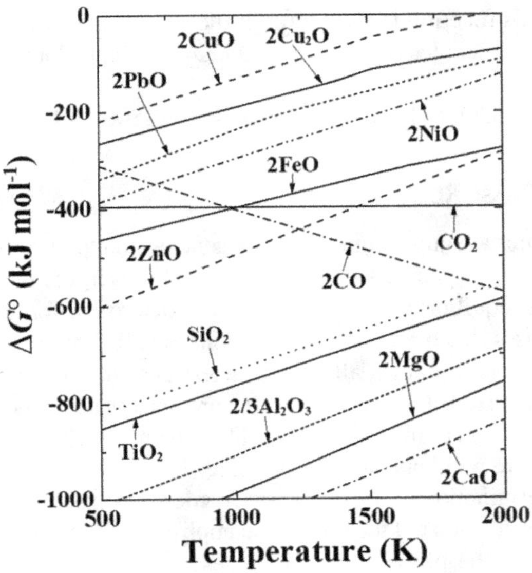

FIGURE 6.6 Ellingham diagram for the formation of selected oxides (per 1 mole of O_2).

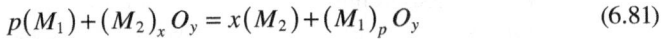

$$p(M_1)+(M_2)_x O_y = x(M_2)+(M_1)_p O_y \tag{6.81}$$

Let us now consider the formation reaction of CO.

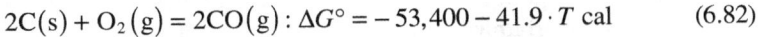

$$2C(s) + O_2(g) = 2CO(g) : \Delta G^\circ = -53{,}400 - 41.9 \cdot T \text{ cal} \tag{6.82}$$

In this reaction, 1 mole of a gas forms 2 moles of a gas and thus $\Delta S > 0$(41.9 cal/K). Thus, the slope of the line is negative, and this fact has an important implication with respect to metal extraction, especially in the production of iron. The reason is that carbon can reduce most metal oxides at a sufficiently high temperature. However, metals with high affinity to oxygen like Al, Ti, and Ca form carbides before being reduced to metal, and thus it is difficult to produce such metals by using carbon as the reductant. Depending on metal oxides, the lines of formation fall below that of CO formation and, thus, their reduction by carbon is possible at relatively moderate temperatures. The production of iron from iron oxides is the most important example. In non-ferrous metal extraction, the production of lead and zinc is an example.

Another consideration is that CO is not as stable at relatively low temperatures and thus CO_2 must be considered as an oxide of carbon.

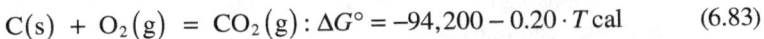

$$C(s) + O_2(g) = CO_2(g) : \Delta G^\circ = -94{,}200 - 0.20 \cdot T \text{ cal} \tag{6.83}$$

In this reaction, 1 mole of gas produces 1 mole of gas and thus $\Delta S \cong 0$ (0.20/calK), and the corresponding line in the Ellingham diagram is essentially horizontal. From reactions (Eqs. 6.82 and 6.83), an important reaction called the Boudouard reaction can be derived. This reaction is important in the reduction of metal oxides (carbothermic reduction) and decides the ratio of CO/CO_2 in the product gas.

$$C(s) + CO_2(g) = 2CO(g) : \Delta G^\circ = +40{,}800 - 41.7 \cdot T \text{ cal} \tag{6.84}$$

6.1.16 Gibbs' Phase Rule

All the characteristics at equilibrium of a homogeneous matter are specified if the pressure and temperature or specific volume are fixed. However, the number of conditions necessary to fix the equilibrium state is difficult to determine for a system containing more than two phases. Such number of conditions is called the degree of freedom. In the case of a homogeneous matter, all the characteristics are decided if two of the above three conditions are fixed. However, at 1 atm pressure water vapor and liquid water (non-homogeneous matter) can coexist at equilibrium only at 100°C, and it is impossible to change temperature without changing pressure. Thus, the degree of freedom, in this case, is 1. The determination of the degree of freedom is difficult to determine simply, but J. Willard Gibbs formulated the phase rule applicable to equilibrium conditions.

Before discussing the phase rule, it is necessary to clearly define a "phase". A phase is a region that contains a physically unique and uniform matter, including a mixture. If the characteristics including the composition of the matter change discontinuously at a plane, the regions on the two sides of the plane are different phases. A gas phase and a condensed liquid or a liquid and a solid are clearly different phases. Different crystal structures form different phases. All the gas mixtures or solutions make one phase, and coexisting water vapor and liquid water form two phases. A very important point here is the fact that the presence and not the amount of a phase determines phase equilibria. Thus, the phase rule applies only when the amount of a phase may change arbitrarily. Broadly speaking, the degree of freedom is the number of conditions that can be varied without changing the constitution (composition), not the amount, of each phase at equilibrium. The conditions used are temperature, total pressure, partial pressures of gas components, and concentrations of condensed phase solution. Density or other conditions could be used, but it is convenient to select from these four parameters that can be directly measured and controlled.

The composition of each phase is denoted by the components of the mixture that make up that phase, and its number is the minimum number of chemical species that make up that phase. When there is no chemical reaction, it is the number of chemical compounds, including elements, that exist in the system.

When the number of components in a system is C and that of phases is P, the conditions that can be varied are T, total pressure P_T, and the concentrations of the components in each phase. Considering that the concentration of the last component is fixed by the concentrations of the rest of the components, $C - 1$ concentrations must be fixed in each phase, and thus $P(C - 1)$ concentrations must be fixed in P phases. Thus, the total number of conditions including T and P_T becomes $P(C - 1) + 2$. However, the phase rule applies to the conditions at equilibrium state and, thus, there is the limitation that the chemical potential of a component is the same in all phases, resulting in $C(P - 1)$ limitations. As a result, the number of independent conditions, namely the total degree of freedom, at equilibrium in a system is given by

$$F = P(C-1)+2-C(P-1)=C-P+2. \quad (6.85)$$

Here, P = number of phases, C = number of components, and F = degree of freedom (number of conditions that can be varied.

In the case in which there are chemical reactions involved in phase equilibria, the equilibria of chemical reactions bring additional limitations, making additional considerations necessary to correctly apply the phase rule. As an example, if in a three-component system the reaction $D(s)=A(s)+B(g)$ takes place, the number of components is three and the number of phases is also three, but in this system, the equilibrium of the chemical reaction must also be satisfied, adding an additional limitation. Thus, the degree of freedom becomes $F = C - P+2 - 1=1$. In other words, only temperature or pressure can be varied: If the temperature is fixed, the pressure of $B(g)$ is also fixed. On the contrary, a certain pressure of $B(g)$ can exist only at a corresponding temperature. Let us consider the following reaction that forms two gas components from the same solid reactant: $D(s)=A(g)+B(g)$. In this case, the number of phases decreases to two, and the degree of freedom might be considered to increase by 1, but here, the composition of the gas, namely the molar ratio of $A(g)$ and $B(g)$, must be fixed by the stoichiometry of the reaction and, thus, the degree of freedom remains to be 1. If $A(g)$ or $B(g)$ is independently added to the system, this stoichiometric limitation is removed, and the degree of freedom is 2, allowing variation of 2 out of temperature, pressure, and gas composition.

The expanded, general Gibbs' phase rule, including the additional restrictions described above, is expressed as follows:

$$F = C - P + 2 - R. \quad (6.86)$$

Here, R is the number of restrictions, which is typically determined by the number of chemical reactions that must satisfy equilibrium relationships and the number of molar ratios of components within the phase according to stoichiometry.

Consider the decomposition of calcium carbonate.

$$CaCO_3(s) = CaO(s) + CO_2(g) \quad (6.87)$$

The number of compounds in this system is three, the number of phases is three, and the number of chemical reactions is one. Thus, $F = C - P+2 - 1=1$, and only one between temperature and $CO_2(g)$ pressure can be varied. In this case, the gas is pure CO_2; thus, even if CO_2 is added separately, the degree of freedom remains the same. If another gas like nitrogen is present, the number of components is increased by one and, thus, the degree of freedom also increases by one. For reference, one could choose Ca, C, and O as the components. In this case, we have a restriction that the C/O ratio in the gas must be ½ and the degree of freedom is the same at 1. For pure solids, the definition of the phase includes the composition, and one cannot consider the variation of the composition. However, in the case of a solid solution or a liquid solution, the composition can vary and, thus, the concentration should be considered in the calculation of the degree of freedom much like in the case of a gas mixture.

6.1.17 Stability Diagram

The Ellingham or $\Delta G° - T$ diagram discussed above conveniently represents the reactions in which one gaseous element participates to produce one compound. When such a reaction may produce several solid products, the $p_{O_2} - T$ diagram shown below is convenient.

Consider the oxidation of a metal as an example.

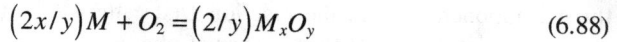

$$(2x/y)M + O_2 = (2/y)M_xO_y \tag{6.88}$$

At equilibrium,

$$p_{O_2} = \exp\left(\frac{\Delta G°}{RT}\right) \tag{6.89}$$

Taking the logarithm of each side,

$$\text{Ln } p_{O_2} = \frac{\Delta G°}{RT} \tag{6.90}$$

Most diagrams use $\log_{10} p_{O_2}$ and $1/T$ as the axes, according to

$$\log_{10} p_{O_2} = \frac{\Delta G°}{2.3R} \cdot \frac{1}{T} \tag{6.91}$$

Considering the Fe-O system, the phases hematite (Fe_2O_3), magnetite (Fe_3O_4), wűstite (Fe_xO), and iron (Fe) can exist depending on temperature and p_{O_2}. The above equation can be plotted by applying it to the equilibrium between the adjacent phases, resulting in Figure 6.7. It is noted that one cannot consider equilibrium between the phases that are not adjacent like Fe_2O_3 and Fe.

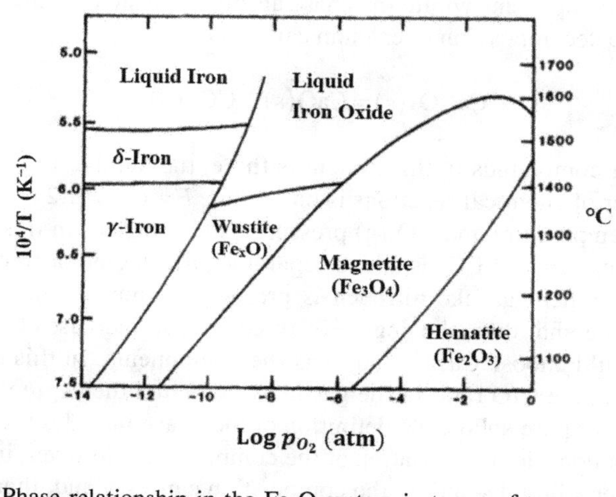

FIGURE 6.7 Phase relationship in the Fe-O system in terms of temperature and oxygen pressure.

6.2 REACTION KINETICS OF FINE SOLID PARTICLES WITH A GAS

6.2.1 INTRODUCTION

The rate processes of gas–solid reactions, in general, involve many factors such as changes in the solid structure, the effects of mass and heat transfer, and the diffusion of gaseous species through pores, in addition to chemical reactions, especially for large pieces of solids or pellets. A comprehensive discussion and derivations of theories involving such general systems of gas–solid reactions can be found in a recent monograph on the subject by Sohn [2020a]. The discussion here will be limited to the reaction of fine particles of the concentrate size, $< 100\,\mu m$, relevant to the flash ironmaking process.

The reaction rates of fine particles in the absence of external mass transfer and intraparticle pore-diffusion effects involve mainly the intrinsic rate of the chemical reactions, although in some cases, solid-state diffusion effects or pore diffusion through a solid product layer covering the solid reactant are lumped with chemical kinetics because the solid geometry of fine particles is intractably complex or undefinable. Furthermore, often the geometry and structure are intrinsic properties of the solid and are not technically or economically feasible to alter. When the small particles have some initial porosity, the reaction process is more complicated because the diffusion of gases through the pores together with a change in morphology may be significant. However, such pore structure and its change are usually difficult to characterize, let alone control. In many cases, pore diffusion effects are negligible compared with the effects of chemical reactions for such small particles. Furthermore, the external mass convection is typically very fast for fine particles of less than a few hundred micrometers, because the mass transfer coefficient is inversely proportional to particle size, among other reasons.

Unlike mass transfer effects, there are no theories that allow one to estimate chemical kinetics to any reasonable degree of accuracy, and thus this information must always be obtained by correctly designed experiments. In such a case we will call it "particle kinetics." Many use the term "intrinsic kinetics" for this term, but it is here broadened to include the cases in which diffusional effects, especially involving solid-state diffusion, are present in the reaction of fine particles in combination with chemical reaction.

The kinetics of gas–solid reactions are often determined by thermogravimetry, but they can also be studied by other methods such as the laminar flow reactor for fast reactions [Elzohiery et al., 2017; Fan et al., 2016, 2017; Chen et al., 2015; Wang and Sohn, 2013; Choi and Sohn, 2010], chemical analysis of product gas [Sohn et al., 1998; Lin and Sohn, 1987; Chaubal and Sohn, 1986; Sohn and Yang, 1985; Shih and Sohn, 1980], chemical analysis of solid at various reaction times [Rhee and Sohn, 1990a, b], differential thermal analysis (DTA) or differential scanning calorimetry (DSC) [Xiao et al., 1997; Khawam and Flanagan, 2006a; Vyazovkin and Dranca, 2005], and X-ray diffraction (XRD) [Kirsch et al., 2004]. For any analytical method, the measurements must be transformed into the extent of solid conversion that can be used in the kinetic equations. A more detailed discussion of experimental methods for gas–solid reactions is presented in Sohn [2020a].

A number of rate expressions describing "particle kinetics" are described below. It is worthwhile to note that these rate equations are in general associated with the geometry of the solid product formed from the reaction, although it is often not possible to predict it without actual experimental observation. Therefore, it is important to perform a microstructural examination of partially reacted particles together with the determination of the reaction rate in the study of 'particle kinetics'.

The reaction rate of fine solid particles with a gas depends on the temperature and concentration of the gaseous reactant. Although the solid is a reactant, the "concentration" or "activity" of the solid reactant often does not vary as it is consumed by the reaction, unlike that of the fluid species. This is because components in solid do not have a very wide miscibility. Rather, solids form distinct "phases" as the contents of the components vary upon reaction. Thus, when hematite (Fe_2O_3) is reduced by a gas (H_2 or CO), a new distinct solid phase is formed instead of the oxygen content in the solid decreasing gradually. In other words, the oxygen "content" in the solid changes essentially as a step function across the solid-solid interphase. The new phase in this case is magnetite, Fe_3O_4, followed by wustite (FeO) and Fe, depending on temperature, as oxygen is removed by reaction. Therefore, familiarity with the phase diagram is very important in the analysis of fluid–solid reactions. The product solid phase is often porous, thus allowing the gaseous reactant to reach the unreacted solid. Since the unreacted solid remains chemically the same as the original solid reactant, its reactivity (or activity) remains the same, except that its surface area for reaction changes with the progress of the reaction. In certain cases, the area could even increase with the consumption of the solid reactant before eventually decreasing. This changing surface area gives rise to the rate dependence on the remaining amount (typically expressed by unreacted fraction) of the solid reactant.

In terms of thermodynamic activity, the activity of a fluid species changes with the diminishing concentration as the reaction progresses. In the case of a solid reactant, the activity of the solid reactant remains the same at the original value as its amount decreases with reaction. Then, the rate dependence on the amount of the solid reactant arises from the change of its surface area with reaction.

There are a number of rate dependences on the fraction of solid reactant consumed, depending on how the fluid–solid interfacial area changes with the progress of reaction. The rate of a solid-state reaction can generally be described by

$$dX/dt = k_o \cdot \exp\left(-E/RT\right) \cdot f(X) \tag{6.92}$$

where k_o is the pre-exponential (frequency) factor, E is the activation energy, T is absolute temperature, R is the universal gas constant, the conversion function $f(X)$ is the rate dependence on solid conversion, and X is the fractional conversion. For the kinetics of reactions involving a gaseous reactant, the right-hand side (RHS) is usually multiplied by an additional term that represents the concentration driving force of the gaseous reactant. A term representing the effect of particle size, if present, may also be included here.

Kinetic parameters are often obtained from isothermal rate measurements by applying Eq. (6.92). Separating variables in Eq. (6.92) and integrating the equation gives the integral forms of the isothermal rate equation:

$$g(X) = k_o \cdot \exp(-E/RT) \cdot t \tag{6.93}$$

where $g(X)$ is the conversion function, defined by

$$g(X) \equiv \int_0^X \frac{dX}{f(X)} \tag{6.94}$$

Alternatively, kinetic parameters can be obtained from non-isothermal rate data. The non-isothermal method routinely adopts the use of a linear temperature increase with time, as follows:

$$T = a \cdot t + T_o \tag{6.95}$$

where a is the heating rate and T_o is the starting temperature.

Equation (6.92) can be transformed to a non-isothermal rate expression describing reaction rate as a function of temperature at a constant heating rate, as follows:

$$dX/dt = (dX/dT)(dT/dt) = a(dX/dT) \tag{6.96}$$

where dX/dT is the non-isothermal reaction rate. Substituting Eq. (6.92) into Eq. (6.96) gives the differential form of the non-isothermal rate law,

$$dX/dT = (k_o/a) \cdot \exp(-E/RT) \cdot f(X) \tag{6.97}$$

and

$$g(X) = (k_o/a) . \int_0^T \exp(-E/RT) \cdot dT \tag{6.98}$$

The closed-form evaluation of the integral on the RHS requires approximation, as will be discussed below. Sohn [2016] proposed a new, non-linear *T-t* program that results in the exact analytical expression of this integral and has other additional advantages.

The form of $f(X)$, and thus of $g(X)$, depends on the mechanism of the reaction and the change in the geometry of the reaction interface with the progress of the reaction.

6.2.2 Chemically Controlled Shrinking-Core Kinetics

Let us consider a fluid–solid reaction of the general type

$$A(f) + bB(s) = cC(f) + dD(s) \tag{6.99}$$

When solid D forms as a layer around the reactant solid B, the former may occupy a smaller (Case 1) or larger (Case 2) volume than the latter.

In Case 1, the product layer is probably porous, and the rate-controlling step is typically the chemical reaction occurring at the interface between the reactant and product solids. This is true for fine solid particles under consideration, in general. Under these circumstances, the rate is determined by the available surface area of the reactant solid under the rate control by chemical kinetics. In Case 2, the product usually forms a protective layer around solid B, and the rate is likely controlled by diffusion through this layer.

The relationship between the rate of consumption of gaseous reactant and that of solid reactant yields

$$\begin{pmatrix} \text{molar rate} \\ \text{of consumption of } A \end{pmatrix} \times \begin{pmatrix} \text{surface area} \\ \text{of solid } B \end{pmatrix} \times \begin{pmatrix} \text{stoichiometric} \\ \text{coefficient} \end{pmatrix} = \begin{pmatrix} \text{molar rate of} \\ \text{consumption of } B \end{pmatrix} \quad (6.100)$$

Following the process discussed by Sohn [2020a], Eq. (6.100) yields

$$-\left(\frac{\alpha_B \rho_B}{b}\right)\left(\frac{dr_p}{dt}\right) = k\left(C_{Ao}^n - \frac{C_{Co}^m}{K^n}\right) \quad (6.101)$$

in which α_B is the volume (and thus area) fraction of the solid occupied by reactant B, ρ_B is the true molar density of the solid reactant B and thus $\alpha_B \rho_B$ represents the number of moles of B contained in a unit volume of the solid; b is the stoichiometry coefficient in reaction (Eq. 6.99); r_p is the length coordinate perpendicular to the solid surface; k is the reaction-rate constant; C_{io} is the fluid species concentration in the bulk fluid; and K is the equilibrium constant.

Thus, the conversion-vs-time relationship is obtained as [Sohn, 2020a]

$$1-(1-X)^{1/F_p} = \frac{bk}{\alpha_B \rho_B r_p}\left(C_{Ao}^n - C_{Co}^m / K^n\right)t \quad (6.102)$$

in which X is the fractional conversion of the solid reactant; and F_p is a shape factor that has the values of 1–3, respectively, for flat plates, infinite cylinders, and equi-dimensional solids (spheres, right cylinders, cubes, other regular polyhedrals, etc.).

6.2.3 Nucleation and Growth Kinetics – Avrami–Erofeev Equation

For many gas–solid reactions, the extent of conversion varies with time in a sigmoidal fashion. When solid particles are very small, typically less than a few hundred micrometers, their reaction with a gas typically goes through the period of nucleation and growth of nuclei up to a significant extent of conversion. The conversion-vs-time curves in these cases are characterized by an initial induction or incubation period, followed by an acceleratory stage in which the rate increases with conversion before falling off in the final stages. This behavior may be interpreted as arising from the production of nuclei at various distinct sites on the solid surface, followed by the growth of these nuclei and eventual merging of the nuclei, thereby reducing the

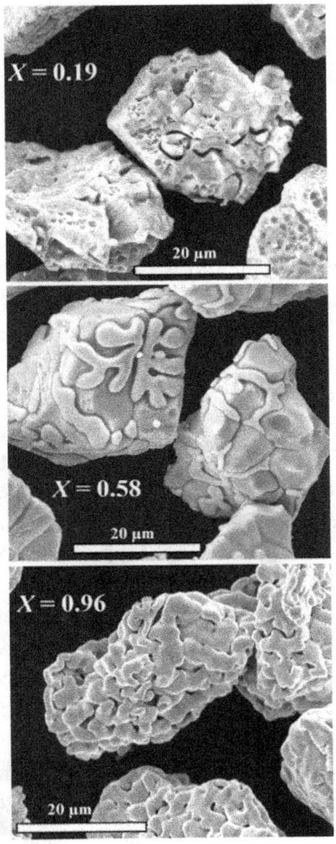

FIGURE 6.8 SEM micrographs of solid magnetite particles with product iron at different degreesof reduction. (Adapted from Elzohiery et al. [2017]).

area for further reaction. This process is illustrated in Figure 6.8 for the case of iron oxide reduction by hydrogen gas [Elzohiery et al., 2017], indicates that iron nuclei are formed from selected points on the surface of iron oxide, grow as the reduction continues, and eventually merge to form a product of iron layer. Crystals have fluctuating local energies from imperfections due to impurities, surfaces, edges, dislocations, cracks, and point defects [Boldyrev, 1986]. Such imperfections are sites for reaction nucleation since the reaction activation energy is minimized at these points.

The kinetics of many solid-state reactions have been analyzed by the nucleation models, specifically the Avrami–Erofeev model. The theoretical derivations of the various nucleation models have been summarized by Khawam and Flanagan [2006b] and, before them, by Sestak [1984]. The reader is referred to these informative references for detail.

Avrami [1939, 1940, 1941] and Erofeev [1946] derived, by following somewhat distinct approaches [Khawam and Flanagan, 2006b], the Avrami–Erofeev equation that describes the processes in which the rate of formation and growth of nuclei dominates the conversion of a solid. This widely used conversion-vs-time relationship is given by

$$\left[-\text{Ln }(1-X)\right]^{1/n} = k_{\text{app}} \cdot t \tag{6.103}$$

This equation is also called the JMAEK model, which stands for Johnson, Mehl, Avrami, Erofeev, and Kholmogorov, in recognition of the researchers who have contributed to its development [Galwey and Brown, 1999].

The value of n should be an integer in the range of 1–4, according to the theoretical derivation by Avrami [1939, 1940, 1941], which reflected the nature of the transformation in question. Sestak [1984] showed that for diffusion-dependent growth of the nuclei, the value of n may become 1.5 (one dimension), 2 (two dimensions), and 2.5 (three dimensions) with the diffusion coefficient (D) included in the reaction rate constant (k) [Khawam and Flanagan, 2006b]. However, it has become customary to regard it as an adjustable parameter that may be a non-integer [IUPAC, 1997], which must usually be determined by experiment. A value of n even less than unity has been found appropriate [Fan et al., 2021; Elzohiery, 2018; Cumbrera and Sánchez-Bajo, 1995; Wunderlich, 1976].

This equation was originally derived to describe the rate of decomposition of solids that produce another solid and a gaseous product [Khawam and Flanagan, 2006b] but has been applied to various other reaction systems, including the reactions of fine solid particles with gases. In such a case, the RHS typically contains the effect of fluid-reactant concentration driving force.

Sohn and coworkers [Sohn and Kim, 1984, 1987; Sohn and Won, 1985; Kim and Sohn, 2001b; 2002a; Wang and Sohn, 2013; Yuan et al., 2013; Yuan and Sohn, 2014; Clayton et al., 2014] considered the apparent rate constant in Eq. (6.103) to be a function of temperature, gaseous reactant partial pressure, and particle size. Thus, k_{app} was expressed as

$$k_{\text{app}} = b.k.f_d\left(d_p\right) \cdot f_p\left(p_A, p_C\right) \tag{6.104}$$

where b is the stoichiometry constant for reaction (Eq. 6.99).

Selected related examples are listed below:

- Solid-state reduction of metal oxide particles:

  Hydrogen reduction of magnetite and hematite concentrate particles; $n=1$ [Elzohiery et al., 2017; Fan et al., 2016, 2017; Chen et al., 2015].

  Hydrogen reduction of nickel oxide; $n=1.5$–4 depending on temperature [Sohn and Kim, 1984].

- Oxidation of metal particles:

  Oxidation of iron particles by water vapor; $n=1.26$ [Yuan et al., 2013].

  Oxidation of iron particles by O_2; $n=1$ [Yuan and Sohn, 2014].

  Oxidation of lower oxides; $n=1$ [Clayton et al., 2014].

6.2.4 Nucleation and Growth Kinetics – Prout–Tompkins Model (Also Called the Autocatalytic Model)

Another equation for the rate of a reaction dominated by the nucleation-and-growth process was developed by Prout and Tompkins [1944], which is given by

$$\frac{dX}{dt} = k_{app}X(1-X) \tag{6.105}$$

Separating variables and integrating, this equation yields

$$\text{Ln}\left(\frac{X}{1-X}\right) + C = k_{app}t \tag{6.106}$$

A disadvantage of this law is that it cannot handle the beginning of the conversion, that is, the time for $X=0$ is negative infinity. In other words, there must be some conversion of solid already at time zero. As a result, integration constant appears in the Prout–Tompkins Eq. (6.106). There is no general criterion for what the integration constant should be; however, Prout and Tompkins used t_{max}, which is the time needed for the maximum rate (i.e., the inflection point) and is approximately the same as $t_{1/2}$ [Khawam and Flanagan, 2006b], resulting in the following equation:

$$\text{Ln}\left(\frac{X}{1-X}\right) = k_{app}(t - t_{x=0.5}) \tag{6.107}$$

6.2.5 Solid-State Diffusion Model

For very fine particles, fluid diffusion through a porous product layer usually does not influence the overall rate, as mentioned earlier; however, if the product forms a dense layer, solid-state diffusion through the product layer, which is slow, often controls the overall rate. Such solid-state diffusion typically involves ionic species and depends on lattice defects [Welch, 1955]. In this case, the diffusion of fast-diffusing species controls the overall rate.

In diffusion-controlled reactions, the growth rate of the product layer is proportional to the flux of the diffusing species at the reaction interface. From this consideration, the conversion-vs-time relationship can be derived for various particle geometries. Such a relationship is mathematically the same as the reaction of a solid forming a porous solid product layer in which the pore diffusion of the gas species controls the overall rate. The reader is referred to Chapters 5 and 6 of a monograph on the subject Fluid–Solid Reactions by Sohn [2020a] for detailed derivations of the relationships. Here, we will simply present the resulting equations representing the conversion-vs-time relationship. Let us consider the following reaction:

$$A(\text{diffusing species}) + b.B(\text{solid reactant}) = d.D(\text{solid product}) \tag{6.108}$$

$$\left(\frac{F_p}{F_p-2}\right)\left\{\frac{Z-[Z-(Z-1)(1-X)]^{2/F_p}}{Z-1}-(1-X)^{2/F_p}\right\}$$
$$=\left(\frac{2bF_pD_A}{\rho_B}\right)\left(\frac{A_p}{F_pV_p}\right)^2(C_{Ao}-C_{Ae})t \tag{6.109}$$

where F_p is the particle shape factor (= 1, 2, or 3 for a flat plate, long cylinder, or an equidimensional particle such as a sphere), z is the volume ratio of the product solid and the reactant solid, X is the fractional conversion of the solid reactant, D_A is the solid-state diffusivity of the diffusing species, ρ_B is the molar density of the reactant solid B, A_p and V_p are the original surface area and volume of the solid reactant, and C_{Ao} and C_{Ae} are the concentration of A in the product layer at the outer surface and at equilibrium with B at the interface between B and D.

This equation was developed by Sohn [1980], based on a similar equation for the case of $z=1$ obtained by Sohn and Sohn [1980], and contains the effects of F_p and z in a single equation, rather than three separate equations for the three basic geometries, as formulated by others. Fine particles involved in gas–solid reactions typically have at least two dimensions that are small, that is, between spherical and cylindrical shapes, seldom plate-like. Thus, F_p has a value typically equal to 2 or 3. Some particles may have a shape between these two, or the progress of the reaction front is such that the above equation may fit the experimental data best with F_p containing a fractional value, which can be accommodated conveniently by the single Eq. (6.109). This approach for diffusion-controlled reaction of solids with non-basic shapes is described in detail in Section 5.3.7 of the abovementioned monograph Fluid–Solid Reactions by Sohn [2020a].

For the three basic geometries considered above, Eq. (6.109) yields the following:

$$ZX^2=\left(\frac{2bD_A}{\rho_BR_p^2}\right)(C_{Ao}-C_{Ae})t \quad \text{for } F_p=1, \tag{6.110}$$

$$\frac{[Z-(Z-1)(1-X)]\,\mathrm{Ln}[Z-(Z-1)(1-X)]}{Z-1}+(1-X)\,\mathrm{Ln}(1-X)$$
$$=\left(\frac{4bD_A}{\rho_BR_p^2}\right)(C_{Ao}-C_{Ae})t \quad \text{for } F_p=2 \tag{6.111}$$

$$3\left\{\frac{Z-[Z-(Z-1)(1-X)]^{2/3}}{Z-1}-(1-X)^{2/3}\right\}.$$
$$=\left(\frac{6bD_A}{\rho_BR_p^2}\right)(C_{Ao}-C_{Ae})t \quad \text{for } F_p=3 \tag{6.112}$$

where R_p is the half thickness of a flat particle and radius of a long cylinder or a sphere (half of the side for a cube and radius of an equidimensional cylinder, etc.). The relationships for $F_p=2$ and 3 when $z=1$ can be obtained from the above equations by applying L'Hospital's rule to yield the following more familiar forms:

$$X+(1-X)\mathrm{Ln}(1-X)=\left(\frac{4bD_A}{\rho_B R_p^2}\right)(C_{Ao}-C_{Ae})t \qquad \text{for } F_p = 2 \qquad (6.113)$$

$$1-3(1-X)^{2/3}+2(1-X)=\left(\frac{6bD_A}{\rho_B R_p^2}\right)(C_{Ao}-C_{Ae})t \qquad \text{for } F_p = 3 \qquad (6.114)$$

For spherical particles reacting under the same conditions, Jander [1927] formulated the following equation by making an assumption that is unnecessary and does not yield any simpler form:

$$\left[1-(1-X)^{1/3}\right]^2 = t/t_{X=1} \qquad (6.115)$$

This equation is only valid at low values of X and, thus, should be avoided in favor of Eq. (6.114) above.

6.2.6 Summary of Various Solid-State Reaction Kinetics Models

The foregoing models are chosen from the vast literature on the kinetics of solid-state reactions but are by no means exhaustive. They merely serve to illustrate the mechanisms involved in the development of some of the more widely used rate expressions. For other models of gas–solid reactions, the reader is referred to a monograph on the subject by Sohn [2020a].

Table 6.1 summarizes some of the most often used equations for the kinetics of reactions involving solids:

Most experimental data on gas–solid reaction systems have scattered even with the most carefully performed experiments. This is because of the nature of solid particles including the intrinsic variability of morphologies and crystal defects. Furthermore, the effect of particle size distribution is always present in a particle assemblage regardless of screening. Therefore, the choice of rate dependence on X out of this list, including fractional values of F_p, n, or m, solely by the goodness of fit is not very meaningful. Rather, the use of an expression that is based on a plausible reaction mechanism is recommended.

6.2.7 Analysis of Rate Data

To obtain the rate parameters, isothermal experiments are normally preferable because rate information at any stage of the reaction is obtained at selected temperatures. However, a non-isothermal method is advantageous for certain reactions and from several viewpoints. Isothermal experiments are difficult for reactions that just require heating because a significant extent of reaction can take place during heating of the reactant to a target temperature. Furthermore, a non-isothermal method can, in principle, yield the activation energy and the pre-exponential factor from a single experimental run. The disadvantages of this method include the increased inaccuracies of the rate parameters because a larger number of parameters (activation energy,

TABLE 6.1
Different Kinetics Models for Gas–Solid Reactions

$$\frac{dX}{dt} = kf(X);\ g(X) = \int_0^X \frac{dX}{f(X)} = kt;\ f(X) = 1/g'(X)$$

Model	f(X) [Differential]	g(X) [Integral]	Note
Topochemical Contraction Model (Shrinking-Core Model)			
$F_p = 1-3$	$F_p \cdot (1-X)^{(1-1/Fp)}$	$1-(1-X)^{1/Fp}$	F_p could be a non-integer
$F_p = 1$; 1 – D (Plates)	1	X	
$F_p = 2$; 2 – D (Cylinders)	$2(1-X)^{1/2}$	$1-(1-X)^{1/2}$	Shrinking cylinder
$F_p = 3$; 3 – D (Equidimensional)	$3(1-X)^{2/3}$	$1-(1-X)^{1/3}$	Shrinking sphere
Nucleation and Growth Kinetics – Avrami–Erofeev			
$n = 1-4$	$n(1-X)\cdot[-\mathrm{Ln}(1-X)]^{(1-1/n)}$	$[-\mathrm{Ln}(1-X)]^{1/n}$	n could be a non-integer
Nucleation and Growth Kinetics – Prout–Tompkins Model			
	$X(1-X)$	$\mathrm{Ln}\left(\frac{X}{1-X}\right)+\mathrm{C}$	
Reaction-Order Models			
0-th order	1	X	
1st order	$(1-X)$	$-\mathrm{Ln}(1\ X)$	
nth order ($n = 1-3$)	$(1-X)^n$	$(n-1)^{-1}[(1-X)^{-(n-1)}-1]$	n could be a non-integer
Power Law			
$m = 1-4$	$m \cdot X^{(1-1/m)}$	$X^{1/m}$	m could be a non-integer
Pore-Blocking Kinetics			
	$\lambda \cdot \exp(-X/\lambda)$	$\exp(X/\lambda)-1$	
Solid-State Diffusion Model			
$F_p = 1\text{–}3$	$\frac{1-F_p/2}{1-(1-X)^{(2/F_p-1)}}$	$1-\frac{F_p(1-X)^{2/F_p}-2(1-X)}{F_p-2}$	F_p could be a non-integer
$F_p = 1$; 1-D (Plates)	$(2X)^{-1}$	X^2	
$F_p = 2$; 2-D (Cylinders)	$[-\mathrm{Ln}(1-X)]^{-1}$	$(1-X)\cdot\mathrm{Ln}(1-X)+X$	
$F_p = 3$; 3-D (Spheres)	$[2(1-X)^{-1/3}-2]^{-1}$	$1\text{–}3(1-X)^{2/3}+2(1-X)$	

pre-exponential factor, and the dependence of the rate on conversion, that is, the form of the conversion function) are determined from fewer experiments. In addition, kinetics parameters obtained from a non-isothermal runs often depend on the temperature–time history used in the experiment and frequently show a lower sensitivity to varying conditions such as gas concentrations and solid conversion.

A number of different methods have been developed for analyzing kinetic data obtained by isothermal or non-isothermal experiments. The rate analysis can be performed by either model-fitting or model-free (iso-conversional) methods [Khawam and Flanagan, 2006a]. The latter methods allow one to calculate E_a values without a specific reaction model but cannot describe the conversion-vs-time relationship. A model, *f*(*X*) or *g*(*X*), is needed for a complete kinetic description of a reaction. Some of the more widely used methods for determining rate expressions are presented in the following subsections.

6.2.7.1 The Isothermal Method

Experimental data from isothermal runs can be analyzed by plotting conversion vs. time according to Eq. (6.93). The form of *g*(*X*) that yields the best straight line over the entire conversion range and the range of temperature of interest is determined. From such straight lines at various temperatures, the kinetic parameters and the complete rate expression can be obtained.

6.2.7.2 The Direct Differential Method – Linear *T-t* Program

In this and subsequent subsections, several methods for analyzing data from non-isothermal experiments are discussed.

Equation (6.97) can be rearranged into the following form:

$$\mathrm{Ln}\left[\frac{dX/dT}{f(X)}\right] = \mathrm{Ln}\left[\frac{k_0}{a}\right] - \frac{E}{RT} \tag{6.116}$$

A plot of the LHS vs. $1/T$ yields the activation energy from the slope and the pre-exponential factor from the intercept. The dX/dT term is best obtained by fitting the conversion data into a polynomial or other suitable function of T and taking the derivative of the function. The function *f*(*X*) that yields the best straight line is chosen as the conversion function. A disadvantage of this method is that the differentiation of experimental data is prone to difficulty and error.

6.2.7.3 Coats–Redfern Integral Method – Linear *T-t* Program

In this case, the conversion vs temperature relationship given by Eq. (6.98) is used, with the assumption that the integral on the right-hand side up to the initial temperature T_o is negligible as is usually the case in most non-isothermal experiments. (This issue has been addressed by Cai et al. [2009].) The integral on the RHS can be expanded into a sum of power series terms by making the substitution $u \equiv E/RT$ and using the following relationship [Coats and Redfern, 1964, 1965]:

$$\int_u^{\infty} e^{-u} u^{-b}\, du \simeq u^{1-b} e^{-u} \sum_{i=0}^{\infty} \frac{(-1)^i (b)^i}{u^{i+1}} \tag{6.117}$$

with $b=2$.

For large values of E/RT, which holds for most chemical reactions, $u \gg 1$ and the series on the RHS converge rapidly, and usually, only one or two terms are sufficient.

The approximate solution of the equation then leads to the following [Sohn and Emami, 2011; Sohn and Kim, 1980; Coats and Redfern, 1964]:

$$\frac{g(X)}{T^2}=\left[\frac{k_0}{a}\cdot\frac{R}{E}\left(1-\frac{2RT}{E}\right)\right]\cdot\exp\left(\frac{-E}{RT}\right) \tag{6.118}$$

Taking the natural logarithm of the above equation, we get

$$\mathrm{Ln}\left[\frac{g(X)}{T^2}\right]=\mathrm{Ln}\left[\frac{k_0}{a}\cdot\frac{R}{E}\cdot\left(1-\frac{2RT}{E}\right)\right]-\frac{E}{RT} \tag{6.119}$$

The appropriate form of $g(X)$ is first selected as the function which results in the best straight line for the plot. For $\frac{2RT}{E}\ll 1$, the plot of $\mathrm{Ln}\left[\frac{g(X)}{T^2}\right]$ vs. $\frac{1}{T}$ will yield E from the slope, and the intercept of the line is $\mathrm{Ln}\left[\frac{k_0}{a}\cdot\frac{R}{E}\right]$. To refine the results, the term $\frac{2RT}{E}$ can be included with the value of T at $X=0.5$ after an initial value of E is obtained without it [Sohn and Emami, 2011; Sohn and Kim, 1980; Coats and Redfern, 1964]. More accurate values of parameters can be obtained by iteration or non-linear regression analysis, but typically this is unnecessary [Sohn and Emami, 2011; Sohn and Kim, 1980]. The difficulty associated with the approximation of the temperature integral as well as the very rapid reaction rate in the upper range of temperature can be improved by using a new non-linear temperature program proposed by Sohn [2016], as discussed in Section 6.2.7.5.

As mentioned earlier, more reliable rate parameters are obtained by carrying out the experiments at various heating rates.

6.2.7.4 Iso-Conversional Methods

Typical kinetic data have enough scatter that more than one conversion functions give a reasonable degree of fit. The most serious consequence of this is that the values of activation energy calculated based on different conversions often vary greatly [Sohn, 2016]. Furthermore, for complex reactions for which the mechanism may vary with the progress of reaction, accompanied by a variation of the rate expression. Thus, it is recommended to determine the activation energy first by a method that does not involve the choice of a conversion function [Khawam and Flanagan, 2006]. Such a method calculates the activation energy at certain conversion values and is, thus, called the iso-conversional method [Zsako, 1996; Simon, 2004].

6.2.7.4.1 Using Isothermal Data

These methods are applied to experimental data collected under isothermal conditions at different temperatures and include the standard and Friedman's methods.

Standard iso-conversional method: This method [Vyazovkin, 2000] is based on taking the logarithm of the isothermal rate equation, Eq. (6.93), which yields

$$\mathrm{Ln}[g(X)]=\mathrm{Ln}[k_o]-\frac{E}{RT}+\mathrm{Ln}[t] \tag{6.120}$$

which can be rearranged and applied to different conversion values, as follows:

$$\text{Ln}[t_X] = \text{Ln}\left[\frac{g(X)}{k_o}\right]_X + \frac{E_X}{RT_X} \tag{6.121}$$

A plot of $\text{Ln}[t_X]$ vs. $1/T_X$ for each X gives activation energy from the slope at that conversion regardless of the type of the conversion function $g(X)$.

Friedman's iso-conversional method: This method [Friedman, 1964] is a differential method based on taking the logarithm of Eq. (6.92), which results in

$$\text{Ln}\left[\frac{dX}{dt}\right]_X = \text{Ln}\left[k_0 \cdot f(X)\right]_X - \frac{E_X}{RT_X} \tag{6.122}$$

A plot of $\text{Ln}\left[\frac{dX}{dt}\right]_X$ vs. $1/T_X$ at each X gives activation energy from the slope at that conversion regardless of the type of the differential conversion function $f(X)$.

This method carries the usual difficulties associated with a differential method of having to obtain derivatives of experimental data.

6.2.7.4.2 Using Non-Isothermal Data – Linear T-t Program

For the Coats–Redfern integral method, the iso-conversional method is applied by rearranging Eq. (6.119), as follows:

$$\text{Ln}\left[\frac{a}{T^2}\right] = \text{Ln}\left[\frac{k_0}{g(X)} \cdot \frac{R}{E}\left(1 - \frac{2RT}{E}\right)\right] - \frac{E}{RT} \tag{6.123}$$

A plot of the LHS vs. $1/T$ at a fixed value of conversion from runs made with different heating rates yields the activation energy for that conversion. This still requires neglecting the variation of the $\frac{2RT}{E}$ term. Activation energy values significantly dependent on conversion indicate variation in reaction mechanism as the reaction progresses.

6.2.7.5 Sohn's Non-Linear Temperature–Time Program

As shown above, the treatment of non-isothermally generated experimental data involve complicated mathematical manipulations because of the nature of the integral in Eq. (6.98) [Pérez-Maqueda et al., 2005]. Furthermore, a non-isothermal method has a disadvantage that information on the rate at lower extents of conversion can only be obtained at the lower range of temperature and information at higher conversion values can only be obtained at the higher range of temperature. This problem is particularly pertinent for the rate measurements of solid-state reactions. Furthermore, with the conventional linear temperature increase, the reaction rate increases very fast in the high-temperature range, and the determination of temperature dependence in this range often contains considerable uncertainty. Another factor to note is that the rate parameters obtained by isothermal and non-isothermal expressions may not be the same and thus may not be applicable interchangeably [Khawam and Flanagan, 2006].

Sohn [2016] proposed a non-linear temperature–time program that alleviates several difficulties associated with the linear temperature program. Firstly, the analysis of kinetic information in the high-temperature range of the measurement by this method is improved by slowing the rate of temperature increase in the high temperature range. Secondly, the new temperature program greatly facilitates the data analysis by providing a closed-form solution of the temperature integral and allows a convenient way to obtain the kinetic parameters by eliminating the need for the approximate evaluation of the temperature integral. Furthermore, this temperature program raises temperature rapidly in the low-temperature range, thus rapidly bypassing this range where the reaction rate is typically very slow. The rate analysis based on the new temperature program is robust and does not appear to be sensitive to errors in experimental measurements.

The new temperature–time program proposed by Sohn [2016] is given by

$$\frac{T_o}{T} = -\mathrm{Ln}\left(\alpha t + e^{-1/\beta}\right)^{\beta} \tag{6.124}$$

where α and β are constants and e is the base of the natural logarithm. This *T-t* program has the overall correct properties of describing $T = T_o$ at $t = 0$ and increasing T with time if $\beta > 0$. Since $T > 0$, this requires that α should satisfy the condition $\alpha t + e^{-1/\beta} < 1$ before the completion of the reaction. Thus, the parameters α and β should be so chosen that the reaction would be completed or reach sufficiently high conversion before the experimental run time reached the value given by

$$t_{max} = \frac{1 - e^{-1/\beta}}{\alpha} \tag{6.125}$$

From Eq. (6.124), we get

$$\alpha \cdot dt = \frac{T_o/\beta}{T^2}\exp\left(-\frac{T_o/\beta}{T}\right) \cdot dT \tag{6.126}$$

Substitution of the above equation into Eq. (6.92) yields

$$\frac{dX}{f(X)} = k_o \frac{T_o/\beta}{\alpha T^2}\exp\left(-\frac{T_o/\beta}{T}\right).\exp\left(-\frac{E}{RT}\right).dT \tag{6.127}$$

Unlike in the case of a linear temperature increase, this equation can be integrated analytically and has the following exact, closed-form integral:

$$\int_0^X \frac{dX}{f(X)} = g(X) = \frac{k_o}{\alpha(B+1)}\left\{\exp\left[-\left(1+\frac{1}{B}\right)\cdot\frac{E}{RT}\right] - \exp\left[-\left(1+\frac{1}{B}\right)\cdot\frac{E}{RT_o}\right]\right\} \tag{6.128}$$

where

$$B \equiv \frac{\beta E}{RT_o} \tag{6.129}$$

The second term on the RHS is negligible when T_o is so chosen that the rate of reaction is negligible until the temperature rises much higher, which is usually applicable in a non-isothermal experiment. (For the example discussed below, the second term on the RHS is smaller than the first by a ratio of $10^{-10} \sim 10^{-14}$ over 1% ~ 99% conversion.) Thus, without any loss of accuracy,

$$g(X) = \frac{k_o}{\alpha(B+1)} \cdot \exp\left[-\left(1+\frac{1}{B}\right)\cdot\frac{E}{RT}\right] \tag{6.130}$$

Taking the logarithm of both sides,

$$\mathrm{Ln}\left[g(X)\right] = \mathrm{Ln}\left[\frac{k_o}{\alpha(B+1)}\right] - \left(1+\frac{1}{B}\right)\cdot\frac{E}{RT} \tag{6.131}$$

Thus, plotting the LHS vs. $1/T$ for any given run will yield

$$E = -R\left(\mathrm{Slope} + T_o/\beta\right) \tag{6.132}$$

For the iso-conversional analysis at $X = X_i$ from runs made with several values of α, Eq. (6.131) is rearranged as

$$\mathrm{Ln}\left[\alpha_j\right] = \mathrm{Ln}\left(\frac{k_o}{B+1}\right) - \mathrm{Ln}\left[g(X_i)\right] - \left(1+\frac{1}{B}\right)\cdot\frac{E}{RT_{ij}} \tag{6.133}$$

where α_j is the jth value of α and T_{ij} is the corresponding temperature at X_i. Thus, a plot of the LHS vs. $1/T$ for any given X_i will yield

$$E_i = -R\left(\mathrm{Slope} + T_o/\beta\right) \tag{6.134}$$

as the activation energy at conversion X_i.

The advantages and usage of this new temperature–time program have been discussed by Sohn [2020a; 2016].

7 Development of Flash Ironmaking Technology – Reduction Kinetics of Magnetite Concentrate Particles

A critical question for the flash ironmaking process is whether iron oxide concentrate can be reduced to a high metallization degree within the few seconds of residence time available in a typical flash furnace. Early research performed using iron ore concentrates (~30 μm size) showed that 90%–99% reduction could be achieved within 1–7 seconds of residence time at temperatures of 1,300°C and higher [Choi and Sohn, 2010; Sohn et al., 2009a]. A much more comprehensive rate measurement was performed by additional laboratory testing.

Hydrogen is the main reducing agent in the flash ironmaking process that takes place at temperatures above 1,473 K, even when the partial combustion of natural gas producing H_2+CO mixtures is used. Most previous investigations were done on the reduction of iron ore particles at temperatures below 1,373 K, which are much lower than the temperature at which the flash ironmaking process is expected to operate [Sohn, 2007; Sohn et al., 2009a; Pinegar et al., 2012]. In addition, none of the rate equations reported in the literature contained all the information needed for reactor design relevant to the flash ironmaking process. Thus, a comprehensive kinetics expression applicable to the flash ironmaking conditions was necessary.

Such reduction kinetics of the gaseous reduction of magnetite concentrates were measured, and rate equations that contain the effects of process conditions in a flash ironmaking reactor were developed by Sohn and coworkers [Fan et al, 2016a, 2021a, b; Elzohiery et al., 2017, 2021; Fan et al., 2017; Chen et al., 2015a, b]. These conditions include temperature, partial pressures of hydrogen and water vapor, and particle size. Furthermore, data covering the entire fractional reduction degree (RD) vs. time, which had not been available for the conditions applicable to the flash ironmaking process, were generated in these investigations. An additional emphasis on the formulation was that such a rate expression be readily applicable to the development of an overall process model for the design and simulation of a reactor for the flash ironmaking technology.

DOI: 10.1201/9781003342199-7

7.1 MATERIALS

Magnetite concentrate is obtained by upgrading the low-grade taconite iron ore by first grinding to > 100 μm particle size and concentrating by wet magnetic separation. Thus, such fine concentrate particles are readily usable in a flash reduction process. Table 7.1 presents the chemical analysis of typical taconite concentrate and XRD analysis of the same. It is indicated that most of the iron ore concentrate is magnetite with a small amount of silica, which is the main gangue content, as shown in Figure 7.1.

The ore concentrate particle size is less than 100 μm with a mass average particle size of about 32 μm, and the total iron content ranges from 68% to 72%. The magnetite concentrate particles are irregularly shaped and non-porous, as shown in Figure 7.2.

TABLE 7.1
Chemical Composition of Magnetite Concentrate Particles

Component	Wt.%
Total iron	70.65
SiO_2	1.87
Al_2O_3	0.13
CaO	0.27
MgO	0.13
MnO	0.11
Cr_2O_3	0.11
K_2O	0.01
Na_2O	0.1
TiO_2	0.01
ZrO_2	0.03
P	0.01
S	0.02
C	0.24
Sr	0.01

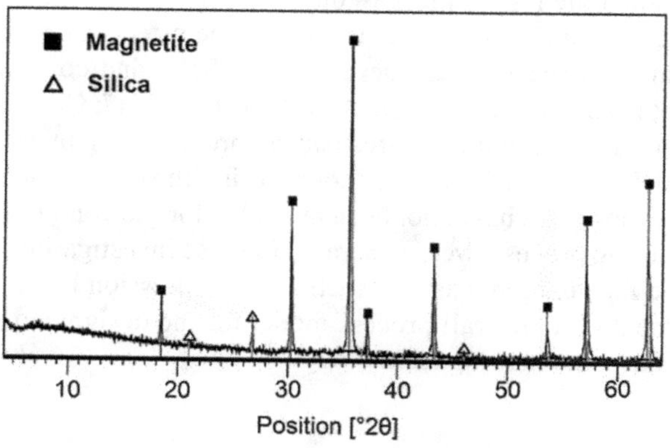

FIGURE 7.1 X-ray diffraction pattern of typical magnetite concentrate.

FIGURE 7.2 SEM micrograph for the unscreened magnetite concentrate particles.

7.2 EXPERIMENTAL APPARATUS

It is extremely difficult to record the weight change of the sample for such a rapid reaction and avoid diffusion resistance among particles in the sample layer in a crucible by using a customary TGA system. Thus, a high-temperature laminar-flow reactor (LFR) also called a drop-tube reactor (DTR) system was used [Choi and Sohn, 2010] for accurate determination of the reduction rate of individual concentrate particles. It uses a dilute particle-gas conveyed system to measure the chemical reaction rate of fine particles entrained in a reducing gas.

The gas composition should preferably be maintained essentially uniform throughout the reactor by a using large excess amount of the reducing gas so that the rate may be measured under known gas composition, rather than under varying concentrations, which complicates the kinetics determination and lowers the sensitivity of the measurements [Elzohiery et al., 2017, 2021; Wang and Sohn, 2012; Chen et al., 2015a, b; Choi and Sohn, 2010]. The LFR consists of five systems: a vertical tubular furnace housing a refractory tube, a pneumatic powder feeding system, a gas delivery system, a particle collection system, and an off-gas outlet system, as shown in Figure 7.3.

Typically, a vertical tubular furnace housing a refractory tube of 5–8 cm of ID and 150–200 cm of length is used.

The pneumatic powder feeding system is shown in Figure 7.4. This system consists of a syringe pump, a vibrator, carrier gas lines, a concentrate container vial, and particle delivery lines. In this system, the gas is fed to the particle container, which then carries particles from the top of the bed in the container, and delivered them to the reactor. The vibration helps in shaking the particles' surface inside the container and along the feed line, making it smoother to carry the particles with the gas. The syringe pump advances up at a pre-set speed, which controls the particle feed rate required for the experimental conditions.

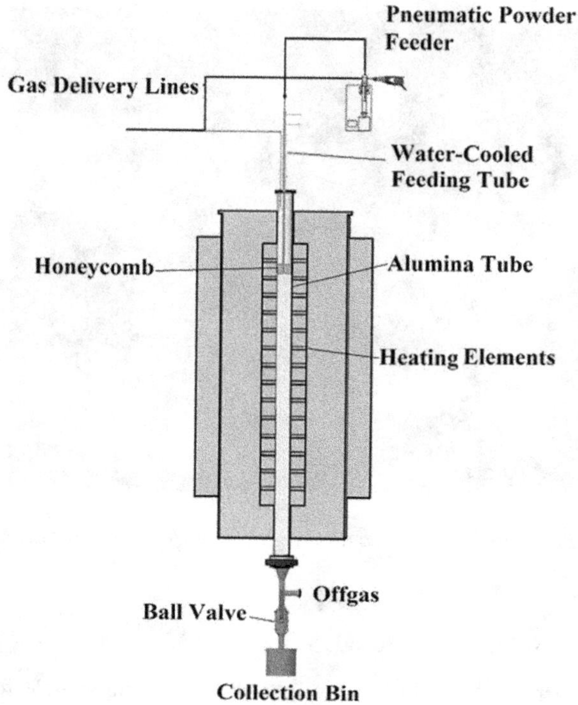

FIGURE 7.3 Schematic of a laminar-flow reactor.

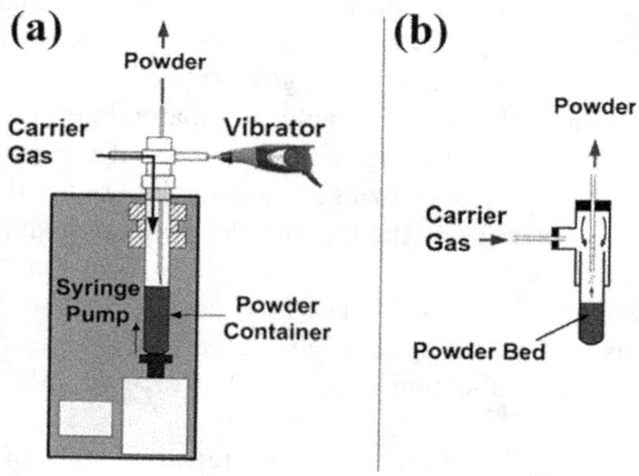

FIGURE 7.4 (a) The pneumatic powder feeding system. (b) Details of sample vial.

The pneumatic powder feeder is pre-calibrated at a constant carrier gas flow rate of 300 mL/min. The excess reducing gas mixture is discharged through a scrubbing system where the off-gas passes through water in a container to capture any particles in the off-gas and avoid blockage of the pipes. The reduced sample is collected in a collection bin that can be closed to keep the sample under an inert atmosphere.

The reactor is set at the target temperature, and the temperature is measured along the centerline of the reactor tube where most particles travel.

As shown in Figure 7.3, the concentrate is fed to the reactor through a water-cooled feeding tube by which the particles are injected directly into the isothermal zone of the reactor to ensure that the reaction starts at that temperature. A mullite honeycomb was fixed around the end of the feeding tube to ensure the mixing and preheating of the gas and straighten the flow into the reactor.

At the end of the isothermal zone, the temperature decreases rather gradually toward the end of the reactor tube and, thus, some additional reduction is expected to take place beyond the isothermal zone. Therefore, corrections to the rate analysis may have to be made to account for this additional amount of reduction according to the procedure described in Chen et al. [2015a] and Elzohiery et al. [2017], preferably corrected by the application of computational fluid dynamics (CFD) modeling of the reactor [Fan et al., 2016a, 2017, 2021].

7.3 EXPERIMENTAL PROCEDURE

Once the system was sealed and all parts were installed, the reactor was heated to the target temperature under a nitrogen flow of 500 mL/min. The heating rate was less than 2°C/minute to avoid any cracking of the alumina tubes. When the target temperature was reached, the reducing gas mixture was fed into the reactor at a predetermined flow rate.

The magnetite concentrate particles were fed into the reactor at a predetermined feeding rate using the pneumatic powder feeder. After running the experiment for a sufficient length of time, the syringe pump was stopped, but the vibrator was kept running for 1–2 minutes to ensure no particles were left in the delivery line. Then, the flow of the reducing gas mixture was stopped, and nitrogen was fed at a high flow rate to purge the system with five times the volume of the reactor tube.

Extra precautions were taken during the kinetics experiments to eliminate any possible source of error in the rate determinations. For the hydrogen reduction experiments, the amount of water vapor generated from the reduction was less than 4% of the total volume of the gas. It was proved that the re-oxidation of the collected sample in flash ironmaking process is quite slow, and the experiments showed that the particles were not pyrophoric [Yuan et al., 2013; Yuan and Sohn, 2014]. In the collector where product iron particles may be kept for up to 1 hour at around 400°C, the expected re-oxidation degree was around 0.02% [Yuan et al., 2013]. However, the stainless-steel collection bin with magnets was heated during the experiment to prevent any water condensation in the collected sample. After the experiment, the collection bin was closed and the collected sample was kept under an N_2 atmosphere. The bin was cooled in water, and then the reduced sample was collected and kept inside a sealed vial for further analyses by ICP, SEM, and XRD.

The reproducibility of the measurements was mostly within $\pm 4\%$ of the average RD [Elzohiery, 2018; Chen et al., 2015a, b].

7.4 FORMULATION OF REDUCTION KINETICS EQUATION

Here, the formulation of the rate equations for the reduction of magnetite by H_2, CO, and H_2+CO gas mixture is presented. The rate equations were formulated in the temperature ranges of 1,150°C–1,350°C and 1,350°C–1,600°C. The formation of wustite, which has a melting point of 1,370°C, and the silica content in the concentrate, cause the particles to fuse at temperatures higher than 1,350°C. Therefore, it is important to determine the kinetics at a temperature lower than 1,350°C separately from that at higher temperatures. The samples produced from the reduction experiments are analyzed by ICP method, which has been developed for the flash ironmaking process to determine the iron content [Mohassab et al., 2016].

7.4.1 Definitions of Parameters

7.4.1.1 Reduction Degree

The RD refers to oxygen that has been removed during the reduction. It was obtained based on the change in the mass of oxygen combined with iron in the particles before and after reduction. The amount of oxygen in the gangue materials that could be reduced and the weight change due to the possible volatile species in the small gangue contents were negligible and, thus, ignored.

The RD is calculated as follows:

$$\text{Reduction Degree}\,(\text{RD}\%) = \frac{W_O^i - W_O^t}{W_O^i} \times 100 = \frac{m_i - m_t}{m_i\,(\%\text{O})_i\,/\,100} \times 100 \qquad (7.1)$$

where W_O^i and W_O^t are the weights of oxygen combined with iron before and after reduction, respectively. m_i is the initial mass of the sample, m_t is the mass of the sample after reduction, and (%O) is the weight percent of oxygen in the ore before reduction.

Since only a small portion of the product is used for the analysis, m_t is unknown and, thus, Eq. (7.1) needs to be converted in terms of Fe contents. As the amount of iron does not change during the reduction, the iron balance yields

$$m_t\,(\%Fe)_t = m_i\,(\%Fe)_I \qquad (7.2)$$

where $(\%Fe)_i$ and $(\%Fe)_t$ are the weight percent of iron in the concentrate and the sample after reduction for time t, respectively. Combining Eqs. (7.1) and (7.2), the RD is obtained from the following:

$$\text{Reduction Degree (RD\%)} = \frac{1-(\%\text{Fe})_i\,/\,(\%\text{Fe})_t}{(\%\text{O})_i\,/\,100} \times 100 \qquad (7.3)$$

Therefore, the RD can be obtained from the iron contents in the samples before and after reduction as well as the removable oxygen content in the sample before reduction. (%Fe) is obtained by ICP analysis with an accuracy of ± 1%. (O%) is the removable oxygen in the concentrate.

The removable oxygen content of the magnetite concentrate (O%) was determined by reducing at least five random samples of the concentrate completely under pure hydrogen at 900°C in a TGA furnace. The amount of weight loss gave the oxygen content in the ore. The $(\%O)_i$ varied with the size fraction and the batch of the ore. Therefore, for each experimental set, $(\%O)_i$ was measured for the accurate determination of RD.

7.4.1.2 The Amount of Excess Reducing Gas

The reduction of magnetite to metallic iron proceeds through two steps where magnetite Fe_3O_4 is first reduced to FeO and then to metallic iron, Fe. The first reaction has a large equilibrium constant, whereas the last reaction is considerably limited by equilibrium.

In the case of reduction by H_2, the % excess hydrogen, defined as the percentage of the hydrogen in excess of the minimum amount required for reduction [Choi and Sohn, 2010], was used for designing the experiment. The minimum amount is the amount of hydrogen required to reduce the iron oxide plus the amount to be present to satisfy the equilibrium of the hydrogen reduction of FeO in the presence of the water vapor produced by the reduction reaction:

$$Fe_3O_4(s) + 4H_2(g) = 3Fe(s) + 4H_2O(g) \tag{7.4a}$$

$$FeO(s) + H_2(g) = Fe(s) + H_2O(g) \tag{7.4b}$$

$$\dot{n}_{H_2,\min} = \dot{n}_O + \frac{\dot{n}_{H_2O}}{K_H} \tag{7.5}$$

$$\%\ \text{Excess}\ H_2 = \frac{\dot{n}_{H_2,\text{supp.}} - \dot{n}_{H_2,\min}}{\dot{n}_{H_2,\min}}\% \tag{7.6}$$

where $\dot{n}_{H_2,\min}$ is the minimum moles of H_2 required for complete reduction, $\dot{n}_O$ is the atoms of removable O in the feed, $\dot{n}_{H_2O}$ is the moles of water vapor produced (from reduction and flame), and K_H is the equilibrium constant for the FeO reduction by H_2, Eq. (7.4b).

In the case of reduction by CO, the % excess CO is calculated based on the following equations:

$$Fe_3O_4(s) + 4CO(g) = 3Fe(s) + 4CO_2(g) \tag{7.7a}$$

$$FeO(s) + CO(g) = Fe(s) + CO_2(g) \tag{7.7b}$$

$$\dot{n}_{CO,\min} = \dot{n}_O + \frac{\dot{n}_{CO_2}}{K_C} \tag{7.8}$$

$$\%\ \text{Excess CO} = \frac{\dot{n}_{CO,\text{supp.}} - \dot{n}_{CO,\min}}{\dot{n}_{CO,\min}}\% \tag{7.9}$$

where $\dot{n}_{CO,min}$ is the minimum moles of CO required to completely reduce the iron oxide to iron including the amount for the equilibrium of FeO reduction, $\dot{n}_O$ is the atoms of removable O in the feed, $\dot{n}_{CO_2}$ is the moles of carbon dioxide produced, and K_C is the equilibrium constant for the FeO reduction reaction by CO, given in Eq. (7.7b).

The kinetics experiments were designed to have a reducing gas amount at more than five times the minimum amount required for reduction to ensure that the gaseous reactant concentration remains essentially constant over the entire reactor length for accurate determination of the kinetics.

7.4.1.3 Excess Driving Force

In the scale-up experiments, the term "excess driving force" (EDF) was used to describe the excess amount of hydrogen present in relation to water vapor compared with the equilibrium condition for the reduction of wüstite (FeO) [Pinegar et al., 2010, 2011]. The EDF in a gas mixture is thus defined by the following equation:

$$\mathrm{EDF} = \frac{\left(\dfrac{p_{H_2}}{p_{H_2O}}\right)_{\mathrm{actual}} - \left(\dfrac{p_{H_2}}{p_{H_2O}}\right)_{\mathrm{equ.}}}{\left(\dfrac{p_{H_2}}{p_{H_2O}}\right)_{\mathrm{equ.}}} = K_H\left(\dfrac{p_{H_2}}{p_{H_2O}}\right)_{\mathrm{actual}} - 1 \quad (7.10)$$

where $p_{H_2,\,\mathrm{actual}}$ and $p_{H_2O,\,\mathrm{actual}}$ are the partial pressures of hydrogen and water vapor in the actual gas mixture, respectively, and $p_{H_2,\,\mathrm{equ.}}$ and $p_{H_2O,\,\mathrm{equ.}}$ are the partial pressures of hydrogen and water vapor at equilibrium in the presence of wüstite and Fe.

7.4.1.4 Particle Residence Time

In the LFR, the duration of the reduction reaction of the iron ore concentrate particles was determined by the residence time (t) of particles in the reaction zone. The value of the residence time was calculated from the length of the reaction zone, which starts from the tip of the powder feeding tube, the linear velocity of the gas, and the terminal falling velocity of particles. Furthermore, Choi and Sohn [2010] that the fully developed state of the flow is reached in less than 5% of the hot zone length by applying the correlation obtained by Durst et al. [2005] given by the following equation:

$$L/D = \left[(0.619)^{1.6} + (0.0567\mathrm{Re})^{1.6}\right]^{1/1.6} \quad (7.11)$$

where L = length needed to reach a fully developed flow pattern, D = diameter of the tubular reactor, and Re = Reynolds number.

The particles' Reynolds number was always less than 0.1 in all the conditions which means that the use of Stokes' law for the terminal velocity calculation is valid. The residence time of the particles is calculated using the following equations [Choi and Sohn, 2010]:

$$u_t = d_p^{\,2} g\left(\rho_p - \rho_g\right)/18\mu \tag{7.12}$$

$$u_p = u_g + u_t \tag{7.13}$$

$$t = L/u_p \tag{7.14}$$

where all in consistent units, d_p is particle size, g is gravitational acceleration, ρ_p is particle density, ρ_g is gas density, μ is viscosity of gas, u_p is particle velocity relative to tube wall, u_g is the linear gas velocity, u_t is the terminal velocity of a falling particle, L is the length of the isothermal zone, and t is the particle residence time.

In the LFR, it was found that more than 90% of the particles flowed around the centerline in most of the length of the reactor [Avila et al., 2012; Lehto, 2007]. Fan [2019] also showed that magnetite particles fed with air traveled mostly in a narrow region near the reactor centerline, as shown in Figure 7.5. This was observed by a particle image velocimetry (PIV) system shown in Figure 7.6. Although most of the particles flow along the centerline, an error analysis was performed by Choi [2010]. It was found that even if the particles were spread over half of the radius of the cross section, the average residence time increased by only about 10%, a tolerable difference. Therefore, using the centerline velocity and temperature of the gas phase is

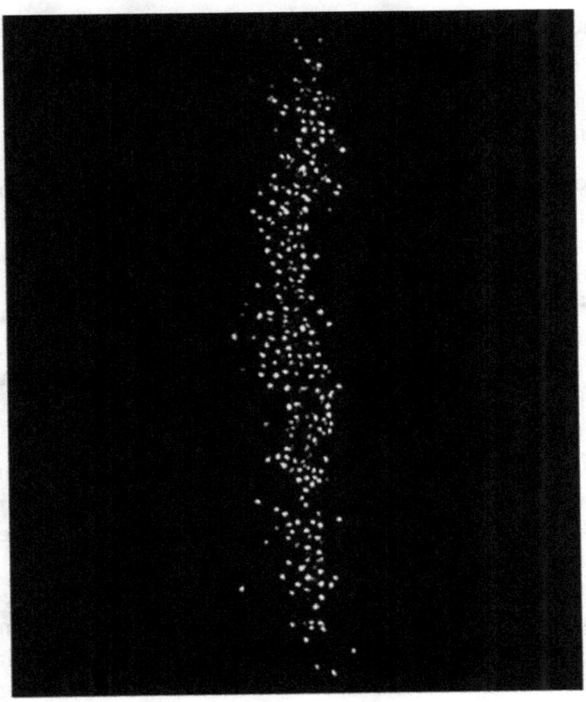

FIGURE 7.5 Image of particle dispersion inside the cold model of drop-tube reactor.

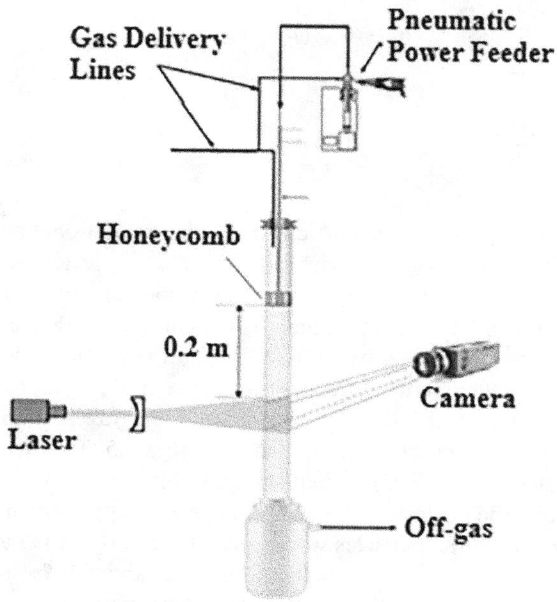

FIGURE 7.6 Schematic representation of the cold model system.

reasonable. Therefore, the centerline gas velocity was used as the linear gas velocity, which is twice the average gas velocity for laminar flow.

7.4.2 Selection of Rate Equation

The "particle kinetics," unaffected by mass transfer effects is the information that is most useful in predicting the rate under any operating conditions because pore diffusion and external mass transfer effects can readily be incorporated, if such phenomena have significant effects in a given operation.

It will be demonstrated by detailed calculation in Section 7.4.6 that the reduction rate would be much faster than the measured rates if the overall rate is controlled by external mass transfer and/or pore diffusion. Therefore, the overall reaction was rate-controlled by the reaction of individual particles and not affected by pore diffusion or external mass transfer.

The reduction rate is affected by reducing gas partial pressure, particle size, and temperature. It is noted that the barometric pressure at Salt Lake City is 0.85 atm (1 atm = 101.32 kPa), which was used as the total pressure in all the experimental design and analysis in this work.

It was determined [Elzohiery et al., 2017; Fan et al., 2017; Chen et al., 2015a, b; Choi and Sohn, 2010] that the nucleation and growth model [Avrami, 1939, 1940, 1941] best describes the reduction of magnetite concentrate particles in all the cases. This is consistent with the features in SEM micrographs shown in Figure 7.7 for the reduction of magnetite concentrate by H_2 gas indicating that iron nuclei form and grow with increasing the RD.

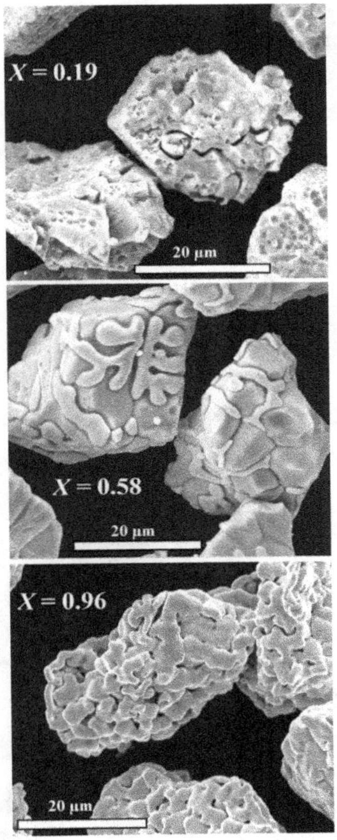

FIGURE 7.7 SEM micrographs of samples with different reduction degrees in the temperature range of 1,137°C–1,324°C.

The reduction of magnetite to iron in the temperature range of this work proceeds through the formation of wustite followed by Fe. However, it is extremely difficult to measure the intrinsic kinetics involving the formation of these phases for small particles going through a rapid reduction. In addition, different parts of a small irregular iron oxide particle react at different rates and thus different oxide phases coexist in the particle at any time, which was verified by the XRD analysis of quenched samples as shown in Figure 7.8.

As the main goal of the new ironmaking technology is to achieve high degrees of reduction, a global nucleation and growth rate expression for the overall reduction process was used in this work where the fractional conversion was expressed by the amount of oxygen removed from the magnetite concentrate. A rate equation that contains the effects of process variables in a flash ironmaking reactor was developed as follows:

$$\left[-\text{Ln}(1-X)\right]^{1/n} = k_{app} \times t \tag{7.15}$$

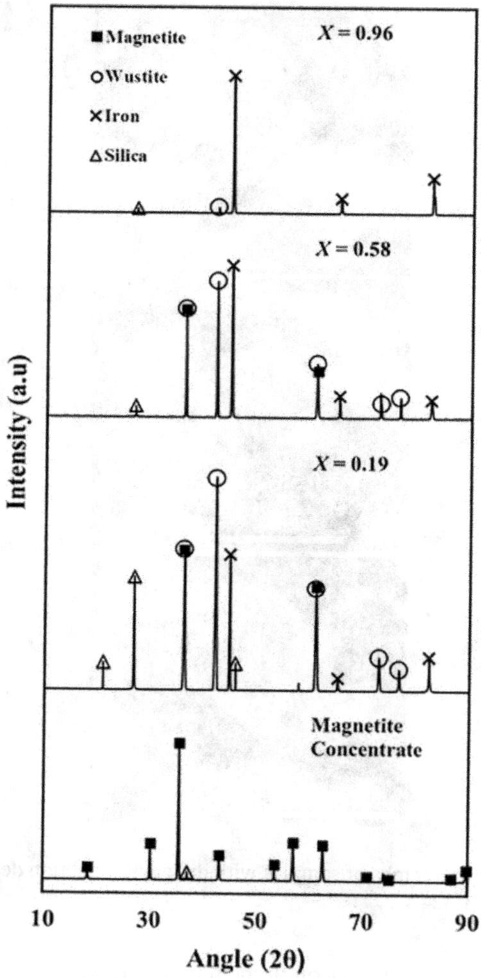

FIGURE 7.8 Coexistence of various iron oxide phases along with iron as indicated by XRDof quenched samples of different reduction degrees.

$$k_{app} = k \times f(p) \times h(d_p) \tag{7.16}$$

$$\left[-\text{Ln}(1-X)\right]^{1/n} = k \times f(p) \times h(d_p) \times t \tag{7.17}$$

where X is the fractional RD, n is the Avrami parameter, k_{app} is the apparent rate constant, k is the rate constant, $f(p)$ is the rate dependence on the gas partial pressures, $h(d_p)$ is the particle size dependence, and t is the particle residence time.

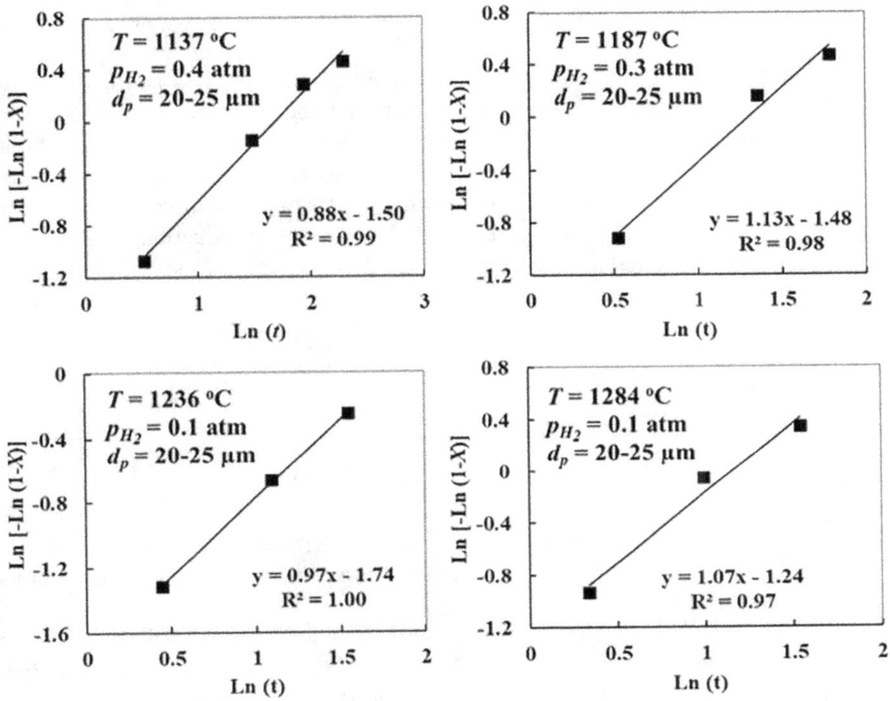

FIGURE 7.9 Determination of Avrami parameter. *t* in seconds.

The Avrami parameter is obtained using the following equation:

$$\text{Ln}\left[-\text{Ln}(1-X)\right] = n\ \text{Ln}(k_{app}) + n\ \text{Ln}(t) \tag{7.18}$$

For the experiments performed at the same temperature, particle size, and reducing gas partial pressure, Ln $\left[-\text{Ln}(1-X)\right]$ is plotted against Ln(t), and the Avrami parameter (n) is obtained from the slope. Some examples are shown in Figure 7.9. After making sure the obtained values of n are within a close range, its average is selected as the single final value and used in the determination of the effects of gas partial pressure and temperature.

7.4.3 Determination of the Reaction Order with Respect to Gas Partial Pressures

To determine the rate dependence on gas partial pressures, the apparent rate constant k_{app} from each set of experimental conditions was first calculated according to the following equation:

$$\frac{\left[-\text{Ln}(1-X)\right]^{1/n}}{t} = k\times f(p)\times h(d_p) = k_{app} \tag{7.19}$$

At the same temperature and particle size, the relation between k_{app} and $f(p)$ should be linear.

As stated previously, the last step in the reduction reaction, that is, the reduction of FeO, is an equilibrium-limited reaction. Therefore, the concentration driving force for the reaction will be lowered in the presence of the product gas. According to Sohn [2014], the exponents on the partial pressure of the reactant and product gases in the rate expression must satisfy the equilibrium relationship. The function representing the partial pressure dependence for H_2 reduction or CO reduction of iron oxide should, thus, be expressed as follows:

$$f(p) = f\left(p_{H_2}, p_{H_2O}\right) = p_{H_2}^{m} - \left(\frac{p_{H_2O}}{K_H}\right)^{m} \quad (7.20)$$

$$f(p) = f\left(p_{CO}, p_{CO_2}\right) = p_{CO}^{z} - \left(\frac{p_{CO_2}}{K_C}\right)^{z} \quad (7.21)$$

where K_H and K_C are the equilibrium constants for the reduction of FeO with H_2 and CO, respectively, and m and z are the reaction orders with respect to the partial pressure of H_2 and CO, respectively.

All experiments were designed to have an excess reducing gas of > 500% to ensure that the reducing gas concentration does not vary much in the reactor. Taking reduction by H_2 as an example, its high excess amount made the partial pressure of water vapor generated from reduction rather low (typically < 5%). However, even its small effect was corrected for by reducing the H_2 driving force. The water vapor partial pressure $\left(p_{H_2O}\right)$ was calculated according to the RD of the sample and the oxygen content of the magnetite concentrate as shown in the following equations:

$$n_{H_2O} = m_{Fe_3O_4} \times (\omega_O)_{(0)} \times X / M_O \quad (7.22)$$

$$p_{H_2O} = \frac{n_{H_2O}}{n_{H_2O} + n_{H_2} + n_{N_2}} \times 0.85 \text{ atm} \quad (7.23)$$

where n_{H_2O} is the number of moles of H_2O produced, $m_{Fe_3O_4}$ is the mass of magnetite fed, X is the RD, $(\omega_O)_{(0)}$ is the fraction of removable oxygen in magnetite before reduction, and M_O is the atomic mass of the oxygen.

7.4.4 Effect of Particle Size

Magnetite concentrate from a taconite ore of the Mesabi Range was used in this study. The concentrate particles were irregularly shaped and angular, as shown in Figure 7.10. The chemical composition for each size fraction is presented in Table 7.2. Most of the iron oxide was magnetite and the total iron content ranged from 68% to 71%. The concentrate was screened to 25–32, 32–38, and 45–53 μm fractions for kinetics measurements

FIGURE 7.10 SEM micrographs – Upper Row: SEM micrographs of screened particles; Lower row: SEM micrographs of polished sections (adapted from Wang and Sohn [2013]).

TABLE 7.2
Chemical Composition (wt%) of Magnetite Concentrate

	Wt% in Samples in the Designated Particle Size Range		
Component	**20–25 μm**	**32–38 μm**	**45–53 μm**
Total Fe	71.0	70.3	67.6
FeO	30.6	30.2	29.7
S	0.02	0.02	0.02
C	0.25	0.28	0.58
SiO_2	1.68	2.08	4.51
Al_2O_3	0.13	0.10	0.17
CaO	0.21	0.35	0.85
MgO	0.11	0.18	0.49
MnO	0.09	0.12	0.24
Cr_2O_3	0.10	0.07	0.11
Na_2O	0.10	0.10	0.10
TiO_2	0.01	0.01	0.01
ZrO_2	0.01	0.03	0.03

in a DTR. These size fractions represent most of the magnetite concentrate that will be used in the flash ironmaking process.

In the temperature range of 1,150°C–1,350°C in which particles remain solid, there was little effect of particle size on the reduction rate within the range tested (20–53 μm), as shown in Figure 7.11. The most likely reason for the absence of size

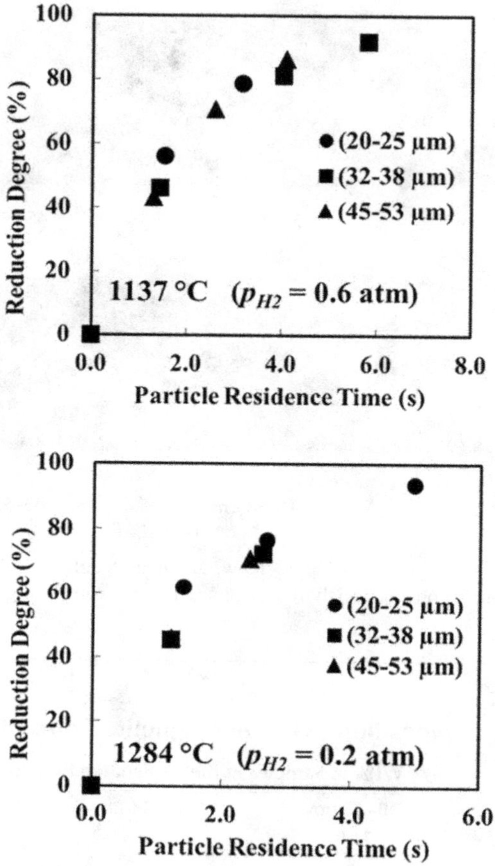

FIGURE 7.11 Determination of particle size dependence.

effect is that the particles experience thermal stress and develop cracks of similar dimensions when rapidly heated as they are fed into the reactor at a high temperature. The reaction rate then depends on the dimensions of the solid between the cracks, assumed similar for all screen sizes, and no longer on the original size.

In the temperature range of 1,350°C–1,600°C in which particles are fused and molten, as shown in Figure 7.12, the reduction rate was inversely proportional to the particle size. The reaction of a molten sphere took place at the surface of the droplet. Therefore, the reduction rate was inversely proportional to the particle size.

7.4.5 Determination of the Activation Energy

After obtaining the partial pressure dependence, the activation energy was obtained using the following equation derived from Eq. (7.19):

$$k = \frac{k_{app}}{f(p) \times h(d_p)} \tag{7.24}$$

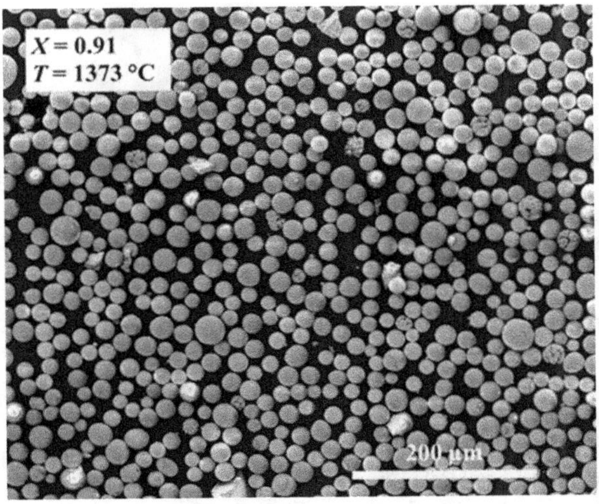

FIGURE 7.12 SEM micrograph for the product particles reduced at T=1373°C and p_{H_2}=0.1 atm to a reduction degree of 91%.

$$k = k_o \times e^{\frac{-E}{RT}} \tag{7.25}$$

$$\mathrm{Ln(k)} = \mathrm{Ln}(k_o) - \frac{E}{RT} \tag{7.26}$$

where E is the activation energy and R is the universal gas constant. Plotting Ln(k) against $1/T$ resulted in a straight line with a slope of $-E/R$ and the intercept with the y-axis equal to Ln(k_o).

7.4.6 Verification of the Absence of the Effects of Mass Transfer and Pore Diffusion

If the external mass transfer presents comparable resistances to the progress of reaction, the reduction rate of magnetite by hydrogen can be expressed as follows:

$$\rho_O V_p \frac{dX}{dt} = k_m A_p \left(C_{H_2} - C^e_{H_2} \right) \tag{7.27}$$

where all in consistent units, ρ_o is the number of atoms of removable oxygen contained in a unit volume of solid, V_p is the volume of the solid, X is the fractional degree of reaction, t is the reaction time, k_m is the mass transfer coefficient, A_p is the surface area of the solid, C_{H2} is the concentration of hydrogen in the reducing gas, and $C^e_{H_2}$ is the concentration of hydrogen at equilibrium with FeO and Fe.

Assuming V_p and A_p remain unchanged by reaction, Eq. (7.27) can be rewritten as

$$\frac{dX}{dt} = \frac{3k_m}{\rho_O r_p}\left(C_{H_2} - C^e_{H_2}\right) \tag{7.28}$$

where r_p is the radius of the solid particle, integration of Eq. (7.28) yields

$$X = \frac{3k_m}{\rho_O r_p}\left(C_{H_2} - C^e_{H_2}\right)t \tag{7.29}$$

In case of full reduction of the particle, $X = 1$, and the time required will be

$$t\big|_{X=1} = \frac{\rho_O \; r_p}{3k_m\left(C_{H_2} - C^e_{H_2}\right)} \tag{7.30}$$

The Re number for the particles based on the slip velocity was less than 0.1, and based on the particle size of ~25 μm, the Sherwood number (*Sh*) can be taken to be its lowest value of 2:

$$Sh = \frac{k_m\left(2r_p\right)}{D} = 2 \tag{7.31}$$

By combining Eqs. (7.30) and (7.31):

$$t\big|_{X=1} = \frac{\rho_O \; r_p^{\;2}}{3D\left(C_{H_2} - C^e_{H_2}\right)} \tag{7.32}$$

where D is the binary diffusion coefficient of the reactant and product gases.

A similar equation is applied in the case of reduction by CO where the concentration of CO is used. Table 7.3 presents the calculated values for reduction by H_2 or by

TABLE 7.3
The Calculated Values for the Effect of Mass Transfer Evaluation

$M_{wt.}$ of $Fe_3O_4 = 231.5$ g/mol	
Density of $Fe_3O_4 = 5.15$ g/cm³	
$\rho_o = 0.089$ atom/cm³	
$r_p = 12.5$ μm	$T = 1{,}673$ K
Reduction by H_2	**Reduction by CO**
$D_{H_2\text{-}H_2O} = 20.91$ cm²/s	$D_{CO\text{-}CO_2} = 4.02$ cm²/s
$p_{H_2} = 0.85$ atm	$p_{CO} = 0.85$ atm
K (H_2-FeO) = 0.829	K (CO-FeO) = 0.261
$p^e_{H_2} = 0.465$ atm	$p^e_{CO} = 0.674$ atm
$\left(C_{H_2} - C^e_{H_2}\right) = 3.19 \times 10^{-6}$ mol/cm³	$(C_{CO} - C^e_{CO}) = 1.28 \times 10^{-6}$ mol/cm³
$\boldsymbol{t_{(X=1)} = 7 \times 10^{-4}}$ **s**	$\boldsymbol{t_{(X=1)} = 9 \times 10^{-3}}$ **s**

CO at a temperature of 1,400°C, which is the average temperature of the tested range in the work. The calculations were made based on a total pressure of 0.85 atm, which is the atmospheric pressure in Salt Lake City, Utah, assuming the reduction occurs under pure reactant gas. The binary diffusion coefficient values were obtained from the literature [Reid et al., 1988].

Therefore, the complete reduction of Fe_3O_4 by either H_2 or CO would take would take several milliseconds if the reaction is limited by external mass transfer, which is much faster than the observation in the experiments where the reduction takes several seconds to progress.

To evaluate the possible effect of the diffusion of gases through the product layer, an estimation for the diffusion-controlled rate is performed using the rate equation for an isothermal, spherical solid with H_2 [Sohn, 2020a]:

$$\left(\frac{K}{1+K}\right)\frac{6D_e\ C_{H_2}}{\rho_O\ r_p^2}\ t = 1-3(1-X)^{\frac{2}{3}}+2(1-X) \quad (7.33)$$

where K is the equilibrium constant and D_e is the effective diffusivity of H_2-H_2O in porous iron. The time required for the complete conversion ($X=1$) can be calculated as:

$$t\big|_{X=1} = \left(1+\frac{1}{K}\right)\frac{\rho_O\ r_p^2}{6D_e\ C_{H_2}} \quad (7.34)$$

The effective diffusivity for H_2-H_2O in porous iron reduced (D_e) needs to be calculated at a temperature of 1,400°C. The value of D_e was obtained by Turkdogan et al. [1971] at 1,200°C for iron reduced by H_2-H_2O, and it was 4 cm²/s. Assuming that Fe_3O_4 was reduced by pure H_2 gas at 1,200°C, Table 7.4 shows the results for these calculations.

The D_e for CO in porous iron is reported to be three to four times lower than that for H_2 [Turkdogan et al., 1971; Moon et al., 1998; Tsay et al., 1976].

The experimentally observed rate of reduction of Fe_3O_4 by CO within the temperature range of 1,200°C–1,600°C is much slower than that under the control of CO diffusion or external mass transfer.

TABLE 7.4
The Calculated Values for the Effect of Pore-Diffusion Control

$M_{wt.}$ of Fe_3O_4 = 231.5 g/mol	
Density of Fe_3O_4 = 5.15 g/cm³	
ρ_o = 0.089 atom/cm³	
r_p = 12.5 μm	T = 1,473 K
Reduction by H_2	
D_e = 4 cm²/s	
p_{H_2} = 0.85 atm	C_{H_2} = 7.04 × 10⁻⁶ mol/cm³
K (H_2-FeO) = 0.787	
$t_{(X=1)}$ = 1.87 × 10⁻³ s	

From these calculations, it is concluded that the reduction of the magnetite concentrate is not controlled by either mass transfer or pore diffusion, as the reduction would be completed in millisecond compared with several seconds observed in this study. Therefore, the rate equations formulated based on the experimental data represent those of the intrinsic chemical reactions unaffected by mass transfer.

7.5 RESULTS OF RATE MEASUREMENTS AND RATE EQUATIONS

The rate determination experiments needed to be divided into two temperature ranges; 1,150°C–1,350°C and 1,350°C–1600°C, around the point of fusion and melting of the concentrate particles, which cover the entire temperature range expected for the flash ironmaking process. No particle agglomeration was experienced at any temperature tested, as shown in Figure 7.13. It is seen in Figure 7.13b that some

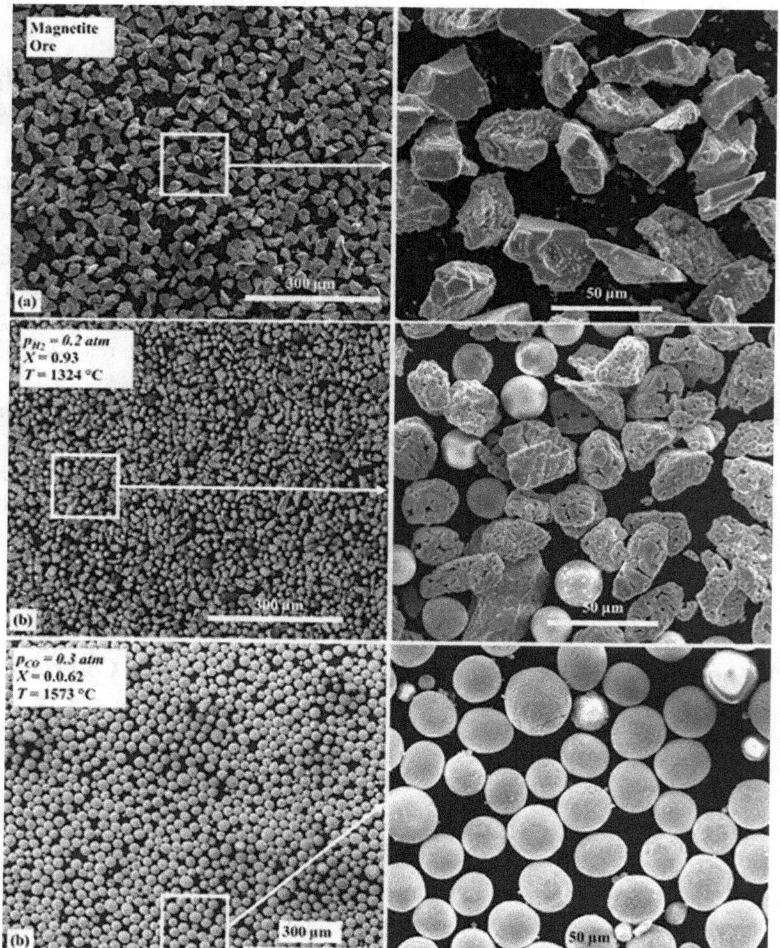

FIGURE 7.13 SEM micrograph of (a) concentrate particles and (b) reduced particles showing no agglomeration even at high temperatures such as 1,324°C and 1,573°C.

particles had spherical shapes, indicating that these particles experienced fusion. These particles have higher oxide gangue contents compared with most of the particles as verified by EDS. These fused particles were observed only at 1,324°C and not at lower temperatures. At higher temperatures, SEM images showed that all the particles were spherical indicating the melting of all the particles. As not all the particles were melted at 1,324°C, 1,350°C was chosen to be the separating temperature between solid-state reduction and reduction in molten state.

7.5.1 Reduction by Hydrogen: *Temperature Range of 1,150°C–1,350°C*

The rate dependence on p_{H_2} was tested in the range of 0.1–0.6 atm by using N_2 and by assuming a half-order or a first-order dependence. The experiments were conducted under a large excess of reducing gas to keep its partial pressure constant throughout the reactor, as mentioned earlier. However, considering the strong equilibrium limitation of the final stage of iron oxide reduction, wustite to iron, the effect of the small amount of H_2O was included by using an average value of the concentration driving force between the inlet and the exit of the reactor. The p_{H_2O} value at the outlet of the reactor was calculated based on the RD of the particles. The appropriate average driving force for a half-order reaction is the arithmetic mean and that for a first-order reaction is the log-mean, as described in Wang and Sohn [2013].

Thus, the function $f(p_{H_2}, p_{H_2O})$ in Eq. (7.20) was tested with $m = 1/2$ and 1, as follows:

$$f\left(p_{H_2}, p_{H_2O}\right)_{\text{avg.}} = \left[p_{H_2}{}^{1/2} - \left(\frac{p_{H_2O}}{K_H}\right)^{1/2}\right]_{am} \tag{7.35}$$

$$f\left(p_{H_2}, p_{H_2O}\right)_{\text{avg.}} = \left[p_{H_2} - \left(\frac{p_{H_2O}}{K_H}\right)\right]_{lm} \tag{7.36}$$

Using the obtained Avrami parameter and plotting k_{app} against $f\left(p_{H_2}, p_{H_2O}\right)_{\text{avg.}}$ at the same temperature and particle size, the value of m that best fits all the experimental data was determined. The results with $m = 1$ best represented the partial pressure dependence for all the experimental data. Figure 7.14 shows an example of the determination of the H_2 partial pressure dependence at 1,187°C and 1,284°C.

The residence time of the particles varied from 1 to 9 seconds with an interval of 2 seconds, which was determined to be enough to obtain the required change in reduction. More than 90% RD was obtained at 1,236°C and residence time of 4 seconds with H_2 partial pressure = 0.3 atm, which are typical conditions that will be available in a flash reactor.

The complete rate equation obtained for the reduction of magnetite with hydrogen gas in the temperature range of 1,150°C–1,350°C is

$$\left[-\text{Ln}(1 - X)\right] = 8.65 \times 10^6 \times e^{\left(\frac{-193{,}000}{RT}\right)} \times \left[p_{H_2} - \left(\frac{p_{H_2O}}{K_H}\right)\right]_{lm} \times t \tag{7.37}$$

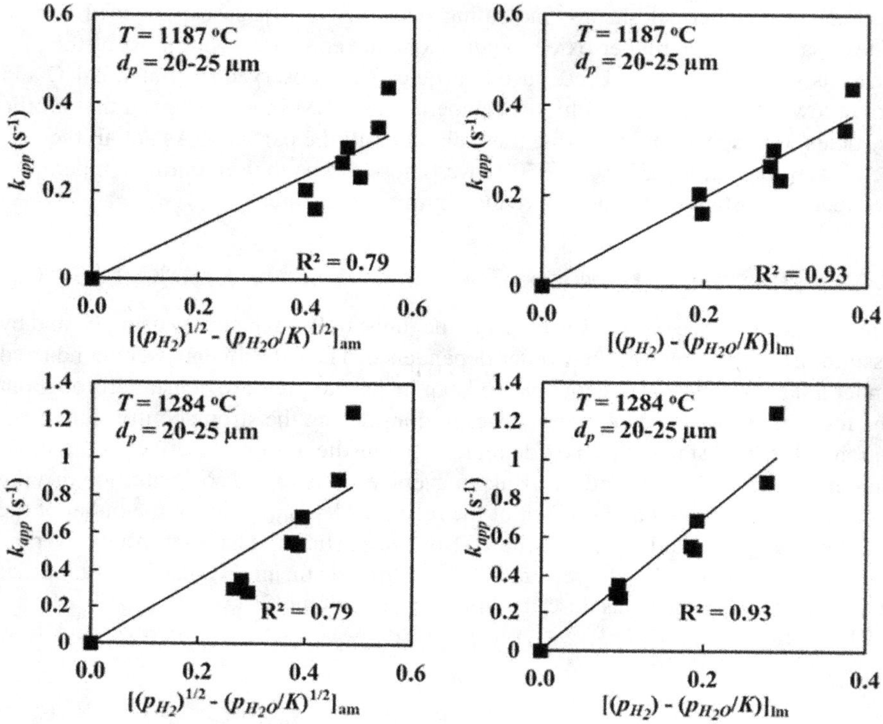

FIGURE 7.14 Relationship between k_{app} and $f\left(p_{H_2}, p_{H_2O}\right)_{avg.}$ at 1,187°C and 1,284°C with particle size fraction 20–25 µm, indicating that $m=1$ best fits the data. p in atm. (Adapted from Elzohiery et al. [2017]).

where R is 8.314 J/(mol·K), T is in K, p is in atm, and t is in seconds. [This kinetics equation is adequate for most purposes. However, this equation has been improved to obtain a more accurate X-t relationship, as will be shown in Section 7.6, by the application of a CFD analysis. It is recommended that the relation presented therein should be used for all purposes.]

This rate equation (Eq. 7.37) was used to calculate the RD of the experiments according to their conditions and compared with the experimental RD. A good agreement was obtained as shown in Figure 7.15a where all the experimental points are plotted. Figure 7.15b shows the change in RD with particle residence time for selected experiments at different experimental conditions with the calculated curve using the developed model. The points are the experimental RD while solid lines represent the RD calculated by the developed rate equation.

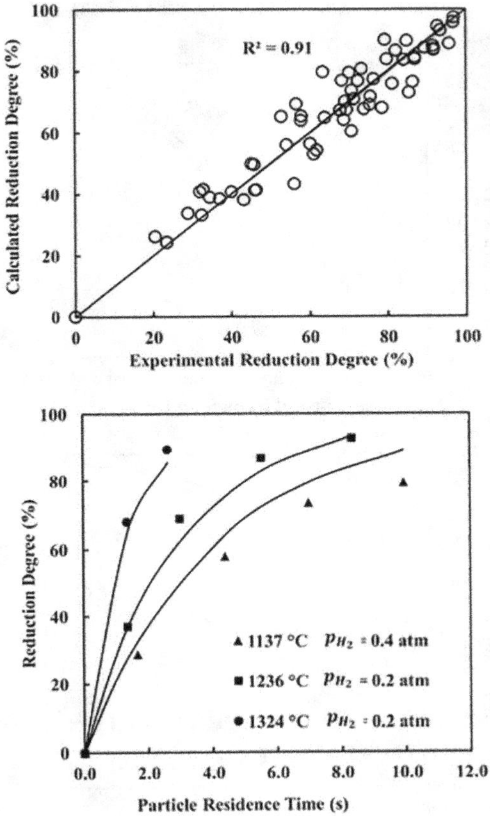

FIGURE 7.15 (a) Comparison between the calculated reduction degree (%) by the developed rate equation and the experimental results. (b) Experimental reduction degree (%) vs. particle residence time where the solid lines represent the reduction degree calculated by the developed rate equation.

7.5.2 Reduction by Hydrogen: *Temperature Range of 1,350°C–1,600°C*

The complete rate equation obtained for the reduction of magnetite with hydrogen in the temperature range of 1,350°C–1,600°C is

$$\left[-\text{Ln}(1-X)\right] = 2.5\times10^{7}\times e^{\left(\frac{-170,000}{RT}\right)}\times\left(d_p\right)^{-1}\times\left[p_{\text{H}_2}-\left(\frac{p_{\text{H}_2\text{O}}}{K_H}\right)\right]_{lm}\times t \quad (7.38)$$

where R is 8.314 J/(mol·K), T is in K, p is in atm, d_p in μm, and t is in seconds. [This kinetics equation is adequate for most purposes. However, this equation has been

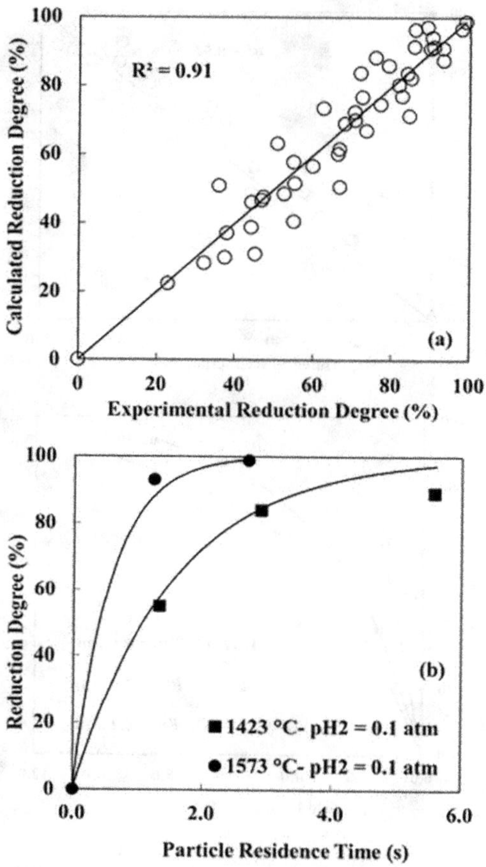

FIGURE 7.16 (a) Comparison between the reduction degree (%) calculated by the developed model and the experimental results. (b) Experimental reduction degree (%) vs. particle residence time where the solid lines represent the reduction degree calculated by the developed rate equation.

improved to obtain a more accurate *X-t* relationship, as will be shown in Section 7.6, by the application of a CFD analysis. It is recommended that the relation presented therein be used for all purposes.]

It is noted that the activation energy in the two temperature ranges was close. This rate equation (Eq. 7.38) was used to calculate the RD of the experiments according to their conditions and compared with the experimental RD. A good agreement was obtained as shown in Figure 7.16.

7.5.3 Reduction by Carbon Monoxide: *Temperature Range of 1,150°C–1,350°C*

The reduction of magnetite concentrate by CO was determined to be much slower than the reduction by H_2 gas. For example, the highest RD at 1,332°C and 8.5 s residence time under pure CO atmosphere was 60% compared with full reduction at the same temperature and residence time and $p_{H_2} = 0.2$ atm.

The complete rate equation obtained for the reduction of magnetite with CO gas in the temperature range of 1,150°C–1,350°C is

$$\left[-\text{Ln}(1-X)\right]^{1/0.5} = 1.21\times10^{13}\times e^{\left(\frac{-430{,}000}{RT}\right)}\times\left[p_{CO}-\left(\frac{p_{CO_2}}{K_C}\right)\right]_{lm}\times t \qquad (7.39)$$

where R is 8.314 J/(mol·K), T is in K, p is in atm, and t is in seconds. (This kinetics equation is adequate for most purposes. However, this equation has been improved to obtain a more accurate X-t relationship, as will be shown in Section 7.6, by the application of a CFD analysis. It is recommended that the relation presented therein can be used for all purposes).

This rate equation (Eq. 7.39) was used to calculate the RD of the experiments according to their conditions and compared with the experimental RD. A good agreement was obtained as shown in Figure 7.17a where all the experimental points

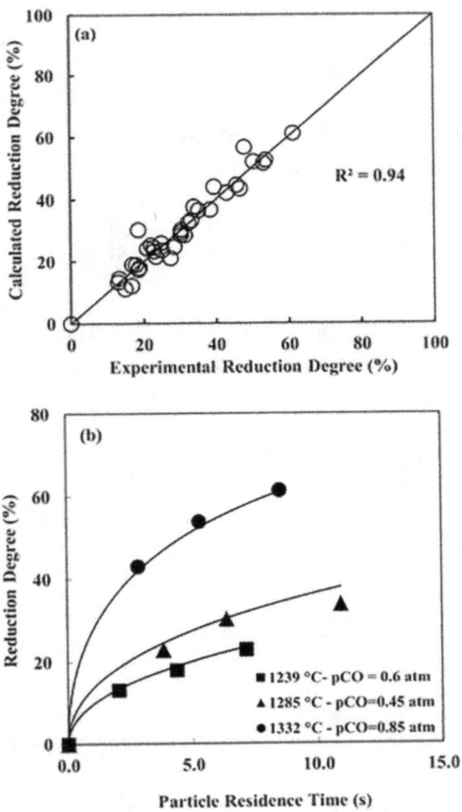

FIGURE 7.17 (a) Comparison between the reduction degree (%) calculated by the developed rate equation and the experimental results. (b) Experimental reduction degree (%) vs. particle residence time where the solid lines represent the reduction degree calculated by the developed rate equation.

are plotted. Figure 7.17b show the change in RD with particle residence time for selected experiments at different experimental conditions.

A note on the low value of the Avrami parameter $n=0.5$ may be in order. The evolution of a new phase in the reaction depends on the nucleation rate and the growth rate. Although the Avrami exponent is usually within 1–4, many researchers have reported an Avrami exponent of less than 1 (mostly 0.5). Wunderlich [1976] showed that the Avrami parameter can have values ranging from less than 1 to above 6. He mentioned that the Avrami exponent can be less than 1 in the case of limiting growth and athermal nucleation. The Avrami parameter (n) depends on the nucleation rate and growth mechanism, and $n=0.5$ was attributed to a fast nucleation rate where all the nuclei form first while the growth is diffusion controlled in one dimension causing thickening of large plates after the edge impingement [Christian, 1975; Altúzar et al., 1991; Cumbrera et al., 1995; Starink, 1997].

7.5.4 Reduction by Carbon Monoxide: *Temperature Range of 1,350°C–1,600°C*

The rate reduction by CO was slower than by H_2, similar to the lower temperature range. RD achieved was 87% at residence time 7.6 seconds, 1,573°C and $p_{CO}=0.6$ atm, whereas complete reduction was achieved at the same temperature and residence time with hydrogen at 0.1 atm.

The complete rate equation obtained for the reduction of magnetite with CO gas in the temperature range of 1,350°C–1,600°C is

$$\left[-\text{Ln}(1-X)\right]^{1/0.5}=2.39\times10^{3}\times e^{\left(\frac{-73{,}000}{RT}\right)}\times\left(d_p\right)^{-1}\times\left[p_{CO}-\left(\frac{p_{CO_2}}{K_C}\right)\right]_{lm}\times t \qquad (7.40)$$

where R is 8.314 J/(mol·K), T is in K, p is in atm, d_p in μm, and t is in seconds. [This kinetics equation is adequate for most purposes. However, this equation has been improved to obtain a more accurate X-t relationship, as will be shown in Section 7.6, by the application of a CFD analysis. It is recommended that the relation presented therein be used for all purposes.]

This rate equation (Eq. 7.40) was used to calculate the RD of the experiments according to their conditions and compared with the experimental RD. A good agreement was obtained as shown in Figure 7.18.

7.5.5 Reduction by H_2 + CO Mixtures: *Temperature Range of 1,150°C–1,350°C*

Experimental conditions were designed in a way to mimic the most used industrial ratios of the syngas: H_2:CO=2:1, 1:1, and 0.5:1 (molar ratio). The rate equations for the gaseous reduction of magnetite concentrate particles by single component gases were presented above. These rate equations were used for obtaining the rate expression of the reduction by H_2+CO mixtures. A simple additive relationship between the reduction rates by the component gases was initially assumed, but under H_2+CO mixtures apparent synergistic effects of the two gases were obtained.

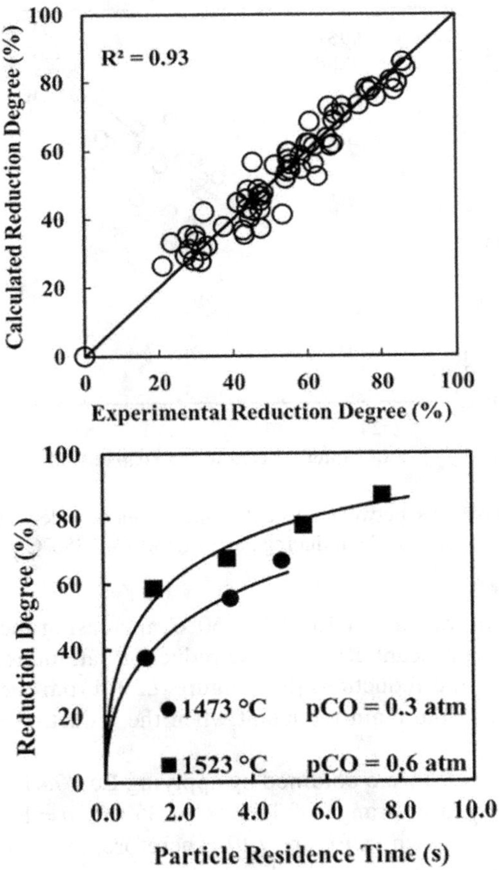

FIGURE 7.18 (a) Comparison between the reduction degree (%) calculated by the developed rate equation and the experimental results. (b) Experimental reduction degree (%) vs. particle residence time where the solid lines represent the reduction degree calculated by the developed rate equation.

Although H_2 was the main contributor to the reduction, the reduction rate was enhanced by the presence of CO when the reduction was under H_2+CO mixtures. A correlation equation in which the contribution of CO is kept the same as in the reduction by CO alone, and a coefficient was multiplied to the H_2 rate was assumed for reduction by an H_2+CO mixture, as follows:

$$\left.\frac{dX}{dt}\right|_{H_2+CO} = a \cdot \left.\frac{dX}{dt}\right|_{H_2} + \left.\frac{dX}{dt}\right|_{CO} \tag{7.41}$$

where $\left.\frac{dX}{dt}\right|_{H_2}$ and $\left.\frac{dX}{dt}\right|_{CO}$ are the instantaneous rate of reaction obtained from the developed reaction rates individually by H_2 and by CO, respectively.

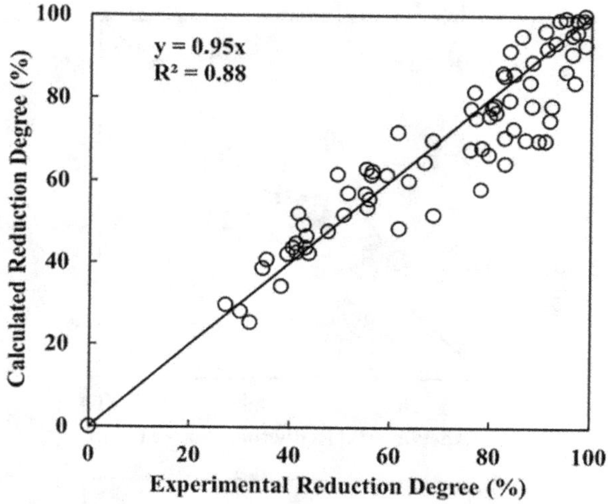

FIGURE 7.19 Comparisons between the calculated reduction degrees vs. experimental results with H_2+CO mixtures as the reducing gas in 1,150°C–1,350°C.

In this temperature range of 1,150°C–1,350°C, increasing the CO partial pressure did not have a significant effect on the reduction rate indicating that the CO did not contribute to the reduction. [See Figure 7.20] From these results, it was concluded that H_2 was the main reductant when the reduction is performed in a mixture gas.

The best-fit values of a were obtained by applying Eq. (7.41) to the experimental data. In the temperature range of 1,150°C–1,350°C, $a=1.3$ was determined to give the best fit as shown in Figure 7.19. Therefore, the rate equation for the reduction by H_2+CO gas mixtures in the temperature range of 1,150°C–1,350°C is given by

$$\left.\frac{dX}{dt}\right|_{H_2+CO} = 1.3\cdot\left.\frac{dX}{dt}\right|_{H_2} + \left.\frac{dX}{dt}\right|_{CO} \quad 1{,}150°C \le T < 1{,}350°C \tag{7.42}$$

$$\left.\frac{dX}{dt}\right|_{H_2} = 8.65\times10^{6}\times e^{\left(\frac{-193{,}000}{RT}\right)}\times\left[p_{H_2}-\left(\frac{p_{H_2O}}{K_H}\right)\right]_{lm}\times(1-X) \tag{7.43}$$

$$\left.\frac{dX}{dt}\right|_{CO} = 6.05\times10^{12}\times e^{\left(\frac{-430{,}000}{RT}\right)}\times\left[p_{CO}-\left(\frac{p_{CO_2}}{K_C}\right)\right]_{lm}\times\frac{-(1-X)}{\mathrm{Ln}(1-X)} \tag{7.44}$$

[This kinetics equation is adequate for most purposes. However, this equation has been improved to obtain a more accurate X-t relationship, as will be shown in Section 7.6, by the application of a CFD analysis. It is recommended that the relation presented therein be used for all purposes.]

7.5.6 Reduction by H_2 + CO Mixtures: Temperature Range of 1,350°C–1,600°C

In the temperature range of 1,350°C–1,600°C, CO enhanced the reduction rate when added to H_2. However, the enhancement did not vary with an increase in the CO partial pressure, as shown in Figure 7.20. Adding CO enhanced the RD at higher temperatures but increasing the CO partial pressure did not have a significant effect on the reduction rate.

The effect of CO in the temperature range (1,350°C–1,600°C) was more significant than in the lower temperature range. Using the same method of fitting the data. The factor a was found to be 2 which gave a good agreement with the experiments, as shown in Figure 7.21. Therefore, the rate equation for the reduction by H_2+CO gas mixtures in the temperature range of 1,350°C–1,600°C is given by

$$\left.\frac{dX}{dt}\right|_{H_2+CO} = 2\cdot\left.\frac{dX}{dt}\right|_{H_2} + \left.\frac{dX}{dt}\right|_{CO} \quad 1{,}350°C \le T < 1{,}600°C \tag{7.45}$$

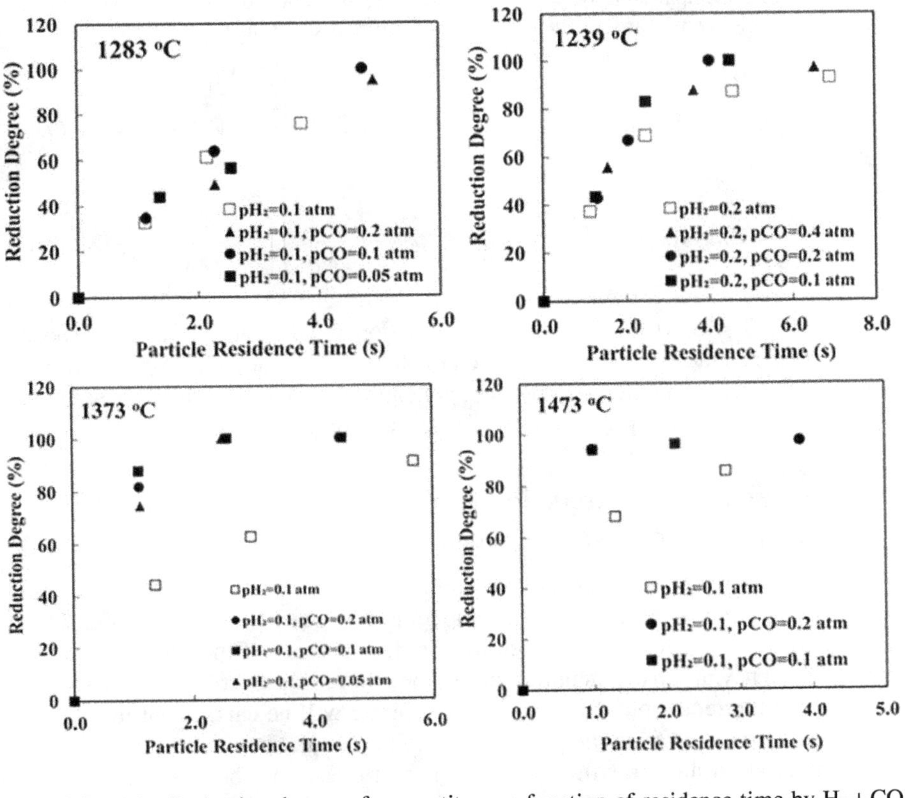

FIGURE 7.20 Reduction degree of magnetite as a function of residence time by H_2+CO mixtures under different conditions.

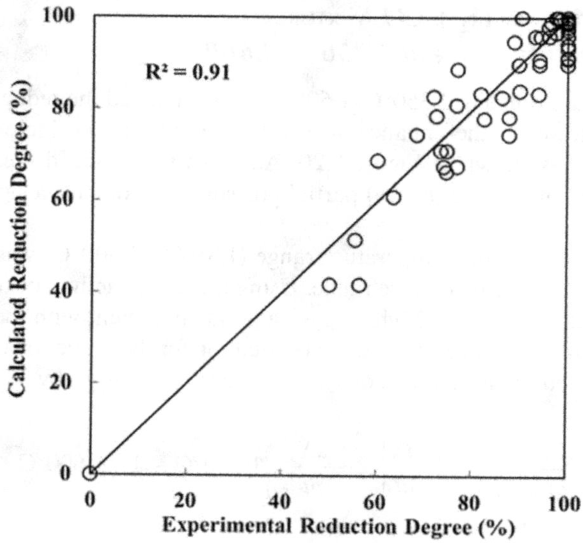

FIGURE 7.21 Comparisons between the calculated reduction degrees vs. experimental results with H_2+CO mixtures as the reducing gas in 1,350°C–1,600°C.

$$\left.\frac{dX}{dt}\right|_{H_2} = 2.5\times10^7 \times e^{\left(\frac{-170,000}{RT}\right)} \times d_p^{-1} \times \left[p_{H_2} - \left(\frac{p_{H_2O}}{K_H}\right)\right]_{lm} \times (1-X) \quad (7.46)$$

$$\left.\frac{dX}{dt}\right|_{CO} = 1.2\times10^3 \times e^{\left(\frac{-73,000}{RT}\right)} \times d_p^{-1} \times \left[p_{CO} - \left(\frac{p_{CO_2}}{K_C}\right)\right]_{lm} \times \frac{-(1-X)}{Ln(1-X)} \quad (7.47)$$

at 1,350°C ≤ T <1,600°C. (This kinetics equation is adequate for most purposes. However, this equation has been improved to obtain a more accurate *X-t* relationship, as will be shown in Section 7.6, by the application of a CFD analysis. It is recommended that the relation presented therein be used for all purposes).

7.5.7 Summary on the Reduction Kinetics of Magnetite Concentrate Particles

The rate equations for the reduction of magnetite concentrate by H_2, CO, and H_2+CO mixture were obtained in the temperature range of 1,150°C–1,600°C. These rate equations have great significance for the development of the flash ironmaking technology. They are instrumental in designing and testing a larger-scale flash reactor in which the reduction of magnetite concentrate will be carried out in a H_2 and CO mixture produced from the partial oxidation of natural gas.

The nucleation and growth model was used to describe the reduction process. The rate of reaction by H_2 was much faster than that by CO. There was no significant effect of particle size in the temperature range of 1,150°C–1,350°C. In the higher temperature range of 1,350°C–1,600°C, where melting and fusion occurred, particle size had an effect on the reaction rate, as increasing the particle size decreased the reduction rate.

These rate equations were used to formulate the rate expression for the reduction by H_2+CO mixtures. Hydrogen was the main contributor to the reduction by the gas mixture. However, the H_2 reduction rate was enhanced by the presence of CO. Therefore, a rate equation was developed where the contribution of CO was kept the same and a coefficient was multiplied to the H_2 rate to describe this enhancement effect. Good agreement was obtained between the experimental results and the developed rate equations.

7.6 REFINEMENTS OF THE RATE EQUATIONS FOR THE REDUCTION OF CONCENTRATE PARTICLES THROUGH COMPUTATIONAL FLUID DYNAMICS MODELING

7.6.1 Approach and Methodology

In situ particle temperature measurement and real-time tracking of particles inside the DTR would require sophisticated and intrusive experimental techniques, for example, PIV and optical access to the DTR interior [Lian et al., 2010; Khatami et al., 2012; Avila et al., 2012; Tolvanen and Raiko, 2014]. Moreover, the use of H_2 atmosphere at high temperature adds an extra layer of complexity to use these techniques. Temperature measurements in the reactor during the experiments in this work as well as CFD simulations found that there was a low-temperature region near the tip of the water-cooled tube through which the carrier gas and particles were injected at room temperature. The presence of this region affects the temperature and flow history of the particles. The use of CFD to incorporate such details allows realistic calculations of particle temperature and residence time as functions of position and improves the accuracy of kinetic analysis. Several efforts of using CFD as an effective tool to investigate the kinetics of fast pyrolysis of biomass and other fuels have been reported in the literature. Alexander et al. [2001] used the CFD approach to predict an accurate time-temperature profile in an entrained flow reactor to study biomass pyrolysis. Simone et al. [2009] developed a procedure coupling experimental results and CFD to evaluate the global biomass devolatilization kinetics in a DTR. Little work has been reported on the CFD simulation of a DTR in the investigation of the kinetics of metal oxide reduction.

A CFD approach, coupled with experimental results, was developed to refine the kinetic parameters of iron oxide particle reduction described above. CFD approach was used in this work to eliminate the small errors caused in the kinetics analysis by the small variation in gas flow and temperature variations near the top and bottom parts of the reactor.

A detailed evaluation of the particle residence time and temperature profile inside the reactor is possible by CFD analysis. This approach eliminates the errors associated with assumptions like constant particle temperature and velocity while the particles travel down a DTR. The gas phase was treated as a continuum in the Eulerian frame of reference, and the particles were tracked using a Lagrangian approach in which the trajectory and velocity were determined by integrating the equation of particle motion. In addition, a heat balance on the particle that relates the particle temperature to convection and radiation was also applied. An iterative algorithm that numerically solved the governing coupled ordinary differential equations was developed to determine the pre-exponential factor and activation energy that best fit the experimental data.

Despite great efforts made to form a uniform temperature zone inside the reactor, there are still temperature variations, particularly in the top and bottom parts of the reactor. Gas stream temperature was measured experimentally only along the centerline over the reaction region. As the wall temperatures of the DTR were not known but essential for setting up boundary conditions in the CFD simulations, preliminary CFD runs were carried out to calculate the suitable wall temperature profiles for different target temperatures using the measured centerline temperature profiles. The wall temperature profiles were first optimized until the calculated gas stream temperature profile agreed with the measured profiles in the isothermal zone and bottom part along the reactor axis. The optimized wall temperature, from the top to near the end of the reactor, was essentially the same as the target gas temperature because the gas comes into the reactor preheated to that temperature by the honeycomb. Toward the bottom, there was some heat loss and thus the wall temperature had to be lowered to reproduce the decreasing temperature. A comparison of the measured and calculated temperature profiles is presented in Figure 7.22. Simone et al. [2009] also used the same approach for setting up the boundary conditions in their DTR simulation work.

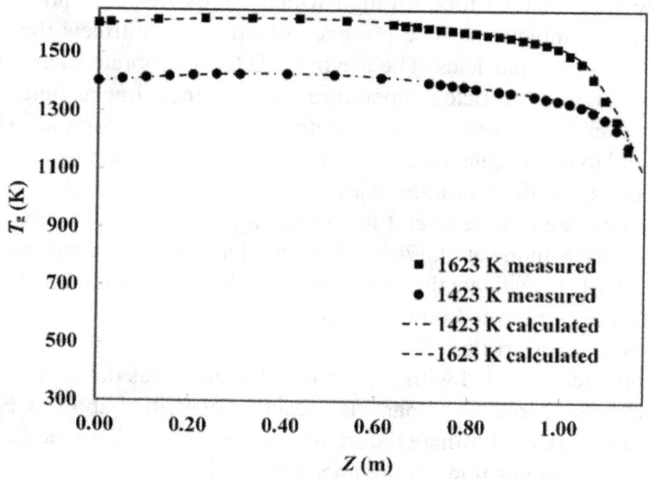

FIGURE 7.22 Comparison of the measured and calculated temperature profiles along the drop-tube reactor axis ($z=0$ corresponds to the exit of particle from the powder feeder injection tube, shown in Figure 7.24 (a)). ■, – – –: experimental and calculated temperature profiles for target temperature of 1,623 K (1,350°C), N_2 flow rate = 1 L/min; •, – · –: experimental temperature and calculated profiles for target temperature of 1,423 K (1,150°C), N_2 flow rate = 1 L/min.

Since the particles are mostly concentrated in the center of the reactor, as mentioned in Section 7.4.1, the central velocity and temperature of the gas phase from the CFD results were used to track the particle in this work. Error analysis of using centerline velocity is further performed below.

The kinetics parameters for magnetite reduction by hydrogen, namely, the pre-exponential factor (k_o) and activation energy (E), are both inputs for the model as well as unknowns to be determined. The procedure used in this study for estimating k_o and E using the experimental data is as follows:

Step 1: Predetermine sufficiently broad ranges of values for k_o and E that contain the actual values for the hydrogen reduction of magnetite and discretize the range into pairs of k_o and E with a specified activation energy increment.
Step 2: The discretized pairs of k_o and E are then used to predict the RD of magnetite particles under each set of experimental conditions.
Step 3: The deviation between each of the predicted RD and the corresponding experimental value is computed. And a summation over all the deviations is performed to calculate the mean of the squared errors.
Step 4: Pick the combination of k_o and E that gives the minimum mean of squared error. The increments of k_o and E close to the optimized values were both < 1% of the adjacent values.

The algorithm used in this study is shown in Figure 7.23.

Detailed thermal and velocity profiles in the gas phase under each set of experimental conditions were first obtained via CFD simulations. The realizable k-ε model [Shih et al., 1995] was chosen for simulating the spread of the carrier gas jet. Radiation was taken into account using the discrete ordinate (DO) model [Chui and Raithby, 1993]. During the drop-tube experiment, the bulk flow of the gas mixture consisted of H_2, N_2,and H_2O. But reducing gas H_2 was usually in more than 500%

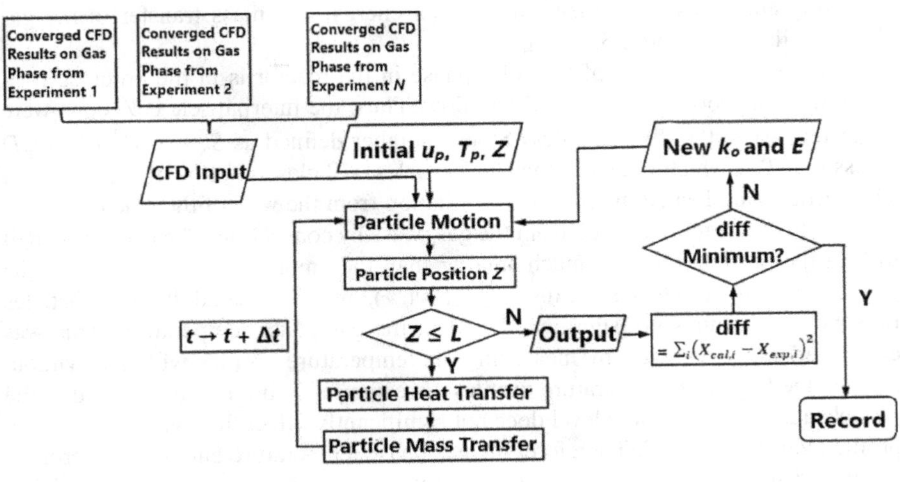

FIGURE 7.23 A block diagram of the algorithm for kinetics parameter optimization.

TABLE 7.5
Governing Equations of Gas Phase

Continuity: $\frac{\partial}{\partial x_i}(\rho u_i) = 0$ (7.48)

Momentum: $\frac{\partial}{\partial x_j}(\rho u_i u_j) = -\frac{\partial p}{\partial x_i} + \frac{\partial}{\partial x_j}\left[\mu\left(\frac{\partial u_i}{\partial x_j} + \frac{\partial u_j}{\partial x_i} - \frac{2}{3}\delta_{ij}\frac{\partial u_l}{\partial x_l}\right)\right] + \frac{\partial}{\partial x_j}\left(-\rho \overline{u_i' u_j'}\right) + \rho g_i$ (7.49)

Energy: $\frac{\partial}{\partial x_i}(\rho u_i h_g) = \frac{\partial}{\partial x_i}\left(\frac{k_{eff}\,\partial T}{\partial x_i}\right) + Q_r$ (7.50)

Species: $\frac{\partial}{\partial x_j}(\rho Y_i u_j) = -\frac{\partial J_j}{\partial x_j}$ (7.51)

Turbulent kinetic energy: $\frac{\partial}{\partial x_i}(\rho k u_i) = \frac{\partial}{\partial x_i}\left[\left(\mu + \frac{\mu_t}{\sigma_k}\right)\frac{\partial k}{\partial x_i}\right] + G_k + G_b - \rho\varepsilon$ (7.52)

Turbulent dissipation rate: $\frac{\partial}{\partial x_i}(\rho \varepsilon u_i) = \frac{\partial}{\partial x_i}\left[\left(\mu + \frac{\mu_t}{\sigma_\varepsilon}\right)\frac{\partial \varepsilon}{\partial x_i}\right] + \rho C_1 S\varepsilon - \rho C_2 \frac{\varepsilon^2}{k + \sqrt{\nu\varepsilon}} + C_{1\varepsilon}\frac{\varepsilon}{k} C_{3\varepsilon} G_b$ (7.53)

Reynolds stress: $-\rho\overline{u_i' u_j'} = \mu_t\left(\frac{\partial u_i}{\partial x_j} + \frac{\partial u_j}{\partial x_i}\right) - \frac{2}{3}\left(\rho k + \mu_t \frac{\partial u_l}{\partial x_l}\right)\delta_{ij}$ (7.54)

Mass diffusion flux: $J_i = -\left(-\rho D_{i,m} + \frac{\mu_t}{Sc_t}\right)\nabla Y_i - D_{T,i}\frac{\nabla T}{T}$ (7.55)

excess of the minimum required amount of H_2 and the amount of water vapor generated in the reduction process only accounted for around 2% [Chen et al., 2015a]. In this scenario, species H_2O is excluded from the gas mixture in this simulation. The governing equations of momentum, thermal energy, and mass transfer in the gas phase are listed in Table 7.5.

As the volume fraction of the solid phase in the reactor is in the order of 10^{-6}, the flow is categorized as a very dilute flow. Thus, the interparticle collisions were neglected. Also, the dimensionless Stokes number defined as $S_{tk} = \rho_p d_p^2 u / 18\mu_g D$ is less than 0.1, which indicates that the particles will closely follow the gas stream. The particles are heated mainly by the radiation from the wall of the reactor tube in addition to the heat transferred from the gas phase by convection. The energy needed to heat the particle phase is much smaller than the amount of heat carried by the gas phase in this dilute system (usually ≤ 5 wt.%), and thus the addition of particles into the reactor does not significantly affect the gas phase temperature. This was confirmed experimentally by measuring the temperature profiles with and without particle feeding. The temperature profiles were similar in both cases. Similarly, the particle addition at this low level does not significantly affect the gas phase velocity profile [Varaksin, 2007]. Therefore, the gas phase temperature and velocity profiles calculated in the absence of particles were used even in the presence of particles. This simplified approach was used to minimize the computational time since a full

CFD simulation of the entire experimental data set that considers coupling between the discrete and continuous phase would have taken a prohibitive amount of computational time.

The flow inside the reactor is a fully developed laminar flow except in the initial part of the reactor where particle laden jet exists. Although most of the particles were believed to flow along the centerline, an error analysis was performed by Choi [2010]. It has been found that even if the particles were spread over half of the radius of the cross section, the average residence time increased by only about 10%, a tolerable difference. Therefore, using the centerline velocity and temperature of the gas phase is reasonable.

The concentrate particle is treated as a discrete phase. It was assumed that there were only three main forces exerting on the particle: drag, gravitational, and buoyancy forces; other external forces like thermophoretic force are negligible compared with them. The particle motion written in a Lagrangian reference frame is described as:

$$\frac{du_p}{dt} = \mathrm{F}_D\left(\vec{u} - \vec{u}_p\right) + \frac{\vec{g}\left(\rho_p - \rho\right)}{\rho_p} \tag{7.56}$$

where, $\mathrm{F}_D\left(\vec{u} - \vec{u}_p\right)$ is the drag force per unit particle mass, and F_D was calculated as

$$\mathrm{F}_D = \frac{18\mu}{\rho_p d_p^2}\frac{C_D \mathrm{Re}}{24} \tag{7.57}$$

The concentrate particle was assumed as a solid sphere, and the drag coefficient was calculated as [Clift et al., 1978]:

$$C_D = \begin{cases} 0.44, & \mathrm{Re} > 1000 \\ \dfrac{24}{\mathrm{Re}}\left(1 + 0.15\mathrm{Re}^{0.678}\right), & \mathrm{Re} \le 1000 \end{cases} \tag{7.58}$$

To account for the turbulent dispersion of particles, the stochastic tracking model [Fan and Zhou, 1998] was used.

A heat balance on the particle that relates the particle temperature to convection and radiation was also applied. The particle thermal energy equation is expressed as

$$m_p c_{p,d}\frac{dT_p}{dt} = hA_p\left(T - T_p\right) - f_h\frac{dm_p}{dt}\Delta H_{reac} + \varepsilon_p A_p \sigma\left(T_s^4 - T_p^4\right) \tag{7.59}$$

in which, the term dm_p/dt was related to particle chemical reaction rate, which is given by [Elzohiery et al., 2017; Chen et al., 2015a]

$$\frac{dX}{dt} = n \cdot k_0 \exp\left(-\frac{E}{RT}\right)\left[p_{\mathrm{H_2}}^m - \left(\frac{p_{\mathrm{H_2O}}}{K_e}\right)^m\right] d_p^s \cdot (1 - X)\left[-\mathrm{Ln}(1 - X)\right]^{1-\frac{1}{n}} \tag{7.60}$$

The particle heat transfer coefficient was evaluated using the correlation of Ranz and Marshall [Ranz and Marshall, 1952].

$$Nu = \frac{hd_p}{k_g} = 2.0 + 0.6\ Re^{1/2} Pr^{1/3} \tag{7.61}$$

A value of 0.8 was chosen as the particle emissivity ε_p, recommended by Hahn and Sohn [1990]. The particle-specific heat was calculated as a mass fraction average of iron and magnetite during the reduction process.

$$c_{p,d} = X \cdot c_{p,Fe} + (1 - X) \cdot c_{p,Fe_3O_4} \tag{7.62}$$

The size of the particles before and after an experiment stays almost the same as SEM micrographs taken in this work revealed (not shown here). The solid particle density equals the particle instantaneous mass divided by the initial particle volume (before reaction) as

$$\rho_p = \rho_{p,0}\left(1 - \omega_O^0 X\right) \tag{7.63}$$

7.6.2 Numerical Procedure

The computational domain includes a section of the injection tube long enough to establish the correct velocity profile at the tube exit ($z = 1.2$ m) and a schematic representation of this is shown in Figure 7.24a. The mesh was generated using ICEM-CFD ANSYS with a total of 150,288 hexahedral cells as shown in Figure 7.24b. Mesh independence was confirmed by halving and doubling the number of cells with the same results.

A mass flow rate boundary condition was imposed at the inlet of the powder injection tube and reactor tube. The operating pressure was kept at 0.85 atm (the barometric pressure at Salt Lake City; 1 atm = 101.32 kPa). A nonslip condition was applied to the reactor wall for the gas flow inside the reactor. At the outlet, the flow was assumed to be fully developed and the gradients for all variables in the exit direction were zero. The temperatures of the gases at their inlets were given a constant room temperature of 298 K (25°C). The wall temperature profile along the longitudinal direction of the reactor to be used as a boundary condition was obtained by the method described in Section 2.3. It is worth mentioning, however, that the water-cooled powder injection tube (see Figure 7.24a) was not installed during the temperature profile measurement due to experimental difficulties. In the CFD simulation of an actual run, the water-cooled powder injection tube and the cold carrier gas were taken into consideration. Although the wall temperature profile as a boundary condition was optimized to match the measured centerline temperature in the absence of the carrier gas, its amount was typically less than 10% of the total gas input. Thus, its effect on wall boundary conditions was neglected. The inner walls were assumed to be gray and diffuse, and a constant value of 0.4 was chosen for the emissivity of the alumina tube wall [Bergman et al., 2011].

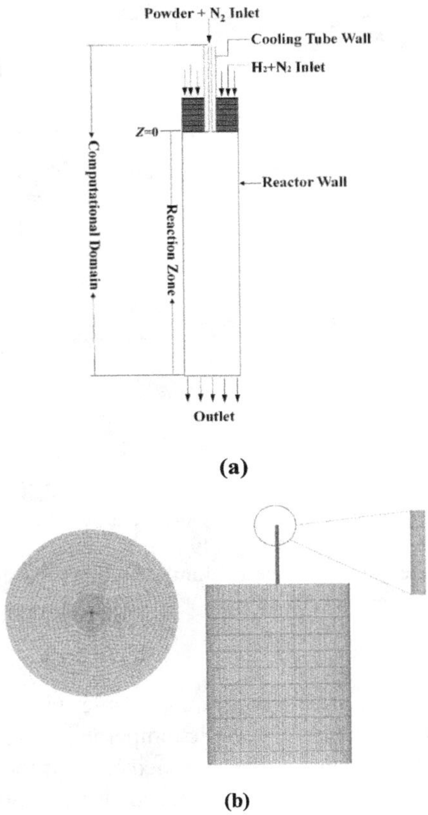

FIGURE 7.24 Schematic representation of the reactor: (a) Computational domain. (b) Mesh for computational domain.

The gas phase governing Eqs. (7.48–7.53) were discretized and solved using the commercial CFD software package ANSYS FLUENT 15.0. The calculation was carried out by a steady-state pressure-based solver and the SIMPLE scheme was chosen for the pressure-velocity coupling. The second-order upwind scheme was chosen for momentum, species transport, and energy equation discretization. A separate program was developed in MATLAB® for solving the Lagrangian particle governing Eqs. (7.56), (7.59), and (7.60) using fourth-order Runge–Kutta method.

7.6.3 Modeling Results

From the computed gas phase temperature inside the DTR under all sets of experimental conditions, it is clear that a low-temperature region does exist adjacent to the water-cooled injection tube in the initial stage of the reactor. The length of this region increases and becomes narrower as the total gas flow rate increases.

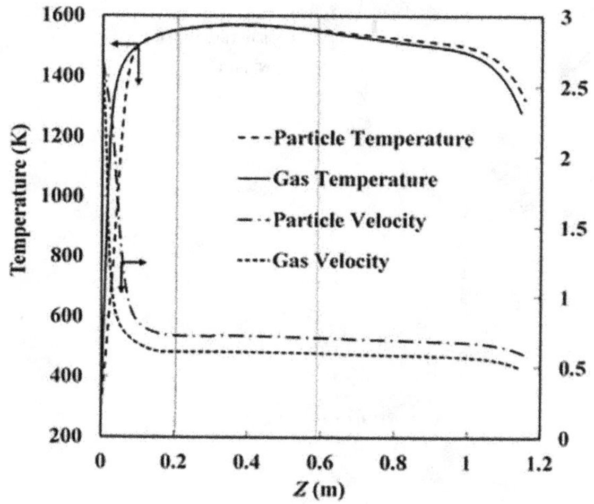

FIGURE 7.25 Particle temperature and velocity profile along the centerline of the reactor for $T_{iso} = 1{,}573$ K (1,300°C), $pH_2 = 0.2$ atm, $d_p = 49$ μm, and total gas flow rate = 8.5 L/min. The two vertical dashed lines represent the beginning and the end of the isothermal zone (target temperature 1,573 ± 20 K (1,300 ± 20°C)).

The typically calculated particle temperature profile within the DTR is shown in Figure 7.25. The position where the particle temperature rises to the temperature of the isothermal zone, marked by the first dashed line on the left in Figure 7.25, is 0.205 m from the tip of the particle injector, which accounts for nearly 17% of the whole reaction zone length in this case. This length can be larger or smaller depending on specific experimental conditions. Within the water-cooled injection tube, temperature remains close to room temperature, and thus no reaction is considered there. It can be seen from Figure 7.25 that the particle temperature stays at the isothermal temperature 1,573 ± 20 K (1,300°C ± 20°C) for some distance (until the second dashed line on the right in Figure 7.23) and then starts to decrease in the bottom part of the reactor. For all cases studied, the gas flow in most of the reactor tube is a fully developed laminar flow except in the region where the carrier gas with the solid particles exits a 2 mm diameter injection tube and forms a jet. The jet spreads and loses its momentum while traveling downward as demonstrated by the velocity profiles of the gas and particles in Figure 7.25.

The velocity vectors shown in Figure 7.26 demonstrate that the gas travels mostly downward along the reactor longitudinal axis and the whole flow transits to laminar flow within a short distance (in the case shown around 0.1 m). The radial velocity was too small to disperse the particles. A similar feature was also computed by other researchers (e.g., Simone et al. [2009]) and was observed with particle tracking velocimetry and optical measurement by Lehto [2007] and Avila et al. [2012] in their study of a DTR.

Since some reaction occurs in the upper low-temperature region, a certain amount of error may occur if this lower temperature zone is ignored. The thermal history and

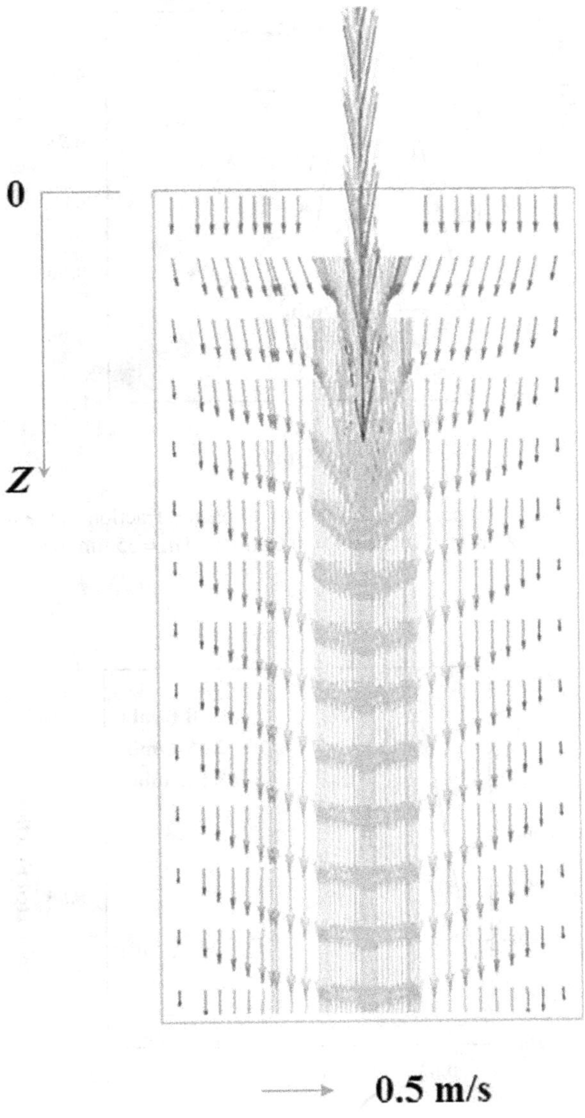

FIGURE 7.26 Gas velocity vector near the top part region of the reactor for T_{iso} = 1,573 K (1,300°C), pH_2 = 0.2 atm, d_p = 49 μm, total gas flow rate = 8.5 L/min.

RD of the particles in the upper low-temperature region for T_{iso} = 1,423 K (1,150°C) and T_{iso} = 1,623 K (1,350°C) are presented in Figures 7.27 and 7.28, respectively. As expected, a lower particle RD is achieved in this region when the target temperature is lower. The length of this zone also affects the RD achieved as can be seen in both figures as the gas flow rate increases.

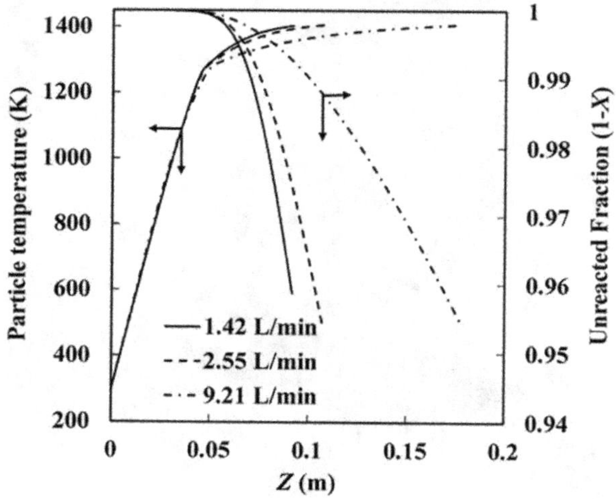

FIGURE 7.27 Particle temperature profile and unreacted fraction in the top part region of the reactor for $T_{iso} = 1{,}423$ K (1,150°C), $pH_2 = 0.6$ atm, and $d_p = 35$ μm, with different total gas flow rate.

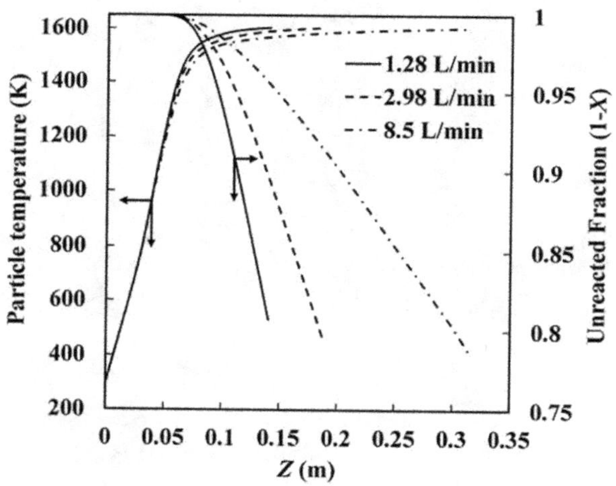

FIGURE 7.28 Particle temperature profile and unreacted fraction in the top part region of the reactor for $T_{iso} = 1{,}623$ K (1,350°C), $pH_2 = 0.6$ atm, and $d_p = 49$ μm with different total gas flow rate.

Typically calculated particle RDs along the reactor length are shown in Figure 7.29 together with particle temperature profiles and the corresponding experimental final values.

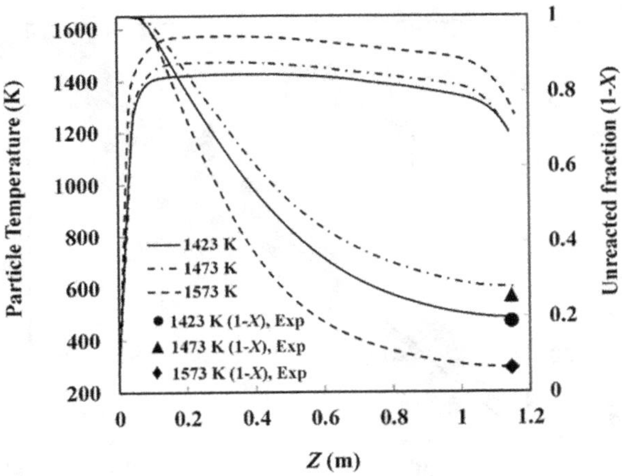

FIGURE 7.29 Calculated profiles of particle temperature and unreacted fraction along the reactor length. ——: $pH_2=0.6$ atm, $d_p=35$ μm, gas flow rate=1.42 L/min; – · –: $pH_2=0.2$ atm, $d_p=22.5$ μm, gas flow rate=2.13 L/min; – – –: $pH_2=0.2$ atm, $d_p=22.5$ μm, and gas flow rate=2.13 L/min.

7.6.4 Kinetics Analysis Procedure

Previous work of Chen et al. [2015a, b] demonstrates that the reduction behavior of iron oxide concentrate particles was affected by the partial pressure of gaseous reducing agents, particle size, particle residence time in the reaction zone, and temperature. The nucleation and growth model was proved to best describe the reduction process of hematite and magnetite concentrate particles by H_2 or CO. In the nucleation and growth kinetics, the reducing agent adsorbs on active sites on the surface and then reacts with oxygen and produces Fe nuclei that grow with time. For small particles, the period of the formation and growth of nuclei occupies essentially the entire conversion range.

A wide range of values of pre-exponential factor and activation energy around the values presented in the previous section were discretized and scanned for searching optimized values. The objective function used in this study was the mean of the squared errors of all experiments as

$$f_{min} = \sum_i \left(X_{cal,i} - X_{\exp,i}\right)^2 / N \tag{7.64}$$

The determination of m and n will be discussed here.

7.6.5 Complete Rate Equations

7.6.5.1 Reduction by Hydrogen: *Temperature Range of 1,150°C–1,350°C*

Combining all the rate parameters obtained through the steps described above, the complete rate equation for the hydrogen reduction of magnetite concentrate

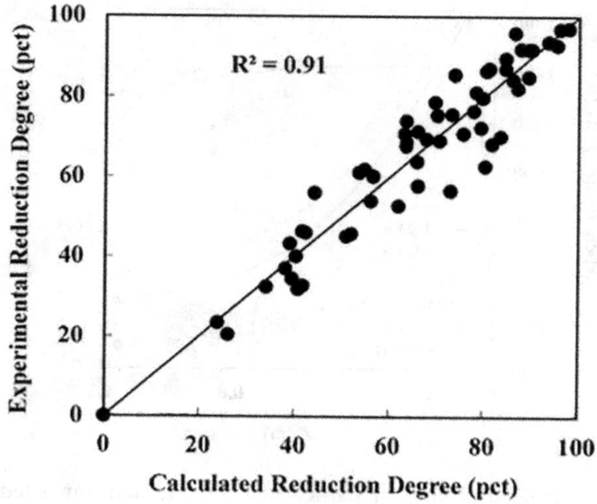

FIGURE 7.30 Comparisons between the calculated reduction degrees vs. experimental results of all runs ($n = 1$, $m = 1$).

particle in the temperature range of 1,423 K (1,150°C) to 1,623 K (1,350°C) is given by

$$\left[-\mathrm{Ln}(1-X)\right] = 1.23\times10^{7}\times e^{\frac{-196,000}{RT}}\cdot\left[p_{H_2}-\left(\frac{p_{H_2O}}{K}\right)\right]\cdot t \tag{7.65}$$

where, R is 8.314 J /mol·K, p is in atm, and t is in seconds.

The comparisons between the RDs calculated using the optimized kinetics parameters and the values obtained experimentally are shown in Figure 7.30.

7.6.5.2 Reduction by Hydrogen: *Temperature Range of 1,350°C–1,600°C*

The complete rate equations for the reduction of magnetite concentrate particles by H_2 in the temperature range of 1,623 K (1,350°C) to 1,873 K (1,600°C) are given, respectively, as

Integral form:

$$\left[-\mathrm{Ln}(1-X)\right] = 6.07\times10^{7}\cdot e^{\frac{-180,000}{RT}}\cdot\left(p_{H_2}-\frac{p_{H_2O}}{K}\right)\cdot\left(d_p\right)^{-1}\cdot t;$$

$$1,623\text{ K }(1,350°\text{C}) < T < 1,873\text{ K }(1,600°\text{C}) \tag{7.66}$$

Differential form:

$$\frac{dX}{dt} = 6.07\times10^{7}\cdot e^{\frac{-180,000}{RT}}\cdot\left(p_{H_2}-\frac{p_{H_2O}}{K}\right)\cdot\left(d_p\right)^{-1}\cdot(1-X) \tag{7.67}$$

where R is 8.314 J/mol·K, p is in atm, K is the equilibrium constant for the reduction of FeO by H_2, d_p is in μm, and t is in seconds.

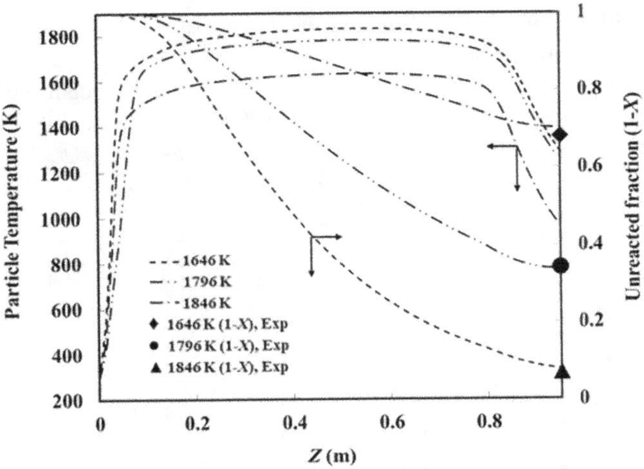

FIGURE 7.31 Calculated profiles of particle temperature and unreacted fractions along the reactor centerline. – – –: pH_2=0.05 atm, gas flow rate: H_2=0.7 L/min, N_2=11.2 L/min, d_p=22.5 μm; – · –: pCO=0.1 atm, gas flow rate: H_2=1.5 L/min, N_2=11.25 L/min, d_p=35 μm; – ·· –: pCO=0.1 atm, gas flow rate: H_2=1.5 L/min, N_2=11.25 L/min, and d_p=22.5 μm. The temperature values in the legend represent the isothermal zone temperature maintained during the experiments.

Typically calculated particle RDs along the reactor centerline obtained using the optimum values of activation energy and pre-exponential factor, together with the corresponding experimental RDs and temperature profiles, are shown in Figure 7.31. It is seen that the simulated RD values match the experimental values very well.

The comparisons between the calculated RDs using the optimum values of activation energy and pre-exponential factor and the experimental values from all the runs are shown in Figure 7.32. It is noted that the optimum kinetic parameters obtained in this work predict the RDs very well.

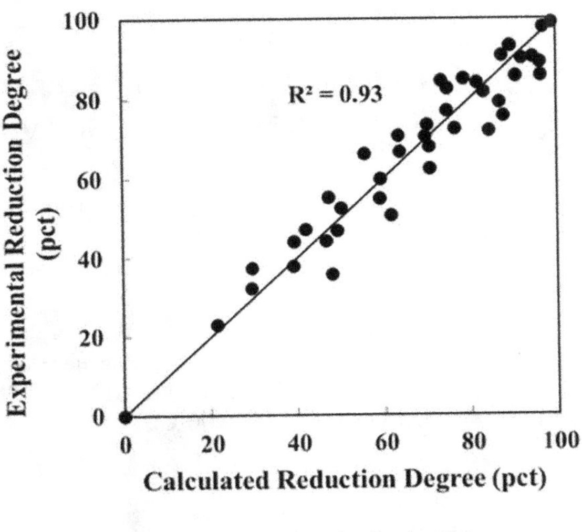

FIGURE 7.32 Comparisons between the calculated reduction degrees vs. experimental results.

7.6.5.3 Reduction by Carbon Monoxide: *Temperature Range of 1,150°C–1,350°C*

The complete rate equations for the reduction of magnetite concentrate particles by CO in the temperature range of 1,473 K (1,200°C) to 1,623 K (1,350°C) and 1,623 K (1,350°C) to 1,873 K (1,600°C) are given, respectively, as:

Integral form:

$$\left[-\text{Ln}(1-X)\right]^{1/0.5} = 1.07\times10^{14}\times e^{\frac{-451,000}{RT}}\cdot\left(p_{\text{CO}}-\frac{p_{\text{CO}_2}}{K}\right)\cdot t \tag{7.68}$$

Differential form:

$$\frac{dX}{dt} = 5.35\times10^{13}\cdot e^{\frac{-451,000}{RT}}\cdot\left(p_{\text{CO}}-\frac{p_{\text{CO}_2}}{K}\right)\cdot(1-X)\cdot\left[-\ln(1-X)\right]^{-1}$$

$$1{,}473\text{ K }(1{,}200°\text{C}) < T < 1{,}623\text{ K }(1{,}350°\text{C}) \tag{7.69}$$

where R is 8.314 J/mol·K, p is in atm, K is the equilibrium constant for the reduction of FeO by CO, d_p is in μm, and t is in seconds. The comparison between the calculated RDs in this temperature range and the experimental values for all the experimental runs are shown in Figure 7.33, together with the comparison for the higher temperature range discussed next.

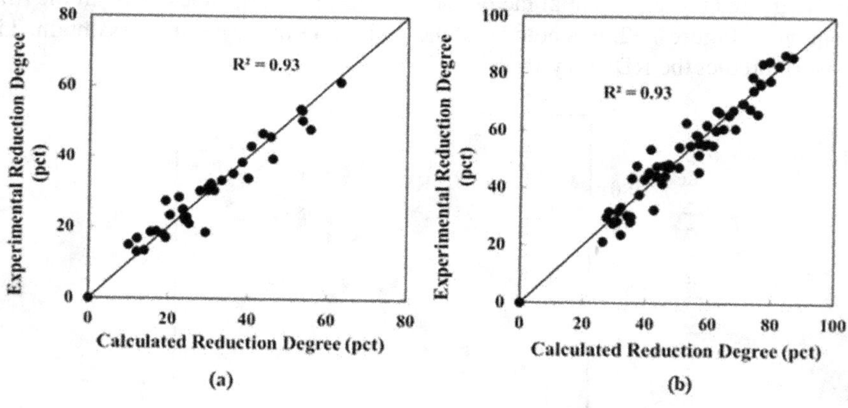

FIGURE 7.33 Comparisons between the calculated reduction degrees vs. experimental results: (a) Lower temperature range of 1,473 K (1,200°C) to 1,623 K (1,350°C). (b) Higher temperature range of 1,623 K (1,350°C) to 1873 K (1,600°C).

7.6.5.4 Reduction by Carbon Monoxide: *Temperature Range of 1,350°C–1,600°C*

Integral form:

$$\left[-\mathrm{Ln}(1-X)\right]^{1/0.5} = 6.45\times 10^{3}\times e^{\frac{-88{,}000}{RT}}\cdot\left[p_{\mathrm{CO}}-\left(\frac{p_{\mathrm{CO_2}}}{K}\right)\right]\cdot d_p^{-1}\cdot t \tag{7.70}$$

Differential form:

$$\frac{dX}{dt} = 3.225\times 10^{3}\cdot e^{\frac{-88{,}000}{RT}}\cdot\left(p_{\mathrm{CO}}-\frac{p_{\mathrm{CO_2}}}{K}\right)\cdot\left(d_p\right)^{-1}\cdot(1-X)\cdot\left[-\ln(1-X)\right]^{-1}$$

$$1{,}623\ \mathrm{K}\ (1{,}350°\mathrm{C}) < T < 1{,}873\ \mathrm{K}\ (1{,}600°\mathrm{C}) \tag{7.71}$$

where R is 8.314 J/mol·K, p is in atm, K is the equilibrium constant for the reduction of FeO by CO, d_p is in μm, and t is in seconds.

The comparisons between the calculated RDs using the optimum values of activation energy and pre-exponential factor and the experimental values for all the experimental runs are shown in Figure 7.33. It is noted that the optimum kinetic parameters obtained in this work predict the RDs very well as the correlation coefficients between the calculated values and corresponding experimental results are high in both cases.

7.6.5.5 Reduction of Hematite Concentrate by H_2 or CO

Similar rate measurements were also carried out with hematite concentrate [Fan et al., 2017]. The abundance of hematite ores in the global market adds to the significance of the information presented here to be highly relevant to potential industrial applications.

The complete rate equations for the reduction of hematite concentrate particles by individual reducing gases H_2 and CO in the temperature range of 1,423 K (1,150°C) to 1,623 K (1,350°C) are given by

$$\left[-\mathrm{Ln}(1-X)\right] = 8.47\times 10^{7}\times e^{\frac{-218{,}000}{RT}}\cdot\left[p_{H_2}-\left(\frac{p_{\mathrm{H_2O}}}{K_{\mathrm{H_2}}}\right)\right]\cdot t \text{ for reduction by } \mathrm{H_2} \tag{7.72}$$

$$\left[-\mathrm{Ln}(1-X)\right] = 5.18\times 10^{7}\times e^{\frac{-241{,}000}{RT}}\cdot\left[p_{\mathrm{CO}}-\left(\frac{p_{\mathrm{CO_2}}}{K_{\mathrm{CO}}}\right)\right]\cdot t \text{ for reduction by CO} \tag{7.73}$$

where R is 8.314 J/mol·K, p is in atm, K is the equilibrium constant, and t is in seconds.

The comparisons between the calculated RDs using the optimum values of activation energy and pre-exponential factor and the experimental values for all the experimental runs are shown in Figures 7.34 and 7.35 for the reduction of hematite concentrate particles by individual reducing gases H_2 and CO, respectively. It is noted that the optimum kinetic parameters obtained in this work predict the RDs very well as the correlation coefficients between the calculated values and the corresponding experimental results are high in both cases.

The activation energy values obtained by Chen et al. [2015a; 2015b] without this CFD-based refinement were 214 and 231 kJ/mol, respectively, for the reduction of hematite concentrate particles by H_2 and CO. The comparisons of the chemical reaction rates under the same reaction temperatures and partial pressures for the previous

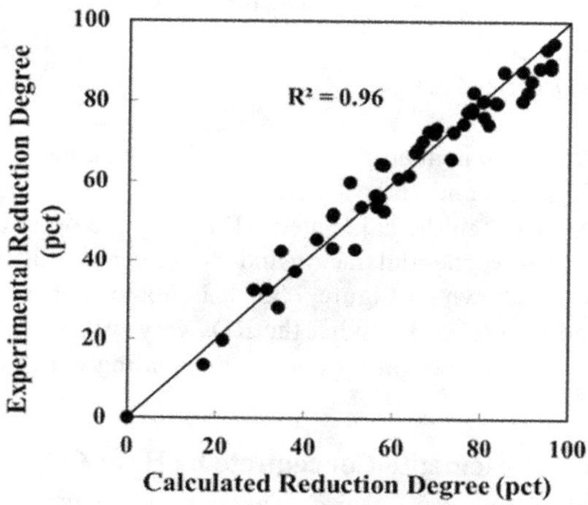

FIGURE 7.34 Comparisons between the calculated reduction degrees vs. experimental results with H_2 as the reducing gas.

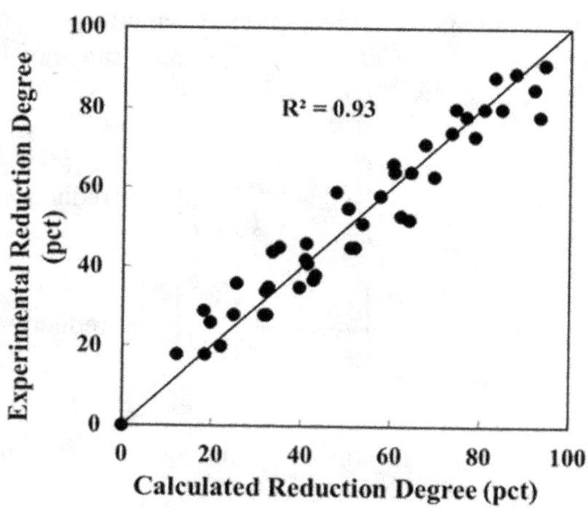

FIGURE 7.35 Comparisons between the calculated reduction degrees vs. experimental results with CO as the reducing gas.

TABLE 7.6
Comparison of the Reaction Rate Constants at Different Temperatures for Hematite Reduction by H_2

Reaction Temp.	$k_{H_2,1}$/(atm·s)	$k_{H_2,2}$/(atm·s)	$k_{H_2,1}/k_{H_2,2}$
1,423 K (1,150°C)	0.84	0.61	1.37
1,473 K (1,200°C)	1.57	1.14	1.39
1,523 K (1,250°C)	2.82	2.02	1.40
1,573 K (1,300°C)	4.88	3.45	1.41
1,623 K (1,350°C)	8.16	5.71	1.43

Note: Subscript under k: 1 represents this work; 2 Chen et al. [2015a, b].

TABLE 7.7
Comparison of the Reaction Rate Constants at Different Temperatures for Hematite Reduction by CO

Reaction Temp.	$k_{CO,1}$/(atm·s)	$k_{CO,2}$ /(atm·s)	$k_{CO,1}/k_{CO,2}$
1,423 K (1,150°C)	0.07	0.06	1.16
1,473 K (1,200°C)	0.15	0.12	1.20
1,523 K (1,250°C)	0.28	0.23	1.23
1,573 K (1,300°C)	0.51	0.41	1.26
1,623 K (1,350°C)	0.91	0.70	1.29

Note: Subscript under k: 1 represents this work; 2 Chen et al. [2015a, b].

rate expressions obtained by Chen et al. [2015a, b] and the current results are listed in Tables 7.6 and 7.7. Because the same rate expressions were used in the two studies in terms of temperature and partial pressure dependences, the comparisons made in Tables 7.6 and 7.7 in terms of the rate constant represent the comparisons of the reaction rates. It is seen that after refinement, the reaction rates under the same reaction temperatures and partial pressures are somewhat higher, even though activation energy values became a little larger. This is due to the fact that the effective particle residence time in the hot zone calculated in this work is shorter as a result of the more accurate description of the low-temperature region in the top part of the reactor.

***Comparison of the reduction kinetics of magnetite equivalent produced from hematite reduction with that of natural magnetite concentrate particles*:**

The rate of the H_2 reduction of hematite particles obtained in this work is also compared with that of natural magnetite concentrate obtained previously [Fan et al., 2016a]. The latter in the temperature range of 1,423 K (1,150°C) to 1,623 K (1,350°C) is given by

$$\left[-\text{Ln}(1-X)\right] = 1.23\times10^{7}\times e^{\frac{-196,000}{RT}}\cdot\left[p_{H_2}-\left(\frac{p_{H_2O}}{K_{H_2}}\right)\right]\cdot t \qquad (7.74)$$

in terms of the same symbols used in Eqs. (7.72) and (7.73).

For this comparison, natural magnetite concentrate was assumed to have been generated from hematite concentrate. The crystal structure and morphology of naturally occurring magnetite minerals are different from those of magnetite equivalent (reduced to the O/Fe ratio of 4/3) produced from the reduction of hematite. This comparison will reveal the effects of these differences. With this assumption, the reduction magnetite equivalent from hematite starts from the time corresponding to $X = 1/9$ (the fraction of oxygen that needs to be removed from the imaginary hematite to produce magnetite), and the RD of natural magnetite is expressed based on the total removable oxygen content in the corresponding imaginary starting hematite. The procedure for converting the X-vs-t relationship for natural magnetite, Eq. (7.74), to one based on the corresponding imaginary hematite is presented in the next section. The comparison is shown in Figure 7.36, in which the RDs were calculated with $p_{H_2} - \left(\frac{p_{H_2O}}{K_{H_2}}\right)$ equal to 0.3 (result from commonly available gas composition obtained from partial combustion of natural gas under the operating conditions of our flash ironmaking process). The curves for hematite were calculated from Eq. (7.72) and that for magnetite, assumed to have been created from the corresponding imaginary hematite, was calculated from an equation for the reduction of the imaginary hematite obtained by converting Eq. (7.74).

It is seen that the reactivity of natural magnetite concentrate is somewhat lower than that of hematite. This difference may be explained by the different crystal structures of hematite and magnetite. The oxygen atoms in hematite are arranged in a

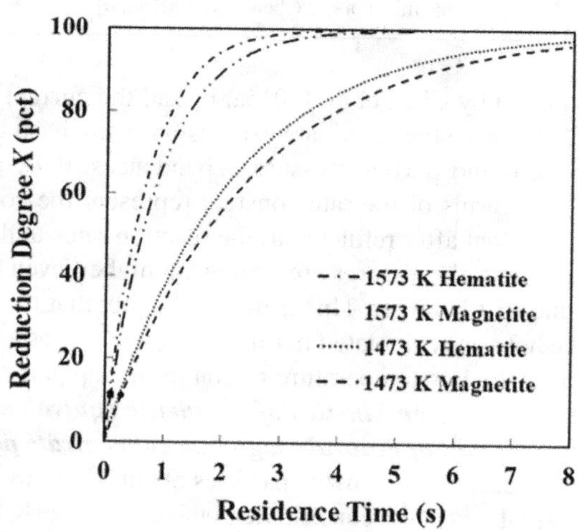

FIGURE 7.36 Comparison of the rate expressions of the reduction of hematite and natural magnetite concentrate particles by H_2 with $p_{H_2} - \left(\frac{p_{H_2O}}{K_{H_2}}\right)$ equal to 0.3; ■: starting point of natural magnetite concentrate particles reduction at $T_{iso} = 1,573$ K (1,300°C); •: starting point of natural magnetite concentrate particles reduction at $T_{iso} = 1,473$ K (1,200°C).

closed-packed hexagonal structure, while they form a face-centered cubic structure in the equilibrium magnetite phase [Chatterjee, 2012]. The crystal structure undergoes significant readjustment when hematite is reduced to magnetite, causing swelling with a volume increase of about 25% [Roller, 1986] due to crack formation, which facilitates the subsequent reduction. However, the oxygen lattice remains unchanged in the reduction of magnetite to wustite, which leads to the lower reducibility when starting with crystalline magnetite.

It is noted that the plot of the RD of the imaginary hematite concentrate particles in Figure 7.36 is exactly the same as Eq. (7.74), except starting at time corresponding to $X = 1/9$. This is particular to the rate expression for nucleation and growth kinetics with $n = 1$, which is the same as a rate equation with first-order dependence on the fraction of solid reactant remaining, i.e., $(1-X)$. The general mathematical derivation for the case of an nth-order dependence is given below.

7.6.5.6 Derivation for Comparison of X-vs-t of a Compound with That of an Intermediate Phase as a Separate Reactant

The proper comparison of the X-vs-t rate equations for hematite (or a general compound) and natural magnetite (or a natural version of an intermediate species formed during the reaction of the compound) will be derived and generalized to other forms of kinetics. We will use, as an example, a rate equation with an nth-order dependence on $(1-X)$. Thus, we want to compare the kinetics of the reduction of hematite (or a compound) given by

$$\frac{dX}{dt} = k_{\text{app}}(1-X)^n \tag{7.75}$$

with that of natural magnetite (or a natural version of an intermediate product) given by

$$\frac{dX_m}{dt} = k_{\text{app}}(1-X_m)^n \tag{7.76}$$

where the fractional RDs X and X_m are defined based on the total removable oxygen contents, respectively, of a hematite (or a compound in general) and a natural magnetite particle (or a natural version of an intermediate product), as follows:

$$X = \frac{\text{Removed Oxygen}}{\text{Removable Oxygen in a hematite Particle}} = 1 - \frac{m_O}{m_{O,H}}, \text{ and} \tag{7.77}$$

$$X_m = \frac{\text{Removed Oxygen}}{\text{Removable Oxygen in a Magnetite Particle}} = 1 - \frac{m_O}{m_{O,M}} \tag{7.78}$$

in which m_O denotes the amount of removable oxygen per unit mass of iron remaining in the sample at any time, and $m_{O,H}$ and $m_{O,M}$ denote the original amounts per unit mass of iron of oxygen, respectively, in hematite and natural magnetite.

When we compare the reduction kinetics of hematite and natural magnetite, the common parameter is the amount of removable oxygen per unit mass of iron. Thus, we need to compare the instantaneous rate of oxygen removal at the same value of m_O whether we start from hematite or natural magnetite. Thus, we first consider the mass change rate from the experimentally determined kinetics of the natural magnetite particles from Eqs. (7.76) and (7.78):

$$-\frac{dm_O}{m_{O,M}dt} = k_{\text{app}}\left(\frac{m_O}{m_{O,M}}\right)^n \tag{7.79}$$

This equation should also describe the mass change of a particle that started from hematite at the same value of m_O. Thus, we can integrate it for the imaginary hematite particle to yield

$$\frac{1}{n-1}m_O^{1-n} = \frac{k_{\text{app}}}{m_{O,M}^{n-1}}t + C \tag{7.80}$$

with the initial condition: at $t = 0$, $m_O = m_{O,H}$, and therefore, $C = \frac{1}{n-1}m_{O,H}^{1-n}$.

Substituting Eq. (7.77) and C into Eq. (7.80) gives,

$$\frac{1}{n-1}\left[\frac{1}{(1-X)^{n-1}} - 1\right] = k_{\text{app}}\left(\frac{m_{O,H}}{m_{O,M}}\right)^{n-1} t \tag{7.81}$$

It is seen that when $n=1$ the rate equation for the reduction of the imaginary hematite, by applying L'Hospital's Rule to the above equation, is identical to Eq. (7.74) for the natural magnetite reduction, as indicated in the discussion above in conjunction with Figure 7.36. However, when $n \neq 1$, the rate equation for natural magnetite to compare with that of hematite will become different due to the modification of the apparent rate constant by $\left(\frac{m_{O,H}}{m_{O,M}}\right)^{n-1}$, as shown by Eq. (7.81).

7.6.6 Reduction by H_2+CO Mixtures

A simple additive relationship was first assumed between the overall oxygen removal rate and the rates by its component gases as $\frac{dX}{dt} = \left.\frac{dX}{dt}\right|_{H_2} + \left.\frac{dX}{dt}\right|_{CO}$, the results of which are shown for experiments in three representative temperatures in Figure 7.37. It is seen that there was a synergistic effect under the H_2+CO mixture as the experimental results were always higher than the calculated results. The reason for this synergistic effect is due to the different morphology of the produced iron by CO reduction, which forms long iron whiskers, unlike H_2 reduction which produces relatively flat islands of iron. This was observed in a previous investigation [Wang and Sohn, 2012] on the reduction of iron oxide compact. This effect is decreased in the presence of H_2 but the structure of iron produced in H_2+CO mixtures has much increased surface areas and can cause swelling of an iron oxide compact [Wang and Sohn, 2012]. The morphological changes are shown in Figure 7.38.

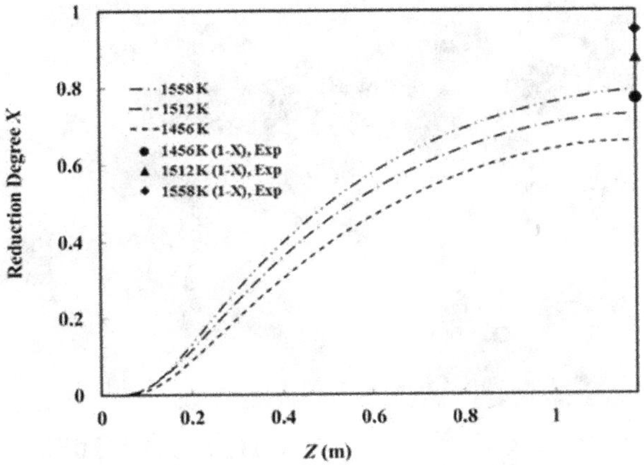

FIGURE 7.37 Comparisons between the calculated reduction degrees by $\frac{dX}{dt}=\left.\frac{dX}{dt}\right|_{H_2}+\left.\frac{dX}{dt}\right|_{CO}$ vs. experimental results. –·–: $pCO=0.4$ atm, $pH_2=0.2$ atm, gas flow rate: CO: 1.6 L/min, H_2: 0.8 L/min, N_2: 1.0 L/min; – – –: $pCO=0.15$ atm, $pH_2=0.3$ atm, gas flow rate: CO: 0.6 L/min, H_2:1.2 L/min, N_2: 1.6 L/min; –·· –: $pCO=0.2$ atm, $pH_2=0.1$ atm, gas flow rate: CO: 0.6 L/min, H_2: 0.3 L/min, N_2: 1.6 L/min. Reactor inner diameter: 5.6 cm, and $d_p=22.5\,\mu m$ for all three cases. The temperature values in the legend represent the experimental temperatures. Each of the experimental conditions was repeated three times and the reduction degree results were found to be within ±5%.

Therefore, to account for the enhanced effect, an enhancement factor was multiplied to the contribution of H_2, as shown below:

$$\frac{dX}{dt}=\left[1+\alpha\left(T,\,p_{CO},\,p_{H_2}\right)\right]\cdot\left.\frac{dX}{dt}\right|_{H_2}+\left.\frac{dX}{dt}\right|_{CO} \tag{7.82}$$

This is based on the reasoning that most of the reduction was done by H_2 in the temperature range investigated and CO mainly enhanced the H_2 reduction rate by modifying the mechanism of hydrogen reduction. The contribution by CO, being rather small, was kept the same as that from CO alone. This can be seen by comparing the reaction rates under the same reaction temperatures, partial pressures, and particle sizes from the individual rate expressions for CO and H_2. The ratios of the reaction rate d*X*/d*t* of the individual gas at $X=0.5$ are listed in Table 7.8. The reaction rate of H_2 under the same experimental conditions is always more than 20 times that of CO. This has not taken into account that the partial pressure of H_2 is almost twice as much as that of CO in a real flash reactor. Moreover, it requires a much higher CO/CO_2 ratio (about 3), compared with an H_2/H_2O ratio of about one, to overcome the equilibrium limitation for the reduction by CO in the temperature range expected for the flash ironmaking process.

FIGURE 7.38 Morphologies of partially reduced iron or concentrated particles showing the effect of CO gas.

TABLE 7.8
Comparison of the Reaction Rate at Different Temperatures with $X = 0.5$ for Magnetite Reduction by CO and H_2 Individually

Temperature	$\left.\frac{dX}{dt}\right\|_{H_2} / \left.\frac{dX}{dt}\right\|_{CO}$
1,423 K (1,150°C)	366
1,523 K (1,250°C)	178
1,573 K (1,300°C)	94
1,723 K (1,450°C)	21
1,773 K (1,500°C)	25
1,823 K (1,550°C)	30

TABLE 7.9
Optimum Values for *a* and *b*

	Lower Temp. Range	Higher Temp. Range
A	0	−0.01
b	1.30	19.65

In Eq. (7.82), $\alpha(T, p_{CO}, p_{H_2})$ is the enhancement factor which is assumed to be of the form

$$\alpha(T, p_{CO}, p_{H_2}) = (a\,T + b) \cdot \frac{p_{CO}}{p_{CO} + p_{H_2}}; \; T \text{ in K} \tag{7.83}$$

with a and b the two parameters to be determined. This correlation factor was formulated by considering that the synergetic effect depends on temperature. Within the relevant temperature range, which is relatively narrow, we linearized the effect of temperature. Secondly, we assumed that the effect of CO is proportional to the fraction $\frac{p_{CO}}{p_{CO} + p_{H_2}}$. This is a simplified correlation but yielded a good result. Different values of a and b are tested, and the optimum values are determined when the calculated RDs most closely match the experimental values.

A wide range of a and b values were discretized and scanned for searching the optimized values. The optimum values are listed in Table 7.9. In the lower temperature range, the temperature effect on this enhancement factor is negligible, whereas in the higher temperature range, temperature has a negative effect on this factor. Given that the H_2/CO ratio is between 0.5 and 2, the value of the enhancement factor is between 0.43 and 0.87 in the lower temperature range. In the high-temperature range, taking $T = 1{,}673$ K (1,400°C) as an example, the enhancement factor is between 0.97 and 1.95 in the same H_2/CO ratio range; while this enhancement factor will drop to only 0.3–0.6 when the experimental temperature is increased to 1,873 K (1,600°C).

Typically calculated RDs along the reactor length obtained using the optimum values listed in Table 7.9, together with the corresponding experimental RDs and temperature profiles, are shown in Figures 7.39 and 7.40. It is seen that the simulated RD values match the experimental values very well.

The results computed based on the CFD method demonstrate that a low-temperature region existed adjacent to the water-cooled injection tube in the top part, and the particle temperature decreased gradually in the bottom part of the reactor. It is seen from Figures 7.39 and 7.40 that it takes about 0.1 m from the tip of the injection tube for the particle temperature to reach the "uniform" isothermal temperatures. The existence of this low-temperature zone in the upper part of the DTR affects the calculation of real particle residence time. From Figure 7.40, it is also seen that larger particles are heated more slowly.

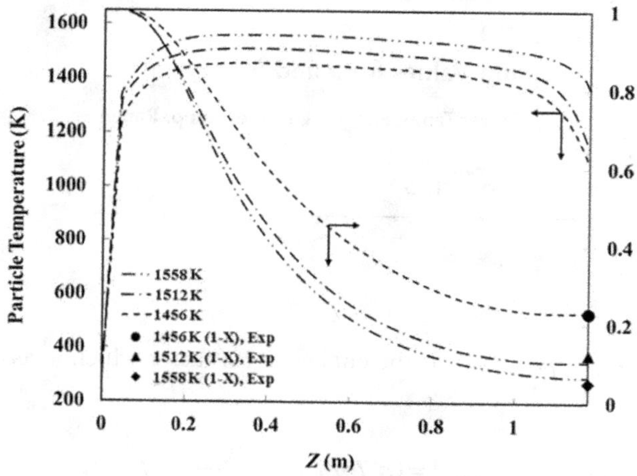

FIGURE 7.39 Calculated profiles of particle temperature and unreacted fraction along the reactor length in the lower temperature range. —· —: $pCO = 0.4$ atm, $pH_2 = 0.2$ atm, gas flow rate: CO: 1.6 L/min, H_2: 0.8 L/min, N_2: 1.0 L/min; — — —: $pCO = 0.15$ atm, $pH_2 = 0.3$ atm, gas flow rate: CO: 0.6 L/min, H_2:1.2 L/min, N_2: 1.6 L/min; —·· —: $pCO = 0.2$ atm, $pH_2 = 0.1$ atm, gas flow rate: CO: 0.6 L/min, H_2: 0.3 L/min, and N_2: 1.6 L/min. Reactor inner diameter: 5.6 cm and $d_p = 22.5$ μm for all three cases. The temperature values in the legend represent the experimental temperatures. Each of the experimental conditions was repeated three times and the reduction degree results were found to be within ±5%.

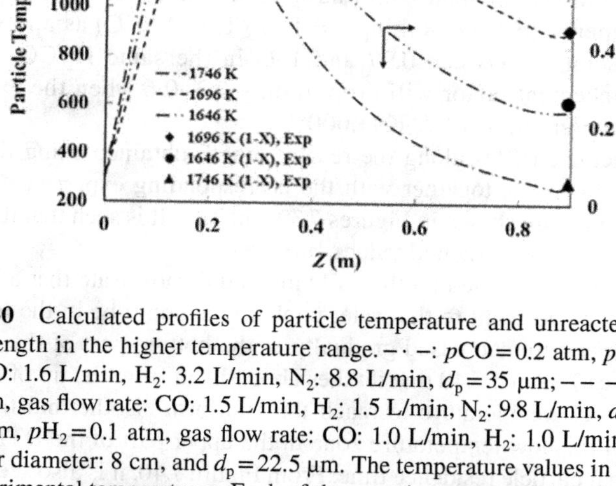

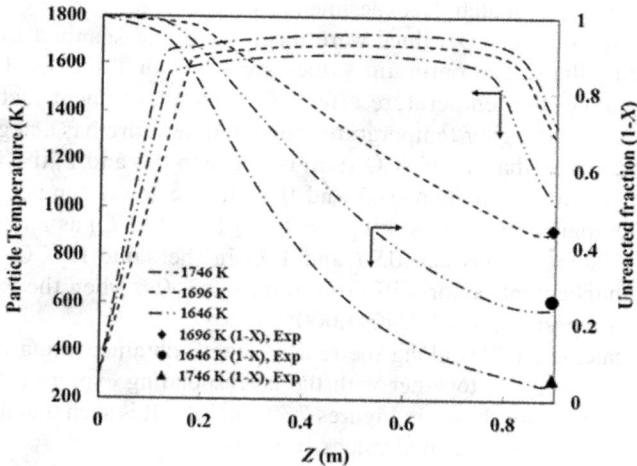

FIGURE 7.40 Calculated profiles of particle temperature and unreacted fraction along the reactor length in the higher temperature range. —· —: $pCO = 0.2$ atm, $pH_2 = 0.1$ atm, gas flow rate: CO: 1.6 L/min, H_2: 3.2 L/min, N_2: 8.8 L/min, $d_p = 35$ μm; — — —: $pCO = 0.1$ atm, $pH_2 = 0.1$ atm, gas flow rate: CO: 1.5 L/min, H_2: 1.5 L/min, N_2: 9.8 L/min, $d_p = 49$ μm; —·· —: $pCO = 0.1$ atm, $pH_2 = 0.1$ atm, gas flow rate: CO: 1.0 L/min, H_2: 1.0 L/min, N_2: 6.3 L/min. Reactor inner diameter: 8 cm, and $d_p = 22.5$ μm. The temperature values in the legend represent the experimental temperatures. Each of the experimental conditions was repeated three times and the reduction degree results were found to be within ±5%.

The CO contributions in the reduction of magnetite concentrate particles are calculated and shown in Figures 7.41 and 7.42. It is seen that the part of contribution that CO plays in the overall reduction is less than 2% in the lower temperature range, while in the higher temperature range, its contribution can raise to around 6%. But overall, the CO contribution to the overall reduction is small.

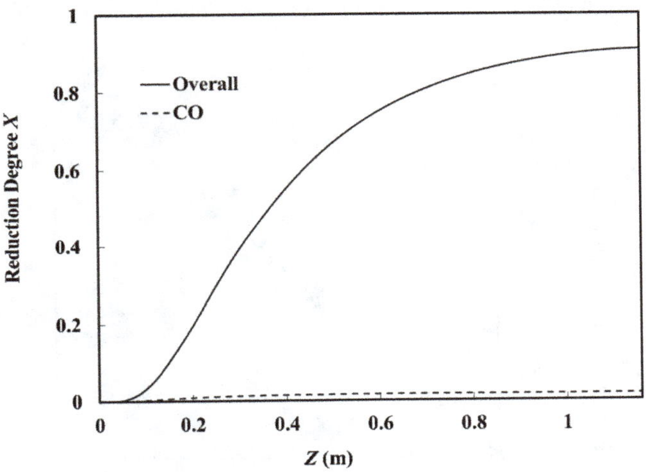

FIGURE 7.41 CO component gas contribution to the overall reduction at the experimental temperature $T = 1{,}512$ K: $p_{CO} = 0.4$ atm, $p_{H2} = 0.2$ atm, gas flow rate: CO: 1.6 L/min, H_2: 0.8 L/min, N_2: 1.0 L/min.

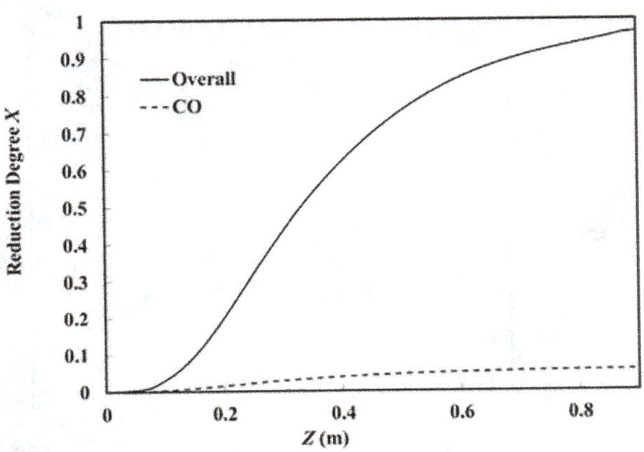

FIGURE 7.42 CO component gas contribution to the overall reduction at the experimental temperature $T = 1{,}746$ K: $p_{CO} = 0.2$ atm, $p_{H2} = 0.1$ atm, gas flow rate: CO: 1.6 L/min, H_2: 3.2 L/min, N_2: 8.8 L/min, and $d_p = 35$ μm.

The comparisons between the calculated RDs using the optimum values of *a* and *b* and the experimental values are shown in Figures 7.43 and 7.44 for the reduction of magnetite concentrate particles by H_2+CO mixtures in the lower temperature range [1,423 K (1,150°C)–1,623 K (1,350°C)] and higher temperature range [1,623 K (1,350°C)–1,873 K (1,600°C)], respectively. It is noted that the optimum values of *a* and *b* obtained predict the RDs very well. Thus, the rate expression proposed as Eq. (7.82) by incorporating an enhancement factor can be used with confidence to describe the reduction of magnetite concentrate particles by H_2+CO mixtures.

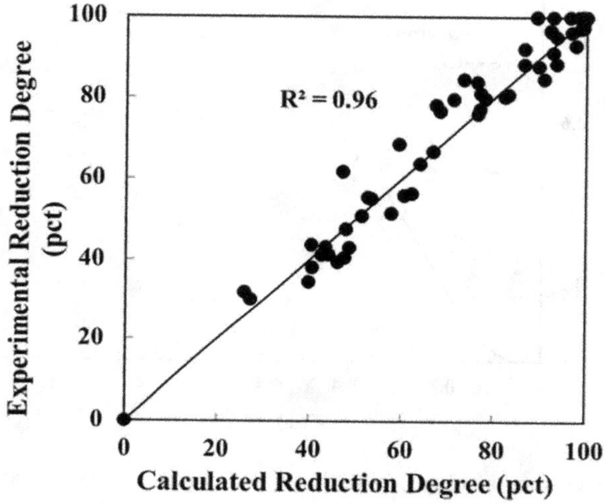

FIGURE 7.43 Comparisons between the calculated reduction degrees vs. experimental results by H_2+CO mixtures in the lower temperature range.

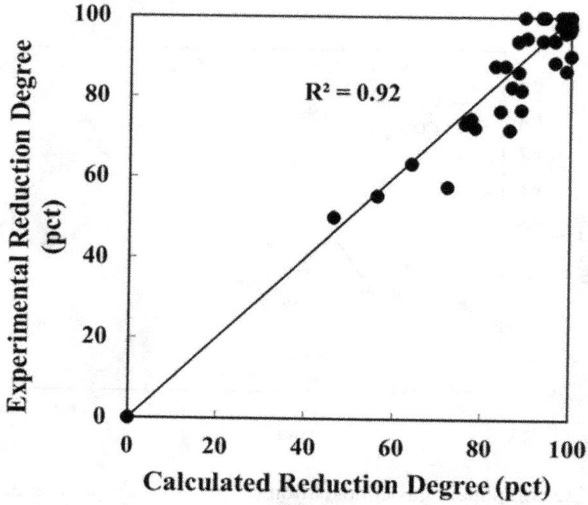

FIGURE 7.44. Comparisons between the calculated reduction degrees vs. experimental results by H_2+CO mixtures in the higher temperature range.

The complete rate equations for the reduction of magnetite concentrate particles by H_2+CO mixtures in the temperature ranges of 1,423 K (1,150°C)–1,623 K (1,350°C) and 1,623 K (1,350°C)–1,873 K (1,600°C) are given, respectively, as:

$$\frac{dX}{dt} = \left(1 + 1.3 \cdot \frac{p_{co}}{p_{co} + p_{H_2}}\right) \cdot \left.\frac{dX}{dt}\right|_{H_2} + \left.\frac{dX}{dt}\right|_{CO} \quad 1{,}423\,\text{K} < T < 1{,}623\,\text{K} \tag{7.84}$$

$$\frac{dX}{dt} = \left[1 + (-0.01T + 19.65) \cdot \frac{p_{co}}{p_{co} + p_{H_2}}\right] \cdot \left.\frac{dX}{dt}\right|_{H_2} + \left.\frac{dX}{dt}\right|_{CO}$$

$$1{,}623\,\text{K} < T < 1{,}873\,\text{K} \tag{7.85}$$

where

$$\left.\frac{dX}{dt}\right|_{H_2} = 1.23 \times 10^7 \exp\left(\frac{-196{,}000}{RT}\right) \cdot \left(p_{H_2} - \frac{p_{H_2O}}{K_{H_2}}\right) \cdot (1 - X)$$

$$1{,}423\,\text{K} < T < 1{,}623\,\text{K} \tag{7.86}$$

$$\left.\frac{dX}{dt}\right|_{H_2} = 6.07 \times 10^7 \exp\left(\frac{-180{,}000}{RT}\right) \cdot \left(p_{H_2} - \frac{p_{H_2O}}{K_{H_2}}\right) \cdot (1 - X) \cdot d_p^{-1}$$

$$1{,}623\,\text{K} < T < 1{,}873\,\text{K} \tag{7.87}$$

$$\left.\frac{dX}{dt}\right|_{CO} = 5.35 \times 10^{13} \exp\left(\frac{-451{,}000}{RT}\right) \cdot \left(p_{CO} - \frac{p_{CO_2}}{K_{CO}}\right) \cdot (1 - X)[-\text{Ln}(1 - X)]^{-1}$$

$$1{,}423\,\text{K} < T < 1{,}623\,\text{K} \tag{7.88}$$

$$\left.\frac{dX}{dt}\right|_{CO} = 3.225 \times 10^3 \exp\left(\frac{-88{,}000}{RT}\right) \cdot \left(p_{CO} - \frac{p_{CO_2}}{K_{CO}}\right) \cdot (1 - X)[-\text{Ln}(1 - X)]^{-1} \cdot d_p^{-1}$$

$$1{,}623\,\text{K} < T < 1{,}873\,\text{K} \tag{7.89}$$

where R is 8.314 J/(mol K), p is in atm, K_j is the equilibrium constant for the reduction of FeO by gas j, d_p is particle size in μm, and t is in seconds. The coefficients in front of the exponential term are the product of n_j k_j. The kinetic parameters for the reduction of magnetite concentrate particles by each component gas are summarized in Table 7.10.

Although the use of the CFD technique should, in principle, allow one to determine the rate expression even from experiments done under spatially varying temperature, velocity, and gas concentrations, the accuracies of the developed rate equations and parameters are greatly enhanced by performing the experiments designed to

TABLE 7.10
Kinetic Parameters for Reduction by Individual Component Gases [Fan et al., 2016a, 2021a, b; Elzohiery et al., 2021; Fan, 2019]

Reducing Gas	Temperature Range	$k_{o,j}$	E_j (kJ/mol)	m_j	N_j	s_j
H_2	1,423–1,623 K	1.23×10^7	196	1	1	0
	1,623–1,873 K	6.07×10^7	180	1	1	1
CO	1,423–1,623 K	1.07×10^{14}	451	1	0.5	0
	1,623–1,873 K	6.45×10^3	88	1	0.5	1

keep these conditions as uniform as possible, in combination with CFD simulation to account for small variations that are difficult to completely eliminate. This is the approach we used in this work.

7.6.7 Summary and Concluding Remarks

Non-uniform temperature regions were found in the upper part of the DTR through CFD simulation and were taken into consideration for rate analysis in this work. It provides a full description of the momentum, heat, and mass transfer of particles during its reduction process, and a detailed evaluation of the particle residence time and particle temperature inside the reactor.

The advantage of this method becomes greater when the reaction temperature varies with position and particle residence time varies with the paths they follow. Although the use of the CFD technique should in principle allow one to determine the rate expression even from experiments done under spatially varying temperature, velocity, and gas concentrations, the accuracies of the developed rate equations and parameters are greatly enhanced by performing the experiments designed to keep these conditions as uniform as possible, in combination with CFD simulation to account for small variations that are difficult to completely eliminate. This is the approach we used in this work.

8 Development of Flash Ironmaking Technology – Tests in a Laboratory Flash Reactor

The ultimate aim of this work was to develop an industrial flash ironmaking process and reactor. Figure 8.1 shows the objectives and the associated scales of test work and facilities for each phase up to the construction of the industrial plant. The kinetics feasibility and rate equations were presented in Chapter 7. These experiments proved that magnetite concentrate is directly reduced to iron within several seconds available in a flash reactor at temperatures as low as 1,250°C. The effects of the flame in the flash reactor and limited excess amounts of reducing gases were investigated in a laboratory flash reactor and will be presented in this chapter. The installation and operation of a pilot-plant-scale flash reactor will be described in Chapter 9.

8.1 LABORATORY FLASH REACTOR

8.1.1 Importance of Testing in Laboratory Flash Reactor

An industrial flash reactor would be significantly different from a laminar-flow reactor (LFR) used for the rate measurement, including the fact that an oxy-fuel burner would be the main source of heat, and the amount of excess reducing gases would be much lower (20%–100%). Furthermore, the pattern of the particle-gas flow inside the reactor would not be as simple as in the LFR, Thus, scale-up experiments in a laboratory flash reactor that is heated by fuel/oxygen flame to test the flash reduction process parameters would be helpful, and at the same time, the obtained rate equations could also be validated.

Therefore, a laboratory flash reactor was built at the University of Utah as a part of developing the novel ironmaking process. In this reactor, iron was produced directly from magnetite concentrate by the partial oxidation of methane and/or hydrogen with pure oxygen producing a non-premixed flame. This flame served as a heating source in addition to electric power that compensated for heat loss. The partial oxidation of methane produced a reducing gas mixture of H_2 and CO that served as the reductant. This was the first experimental realization of the flash ironmaking technology as a part of developing the process and designing a larger bench reactor in which process heat was supplied completely by a natural gas flame. The factors affecting the extent of the reduction of magnetite concentrate which need to be tested are the

DOI: 10.1201/9781003342199-8

Project Objectives	Kinetic Feasibility	Proof of Concept at Lab Scale	Process Validation/ Scale-up	Industrial Pilot TBD (2018+)
Experimental Apparatuses				Approaches 1. Large scale 75-100k tpy 2. Modest scale 10-25k tpy
Funding	Federal: \$350k Industry: \$150k Total: \$500k	Federal: \$0 Industry: \$4.8M Total: \$4.8M	Federal: \$8.0M Industry: \$2.6M Total: \$10.6M	\$10 - 75M Funding TBD

FIGURE 8.1 The phases and objectives of the overall project and associated funding. (Phase 1 was from February 23, 2005, to December 31, 2007; Phase 2 from January 1, 2008 to December 31, 2011; Phase 3 from June 2012 to August 2018.)

temperature, particle residence time, and amount of excess reductant. Sohn et al. [2021] further tested the effects of different flame configurations and concentrate feeding modes. These tests are important in understanding the process as well as developing pilot and industrial reactors.

8.1.2 Apparatus

Figures 8.2 and 8.3 show the laboratory flash reactor used for such test work by Sohn et al. [2021]. The apparatus consisted of six systems: a vertical electrical furnace housing a stainless-steel tube, an electric power control system, a gas delivery system, a pneumatic powder feeding system, a gas scrubbing system, and an off-gas burner. The vertical electrical furnace housed a 316 stainless-steel tube with 19.5 cm of ID and 213 cm of length. This furnace was electrically heated by six SiC heating elements, which were grouped into two series and managed by two silicon-controlled rectifier (SCR) controllers, semiconductor or an integrated circuit (IC) which allow the control of electrical current.

The maximum temperature that could be obtained inside the reactor with this set-up was 1200°C ± 10°C, measured at a depth of 76 cm from the top flange of the reactor. Temperature was monitored inside the reactor tube at different depths (25, 51, 76, 102, and 152 cm from the upper flange) and outside the tube. Temperature and power were monitored and controlled using a computer connected to the SCR power controllers. The concentrate feeding system consisted of two pneumatic powder feeding systems similar to the one described in Section 7.2 to feed up to 0.12 kg/h.

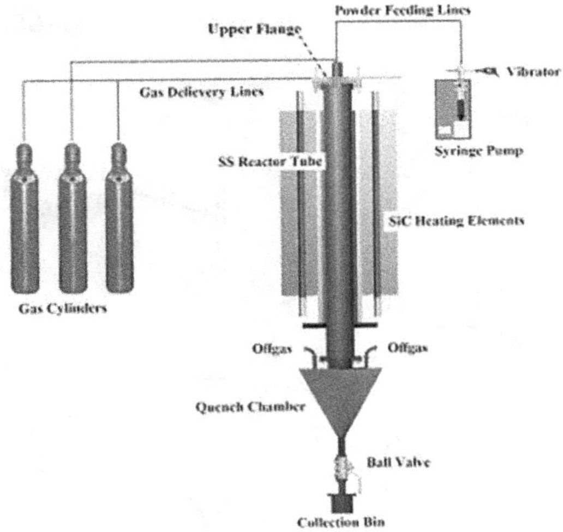

FIGURE 8.2 Schematic representation of the laboratory flash reactor.

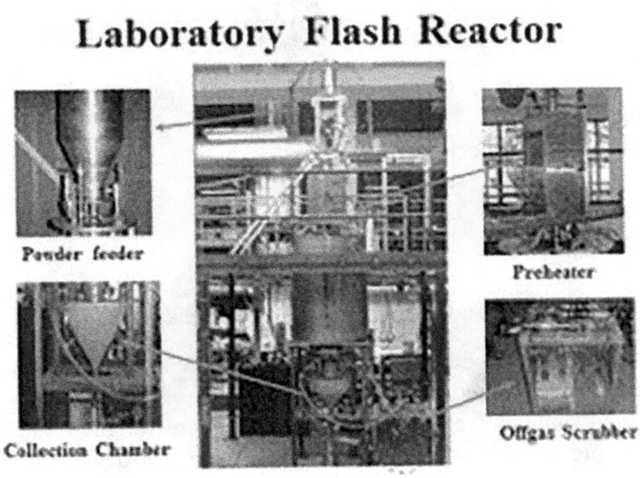

FIGURE 8.3 Photo of the Utah Laboratory flash ironmaking reactor (I.D. 0.19 m and height 2.13 m).

Figure 8.4 shows the bottom part of the reactor where the reduced samples are collected. The quench chamber was designed to quench the hot gases coming from the reactor. The particles settled and stayed in the chamber and the collection bin to minimize particle loss in the off-gas. To increase the amount of the collected sample, a stainless-steel bowl was placed at the end of the reactor tube where most of the particles fall. A magnet was placed under the plate to enhance the amount of collection and capture the particles in the off-gas around the plate. Magnets were installed also on the off-gas elbows to capture more particles that escaped. In most

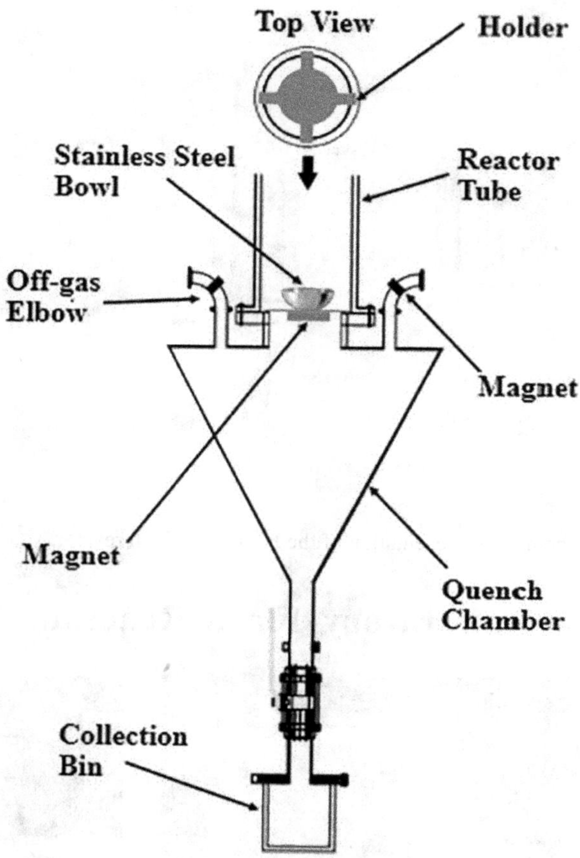

FIGURE 8.4 Collection locations in the laboratory flash reactor.

of the experiments performed in the laboratory flash reactor, more than 60% of the reduced sample was collected.

The collected samples from all these positions were weighed and analyzed for iron content using an ICP. A weighted average reduction degree was determined based on the weight fraction of each sample to the total weight and its reduction degree to determine the overall reduction degree. To verify the results obtained from this method, the samples collected from different locations from the same experiment were mixed, and the analysis of the mixed sample was determined. The difference between the reduction degree obtained from the mixed samples and that from the weighted average was within ±2% of the average reduction degree.

8.1.3 Experimental Procedure

The test variables in this work included the particle residence time, the extent of excess reducing gases (EDF), particle feeding mode, and flame configuration. The system was heated up under a nitrogen flow using electric power. When the temperature reached 1,423 K (1,150°C) measured at a depth of 76 cm in the reactor tube, nitrogen was switched to hydrogen with a pre-determined flow rate, and then oxygen was introduced to the reactor to start the flame. No external igniter was needed to start the flame as the local temperature in the burner was higher than the auto-ignition temperature of hydrogen. Nitrogen flowed through the powder feeder lines all the time to prevent the backflow of the gas. The flame ignition was detected by the sudden increase in temperature of a K-type thermocouple placed 2.5 cm to the side of the burner nozzle where the temperature increased by 100°C–200°C once the flame ignited. The system was then heated by both the flame and the electrical power until the measured temperature at a depth of 76 cm reached 1,473 K (1,200°C).

In the case of experiments with methane, H_2 was replaced with CH_4 gas gradually, and the O_2 flow was adjusted to the required experimental conditions. The flame extinguished after a short time when methane alone was used as fuel. It was found by inspecting the system when extinguishment occurred that serious soot formation was observed at the tip of the burner nozzle as well as the inner wall of the reactor tube. A small amount of H_2 (300 mL/min) was flowed with the CH_4 gas to help maintain a stable flame throughout the experiment and to mimic the H_2 recycled from the off-gas in the operation of an industrial reactor. During the experiment, the electrical power was controlled to maintain the measured temperature at $1{,}473 \pm 5$ K ($1{,}200°C \pm 5°C$).

The methane flow rate was varied to control the particle residence time inside the reactor, and the amount of oxygen fed was varied according to the desired amount of excess reducing gas for the reduction reaction. The flow rates of methane and oxygen gases were controlled using mass flow controllers (AALBORG, Orangeburg, NY) with an accuracy of ± 1% of full scale, while the hydrogen and nitrogen flow rates were controlled using flow meters with an accuracy of ± 2% of full scale.

A nitrogen flow was maintained through the system at the end of a run till it cooled down to 400°C the next day. The collection bin was heated to avoid any water condensation which may mix with the reduced sample. Then, the bottom part of the reactor was detached, and the reduced sample was collected and stored in closed vials for further analyses, including ICP-OES analysis, and SEM. To verify the results obtained from this method, the samples collected from different locations (collection bin, quench chamber inner wall, and bottom part of the reactor) of the same experiment were mixed, and an analysis of the mixed sample was also performed. The difference was found to be within ± 2% of the reduction degree of samples in the collection bin.

8.2 FACTORS AFFECTING THE EXTENT OF REDUCTION

The aim of the experiments performed on the flash reactor was to investigate the conditions that yield the highest reduction degree of the magnetite concentrate particles for the design of the industrial flash reactor. The iron ore concentrate particles used in this work were un-screened magnetite concentrate with an average size of 32.5 μm.

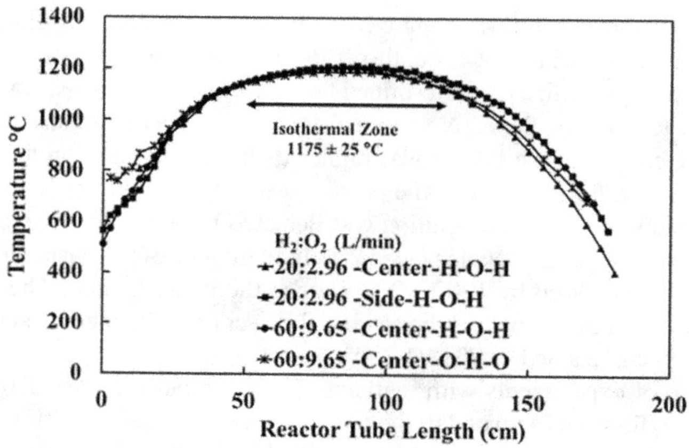

FIGURE 8.5 Temperature profiles measured in the laboratory flash reactor at different conditions.

The length of the reaction zone is required to calculate the particle residence time. Temperature was measured near the wall and at the center of the reactor tube during representative runs under different hydrogen and oxygen flow rates, using a K-type thermocouple at 5 cm intervals. Despite the different amounts of fuel and oxygen in these experiments, the temperature profile was essentially the same, and the isothermal zone defined as the zone where the temperature was 1,175 ± 25°C, was 70 cm long.

In the isothermal zone, the temperature difference between the center and the wall varied from 5°C to 10°C. This temperature difference was small relative to the uncertainties associated with other experimental conditions in this reactor. Therefore, the radial temperature variation in the reactor was considered negligible. The measured temperature profiles under all the different conditions are shown in Figure 8.5.

Various combinations of particles' residence time, excess driving force (EDF), feeding ports, and flame configuration were investigated. In the following sections, the effect of these factors will be discussed in detail with the results of different experiments.

8.2.1 Particle Feeding Modes

The concentrate particles were fed into the reactor through openings in the upper flange installed on the top of the reactor tube. Two different types of feeding modes were tested, which are shown in Figure 8.6: (a) feeding through the center of the fuel/oxygen burner and (b) feeding through two ports on the opposite sides of the burner. The two-side ports were bent toward the center of the tube with an angle of 22.5° from the longitudinal axis to avoid powder accumulation on the inner walls of the reactor tube.

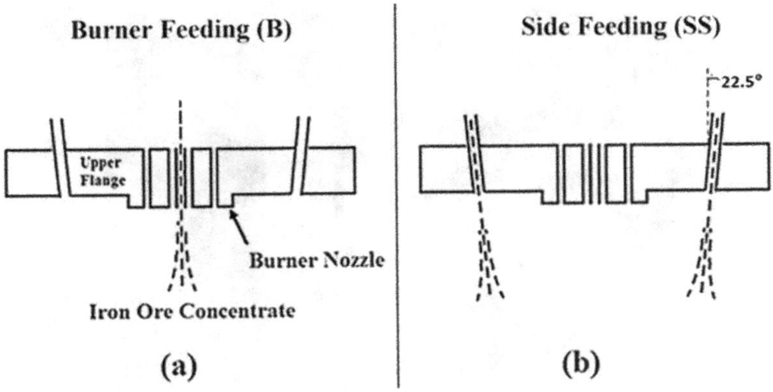

FIGURE 8.6 Powder feeding modes: (a) Burner feeding (through the center of the burner). (b) Side feeding (through two ports on opposite sides of the burner). (Adapted from Elzohiery et al. [2020].)

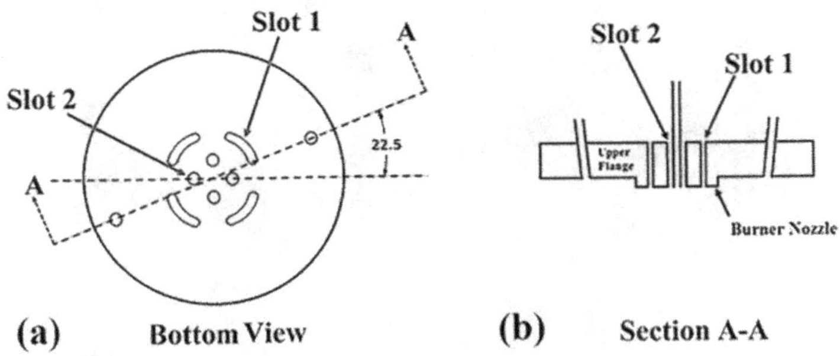

FIGURE 8.7 Sketch of the fuel/oxygen burner: (a) Plan view showing the configuration of the slots. (b) Injection ports at section (A-A). (Adapted from Elzohiery et al. [2020].)

8.2.2 Flame Configuration

A non-premixed burner nozzle made of Inconel with four feeding crescent-shaped inlets (Slot 1) and four cylindrical inlets (Slot 2), shown in Figure 8.7, was used as the burner. Two flame configurations were tested by swapping the fuel and oxygen injection slots. F-O-F (M-O-M and H-O-H when the fuel was methane or hydrogen, respectively) is the first configuration where the fuel was injected through Slot 1 and surrounds the oxygen injected through Slot 2. O-F-O is the second configuration where the oxygen was injected from Slot 1 surrounding the fuel injected from Slot 2.

This variation in the burner configuration changed the temperature distribution in the upper part of the reactor. Figure 8.8 shows the different temperature distributions in the case of hydrogen as the fuel in (a, b) and methane as the fuel in (c, d) simulated using CFD analysis [Elzohiery et al., 2020]. This variation in the flame configuration affected the feeding modes as particles experience different temperatures in different flame configurations.

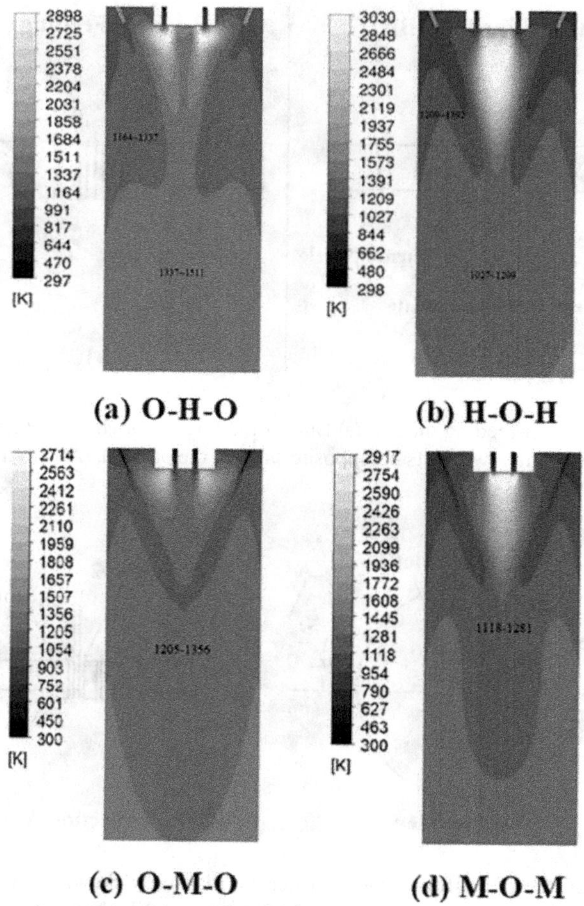

FIGURE 8.8 Temperature profile in the reactor with hydrogen as the fuel in (a) O-H-O, (b) H-O-H; and with methane as the fuel in (c) O-M-O and (d) M-O-M. (Adapted from Elzohiery et al. [2020].)

A non-premixed burner made of Inconel with four crescent-shaped inlets (Slot 1) and four round inlets (Slot 2) was used, as shown in Figure 8.7. Two flame configurations were tested in this study by swapping the inlets for fuel (F) and oxygen (O). In the first flame configuration (designated as F-O-F), the fuel (CH_4+H_2) and oxygen were fed through Slots 1 and 2, respectively. In the second flame configuration (designated as O-F-O), the fuel and oxygen slots were swapped so that the fuel was surrounded by oxygen streams.

8.2.3 Excess Driving Force – for H_2+CO Mixtures

It was determined in Section 7.5.6 that in the temperature range of 1,150°C–1,350°C, the reduction of magnetite concentrate was performed mainly by hydrogen in the case of reduction by H_2+CO gas mixture. In the laboratory flash reactor, the reduction temperature was 1175°C ± 25°C. Therefore, the main reductant in this reactor is hydrogen even when methane was used to produce the reductant gas mixture.

Here, EDF denotes the excess H_2 present in the gas above the equilibrium level, defined as follows:

$$EDF = \frac{\dfrac{p_{H_2}}{p_{H_2O}} - \dfrac{p_{H_2,eq}}{p_{H_2O,eq}}}{\dfrac{p_{H_2,eq}}{p_{H_2O,eq}}} \tag{8.1}$$

The EDF was varied in this reactor from 0.05 to 1.4 when hydrogen was used as the fuel. When methane was used, the tested H_2 EDF values were 0.5 and 1. These factors had a significant effect on the extent of the reduction.

8.2.4 Nominal Particle Residence Time

The residence time of the particles in the reactor is defined as the length of time during which the reduction of particles takes place. As shown in Figure 8.8, the flame configuration changed the temperature distribution in the upper region of the reactor where the particles were fed. Although the temperature profiles showed similar reaction zone in all the measured conditions, the particle temperature changed while traveling in the flame region. The temperature in the flame region was not recorded when the temperature profile was measured at the center as the limit for using the K-type thermocouple is 1,350°C and the flame temperature was much higher. The nominal particle residence time was used to present the data obtained at different flow rates. The nominal particle residence time was calculated using the method described in Section 7.4.1.

The volume of the gases produced from the partial oxidation was considered in the calculation of the gas velocity. The average gas velocity was used as the particles were dispersed in this reactor over the entire area. The reaction zone length was assumed to be the length from the particle inlet to the end of the isothermal zone, which was 117 cm and used for the nominal residence time calculation.

8.3 EXPERIMENTS WITH HYDROGEN

The partial oxidation of hydrogen with pure oxygen produced a flame, which was utilized as a heating source besides the electric heat. This flame produced a hot mixture of hydrogen and water vapor. Changing the flow rate of oxygen changed the ratio of hydrogen and water vapor produced to change the EDF of the gas mixture. A reduction degree >90% with <100% excess hydrogen at a temperature as low as 1,175°C was achieved in several seconds of residence time. There was a preheating system for heating the gases to 500°C before entering the reactor. The excess hydrogen used in that experiment ranged from 0% to 860%. The experimental data are presented in Table 8.1.

The H-O-H flame configuration had a high center temperature (~ 2,624°C) as shown in Figure 8.8, which is higher than the melting temperature of the particles. When fed through the burner in the H-O-H flame configuration, the particles pass

TABLE 8.1
The Conditions of the Experiments Performed with Hydrogen in the Laboratory Flash Reactor

Gas Flow Rates (L/m)			Conc. Feeding Rate (g/m)	Feeding Mode[a]	Flame Config.	EDF	Particles Residence Time (s)	RD (%)
H_2	O_2	N_2						
15.3	2.16	2.80	1.9	SS	H-O-H	0.5	10.6	82
15.3	2.16	2.80	1.8	SS	O-H-O	0.5	10.6	99
15.3	2.36	2.80	1.7	SS	H-O-H	0.4	10.7	76
15.3	2.48	2.80	2.0	SS	H-O-H	0.2	10.7	70
15.3	2.50	2.80	1.9	SS	H-O-H	0.2	10.7	70
15.3	2.72	2.80	2.1	SS	H-O-H	0.06	10.8	57
20.0	2.20	2.80	2.2	SS	H-O-H	1.1	9.5	96
20.0	2.90	1.85	2.0	B	H-O-H	0.5	9.6	26
20.0	2.90	2.00	1.7	B	O-H-O	0.5	9.58	44
20.0	2.96	2.80	1.8	SS	H-O-H	0.5	9.5	84
20.0	2.96	2.80	2.2	SS	H-O-H	0.5	9.5	80
20.0	3.22	2.80	1.9	SS	H-O-H	0.3	9.5	80
20.0	3.70	2.80	2.0	SS	H-O-H	0.1	9.6	63
36.0	5.60	1.85	2.0	B	H-O-H	0.5	7.0	46
40.0	4.32	2.80	2.2	SS	H-O-H	1.4	6.4	92
40.0	6.40	2.80	2.1	SS	H-O-H	0.5	6.5	77
40.0	6.40	2.80	1.9	SS	O-H-O	0.5	6.5	90
40.0	6.40	2.80	2.3	SS	O-H-O	0.5	6.52	87
60.0	7.83	1.85	1.9	B	H-O-H	1.0	5.7	59
60.0	9.65	2.80	1.9	SS	H-O-H	0.5	5.0	74
60.0	9.65	1.85	2.0	B	H-O-H	0.5	5.0	53
60.0	9.65	1.85	2.0	B	H-O-H	0.5	5.0	59
60.0	9.65	1.85	1.5	B	O-H-O	0.5	5.0	75
60.0	9.65	2.80	2.1	SS	O-H-O	0.5	5.0	83
60.0	9.65	2.80	2.2	SS	H-O-H	0.5	5.0	74
60.0	10.65	2.80	2.3	SS	H-O-H	0.3	4.97	72
60.0	11.2	2.80	2.0	SS	H-O-H	0.2	5.0	64
60.0	11.2	2.80	2.0	SS	H-O-H	0.2	5.0	65
60.0	11.2	2.00	2.0	B	O-H-O	0.2	5.0	58
60.0	12.2	2.80	2.0	SS	H-O-H	0.06	5.1	49

[a] B stands for Burner and SS stands for Side Slots.

through the center of the flame and melt due to the high temperature at the core of the flame and then solidify again after the flame to spherical particles. Melting decreases the surface area compared with side feeding where the particles retain their irregular shape and reactivity that produced higher reduction degrees compared to side feeding as shown in Figure 8.9.

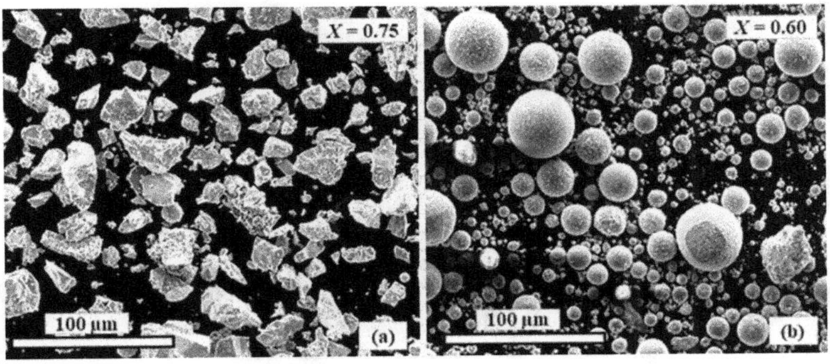

FIGURE 8.9 SEM micrographs of the samples from runs with 60 L/min H_2 and 9.65 L/min O_2 and EDF = 0.5: (a) fed from the side and (b) fed through the burner. [*X* represents the fraction reduced]. (Adapted from Sohn et al. [2021].)

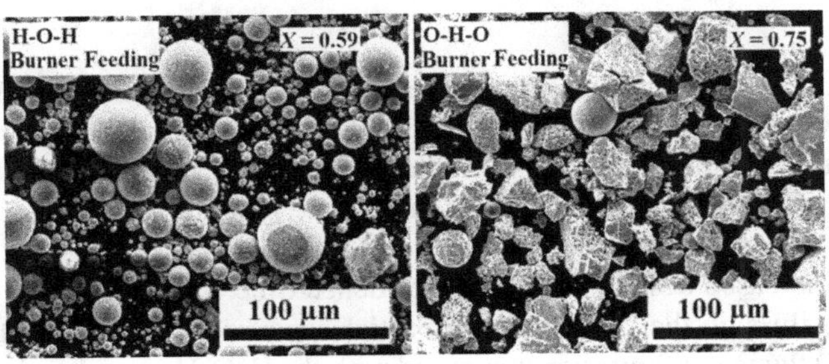

FIGURE 8.10 SEM micrographs of the particles collected from experiments with 60 L/min H_2 and 9.65 L/min O_2, and particles fed through the burner in (a) H-O-H (RD% = 59%) and (b) O-H-O (RD% = 75%) flame configurations. (Adapted from Sohn et al. [2021].)

The amount of oxygen added to hydrogen affected the EDF as well as the flame power. The more the oxygen flow rate, the higher the flame power (directly linked to the flame temperature and length) under the same hydrogen flow rate was. This means that the flame length and temperature increase with a decrease in residence time under the same EDF and temperature.

The experimental results showed that in the case of burner feeding of the particles, the reduction degree decreased with decreasing the overall flow rates of H_2 and O_2 although the residence time was longer. That was attributed to the change in flame length and temperature with the flow rates of H_2 and O_2. In the case of burner feeding, the higher flame temperature and the longer flame length resulted in increasing the particle temperature in the reaction zone and subsequently increased the reduction degree of the particles (RD = 54%), although the residence time was shorter at 5 s compared with the latter case with a lower flame temperature (RD = 26%) with a longer residence time at 9.5 second.

The temperature in the center of the flame with the O-H-O flame configuration is lower than in the case of H-O-H flame configuration, 1,234°C and 2,624°C,

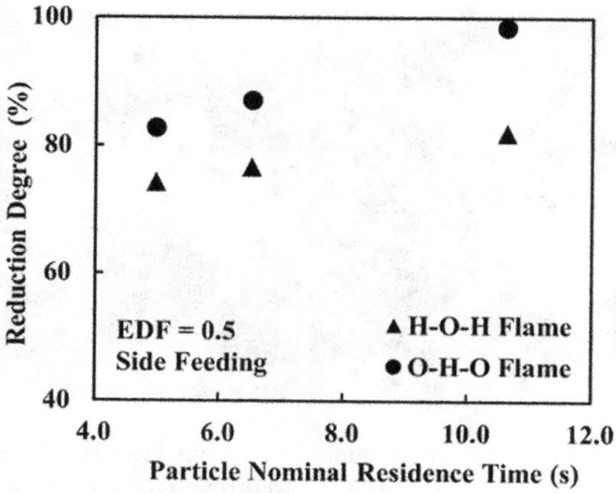

FIGURE 8.11 Effect of flame configuration on the reduction degree at EDF = 0.5 and with particles fed through the side feeding ports. (Adapted from Sohn et al. [2021].)

respectively. Therefore, particles did not melt and retained their irregular (similar to ore) shape even when passing through the flame, which resulted in a higher reduction degree. Figure 8.10 shows the change in the particle shape and reduction degree with the flame configuration under otherwise the same conditions. The effect of flame temperature was still significant with increasing flow rates of H_2 and O_2 as the reduction degree increased despite the shorter residence time.

When fed through the side ports, the particles experienced a higher temperature in the O-H-O flame configuration compared with the H-O-H configuration, as shown in Figure 8.8, which resulted in a faster reaction, as shown in Figure 8.11. However, at this temperature, the residence time effect was not significant, especially with the H-O-H flame.

8.4 EXPERIMENTS WITH METHANE

The flash ironmaking process may be adopted with natural gas as the reductant and fuel. Thus, the laboratory flash reactor was also operated by partially combusting methane gas with pure oxygen in a non-premixed flame, which produced heat and a gas mixture of $H_2 + CO + H_2O + CO_2$ [Elzohiery et al., 2020]. Hydrogen was fed with methane gas to help stabilize the flame. When only methane was injected, the flame was unstable under partial oxidation conditions, and soot was formed at the burner nozzle and the top part of the reactor. Furthermore, in an industrial flash furnace, the injected fuel and reductant gas should be natural gas mixed with the recycled hydrogen recovered from the off-gas, considering the fact that the off-gas must contain a significant amount of hydrogen even at the equilibrium condition. The tested flow rate of hydrogen was varied from 0.1 to 2 L/min.

To set the experiment conditions, HSC 5.11 thermodynamics software [Roine, 2002] was used to calculate the equilibrium product composition. Using the "Equilibrium Composition" module, the input gases' (H_2, CH_4, O_2, N_2) flow rates and the magnetite concentrate feeding rate were added, and the equilibrium composition

TABLE 8.2
The Conditions and Results of the Experiments with Methane in the Laboratory Flash Reactor

Gas Flow Rates (L/min)				Conc. Feeding Rate (g/min)	Feeding Mode[a]	Flame Config.	H_2 EDF	Particle Residence Time (s)	RD (%)
CH_4	H_2	O_2	N_2						
5.0	2.0	4.0	2.8	2.0	SS	O-F-O	1.1	10.5	83
		4.0	2.8	2.4	SS	F-O-F	1.1	10.5	76
		4.5	2.8	2.0	SS	F-O-F	0.5	10.5	64
		4.5	2.8	1.8	SS	O-F-O	0.5	10.5	72
		4.0	2.0	1.9	B	O-F-O	1.1	10.5	83
		4.0	2.0	2.1	B	O-F-O	1.0	10.5	82
		4.5	2.0	2.2	B	O-F-O	0.5	10.5	65
		4.2	2.0	2.1	B	O-F-O	0.5	11	59
		3.75	2.0	1.9	B	O-F-O	1.0	11	78
		3.75	2.8	2.2	SS	O-F-O	1.0	11	81
	0.13	3.75	2.0	2.2	B	F-O-F	1.0	11	33
		3.75	2.0	2.4	B	O-F-O	1.0	11	81
		4.2	2.8	2.5	SS	O-F-O	0.5	11	68
10.0	2.0	8.2	2.8	2.0	SS	O-F-O	1.1	7.6	46

[a] B stands for Burner and SS stands for Side Slots.

of the output was calculated assuming a full reduction of the magnetite to iron at 1,175°C and total pressure of 0.861 bar (the atmospheric pressure in Salt Lake City, Utah). The output gas consisted of H_2, CO, CO_2, H_2O (representing 99.99% of the total composition), and a very small amount of CH_4.

Using this calculated equilibrium composition, the nominal residence time of the particles and EDF for H_2 were calculated. The measured temperature profile in the case of using hydrogen as the fuel was used for the residence time calculation in these experiments where methane is the fuel as it was more dangerous and difficult to measure the temperature profile while using a methane flame. Therefore, a more comprehensive analysis needs to be performed using the CFD simulation for the accurate determination of the residence time, as described subsequently in Chapter 10. The temperature in the flame in the methane experiments was lower than in the hydrogen experiments, as shown in Figure 8.8, and this also contributes to the lower reduction degree with methane.

Table 8.2 lists the experimental conditions. A reduction degree of 80% ± 5% was obtained at EDF = 1 based on hydrogen content and particle residence time of 8 seconds even at the low temperature of 1,175°C.

Due to the limitation of the reactor size and the stainless-steel tube used, it was hard to widely vary the flow rate of natural gas, and this resulted in a narrow range of tested residence time. The reproducibility of the experiments was tested, and it was found to be within ± 5%. A reduction degree of 80% ± 5% was achieved with EDF = 1 of hydrogen and particle residence time of 8 seconds.

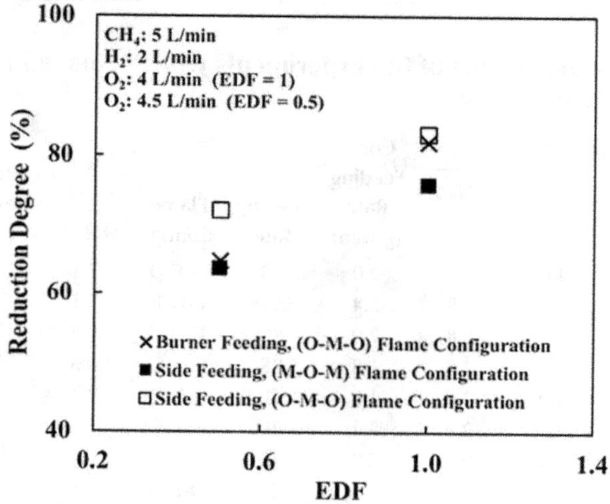

FIGURE 8.12 % Reduction vs. EDF for different feeding positions and flame configurations. (Adapted from Sohn et al. [2021].)

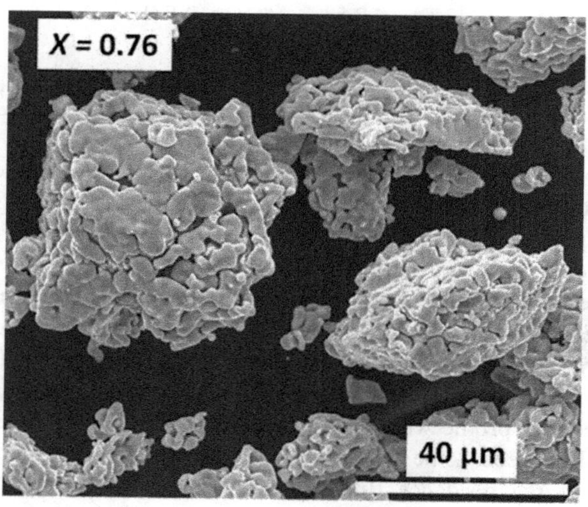

FIGURE 8.13 SEM micrograph of a sample obtained using M-O-M flame configuration, two-side feeding, particle residence time of 8.3 s, and EDF of 1. (Adapted from Elzohiery [2018].)

The effect of flame configuration was investigated by swapping the methane and oxygen slots as described earlier. In the O-M-O configuration, oxygen flow surrounded the methane stream, whereas in the M-O-M configuration, methane surrounded the oxygen flow, as shown in Figure 8.8.

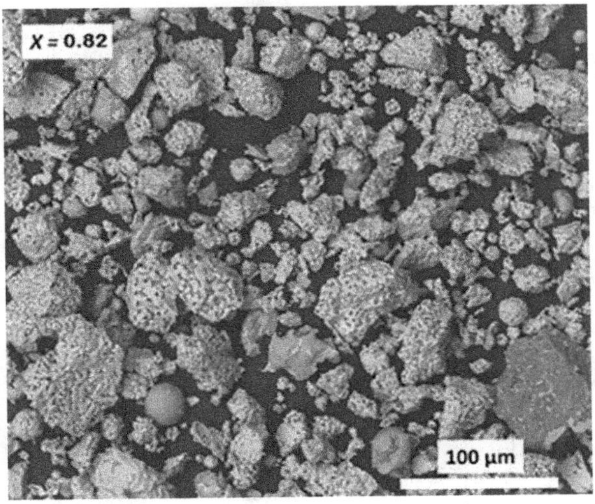

FIGURE 8.14 SEM micrograph of burner feeding sample with the O-M-O flame configuration showing no melting of the particles. (Adapted from Elzohiery [2018].)

Based on the results from the experiments with hydrogen, the O-M-O configuration was used when the solid was fed through the burner to avoid the melting of the particles. Hydrogen is the main reductant at 1,175°C temperature, and CO contributes much less to the reduction, especially in the presence of CO_2. The tested hydrogen EDF values were 0.5 and 1. This configuration yielded reduction degrees higher than with the M-O-M configuration, as shown in Figure 8.12. The samples collected from these experiments showed no melting.

Figure 8.13 shows the SEM micrograph for the sample collected for an experiment performed with the M-O-M flame configuration at EDF = 1. The reduced particles were angular, irregularly shaped, and porous, resulting in a high reduction degree.

At EDF = 1, the burner feeding with the O-M-O flame configuration and the two-side feeding resulted in almost the same reduction degree. Figure 8.14 shows SEM micrographs for the samples collected when feeding was through the burner in the O-M-O flame configuration and EDF = 1 showing no melting of the particles.

Hydrogen gas was flowed with natural gas to help maintain a stable flame throughout the experiment and to mimic an industrial flash reactor operation in which hydrogen in the off-gas is expected to be recycled to the flash reactor. When only methane was used as the fuel, the flame extinguished after a short time, and an inspection of the system revealed that methane cracking occurred and soot was formed at the burner nozzle and the top part of the reactor tube. This may be the reason for not having a stable flame with methane only under partial oxidation conditions in the reactor. The tested flow rate of hydrogen varied from 0.1 to 2 L/min. It was found that under the same EDF, methane flow rate, and temperature, the flow rate of hydrogen did not affect the reduction degree much, as shown in Figure 8.15.

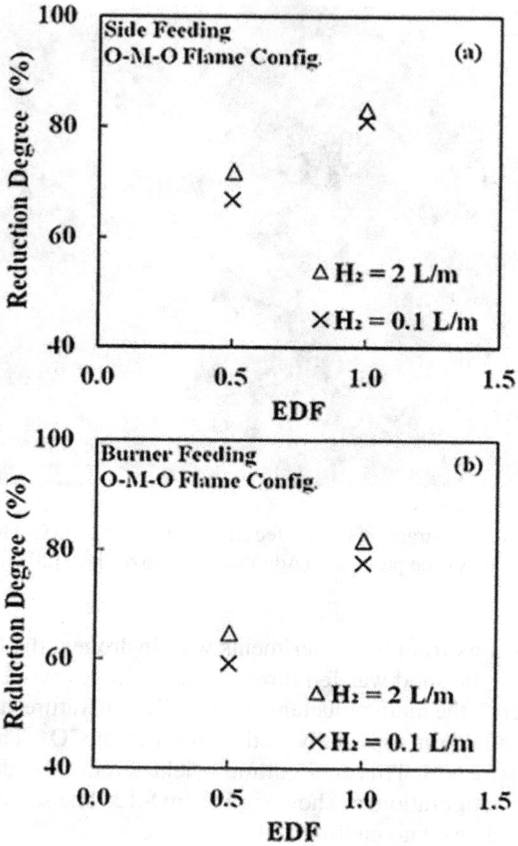

FIGURE 8.15 Effect of EDF on reduction degree (%) with burner feeding, O-M-O flame configuration, and different hydrogen flow rates when (a) feeding through side ports and (b) feeding through the burner port.

It was noted that the reduction degree in the case of methane experiments was less than that obtained in the hydrogen experiments under the same H_2 gas EDF, and the nominal residence time in the methane experiment was longer. As mentioned earlier, a more accurate determination of the realistic residence time needs to be performed. Moreover, a small amount of soot was formed on the stainless-steel reactor tube during the experiments performed with methane. The soot formation was attributed to the methane cracking and therefore, the actual amounts of H_2 and CO produced were less than that obtained by the equilibrium calculations. This soot formation was because the large metal surface of the reactor tube catalyzed the soot formation reaction and, thus, is not expected in larger reactors where a ceramic surface would be in contact with the gas. This was verified with the experiments performed on the mini-pilot flash reactor presented in Chapter 9. Therefore, the soot formation was not studied in more detail in this work.

8.5 CONCLUDING REMARKS

This reactor was the first experimental realization of the flash ironmaking technology in which methane and/or hydrogen was partially oxidized producing heat and reducing gas mixture for direct reduction of magnetite concentrate. The results obtained from this reactor proved that iron oxide can be reduced directly in a flash reactor using natural gas. More than 80% reduction was achieved in this reactor despite its low maximum temperature possible. Different variables affecting the flash ironmaking process were tested such as residence time, excess reducing gas, feeding mode, and flame configuration. The results obtained were used to determine the best feeding modes and the flame configuration to be implemented in a larger-scale reactor as well as an industrial reactor.

Good agreement was achieved between the experiments performed on this reactor and the developed rate equations when hydrogen was used as the reductant and fuel. Magnetite reduction degree > 90% was achieved at a temperature as low as 1,175°C with 100% excess hydrogen driving force or less and a feeding rate of ~ 0.12 kg/h in a few seconds of residence time.

9 Development of Flash Ironmaking Technology – Operation of a Pilot-Plant-Scale Flash Reactor

Author's note: This chapter is based on parts of a published article by H. Y. Sohn, M. Elzohiery, and D.-Q. Fan [2021].

9.1 INTRODUCTION

As one of its kind in the world, a mini-pilot reactor has been operated at the University of Utah to develop a novel flash ironmaking process. In this chapter, we will describe the operation of the mini-pilot flash reactor (MPFR), shown in Figure 9.1, capable of operating at 1,200°C–1,600°C with a concentrate feeding rate of 1–7 kg/h. It is called "mini" only because the industrial production rate of iron is extremely large compared with other metallurgical industries. The reactor installed in this work is no smaller than most pilot plants in the metallurgical industry. The reactor was fabricated and installed by Berry Metal Co. (Harmony, PA, U.S.A.). In this reactor, natural gas was used as the fuel/reducing gas.

This reactor was the first flash ironmaking reactor where the heat and reductants are produced by the partial oxidation of natural gas with pure oxygen.

Commissioning of the reactor with an emphasis on the preheating of the reactor, production of reducing gas mixtures, concentrate feeding, and product particle collection was first completed. The results of experimental runs in this facility, together with other aspects of the overall project such as the CFD simulation, would be used in the design of an industrial-scale pilot plant. This chapter deals with the description of the reactor, its operation, and the difficulties encountered during the operation.

9.2 FACILITY

The MPFR consists of a reactor vessel, a vessel roof, burners, a quench tank, off-gas piping, a flare stack, an off-gas analyzer, a gas valve train, a water-cooling system, gas leak detectors, a concentrate feeding system, and human–machine interface (HMI). A large number of safety precautions were included in the reactor system and its operation. The off-gas was fed into a flare stack where excess oxygen incinerates it to ensure the combustion of the harmful components. A gas analyzer was used to measure the composition of the off-gas stream.

DOI: 10.1201/9781003342199-9

FIGURE 9.1 The mini-pilot flash reactor installed at the University of Utah.

9.2.1 Reactor Vessel and Roof

The reactor vessel, shown in Figure 9.2, is made of a carbon steel shell, which can withstand pressures of up to 3.4 atm. The vessel is lined with three wall layers: 0.3 cm of high-quality, alumina-silica fiber blanket, 8 cm of an insulating layer of high-fired,

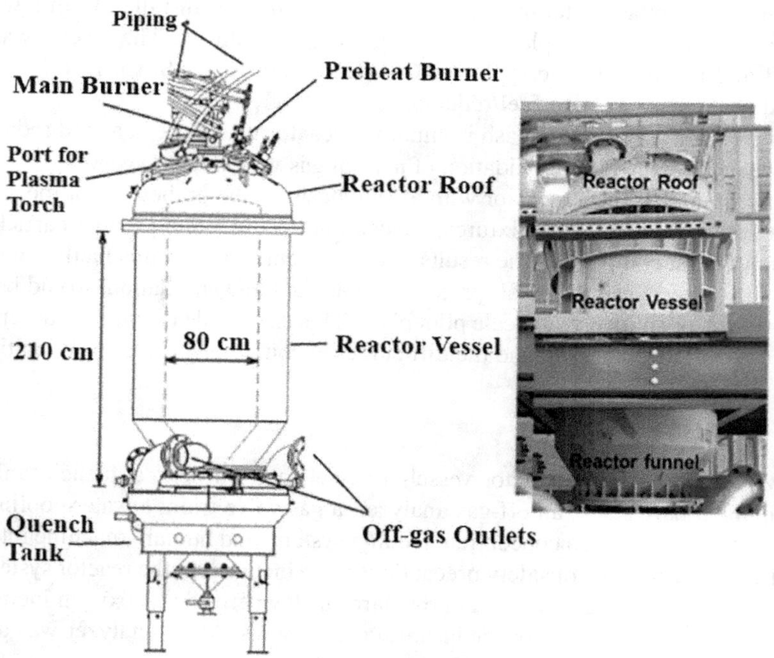

FIGURE 9.2 Schematic diagram and photo of the mini-pilot flash reactor.

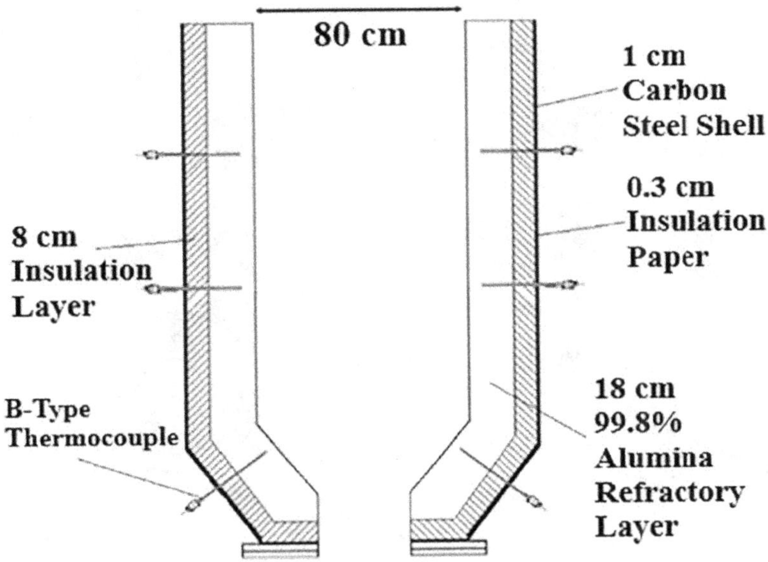

FIGURE 9.3 Schematic diagram of the reactor vessel showing the wall layers.

pressed-shape low iron, lightweight crystalline silica with a special formula to help resist alkali hydrolysis, and 18 cm of 99.8% alumina-castable refractory layer with high hot strength. The materials were supplied and fabricated by ONEX (Crescent, PA, U.S.). These wall layers significantly reduced the heat loss to the surroundings. As an example, the temperature of the steel shell surface was 220°C when the temperature of the inside alumina surface was 1,450°C. The inner diameter of the vessel was 80 cm, and the height was 200 cm, as shown in Figure 9.3.

The reactor roof consisted of the same shell and insulation materials. The roof had an opening for the different burners as well as an emergency off-gas conduit equipped with a rupture disk, which extended through the building roof all the way to building exterior. This rupture disk was designed to burst when the pressure inside the reactor exceeded 1.5 atm above the external pressure for any reason.

B-type thermocouples were embedded in the reactor vessel and roof such that the measuring junctions protruded 2.5 cm into the reactor from the inside surface of the wall. These thermocouples were capable of measuring temperatures up to 1,860°C. The thermocouple signals were used to control the temperature inside the reactor vessel.

9.2.2 Burners

The MPFR was equipped with three burners: a preheat burner, the main burner, and a plasma burner. Figure 9.4 shows a schematic diagram for the cross-section of the preheat burner and the main burner.

The preheat burner used a natural gas-oxygen flame. The preheat burner contained a pilot burner that generates a small flame that ignites the preheat burner.

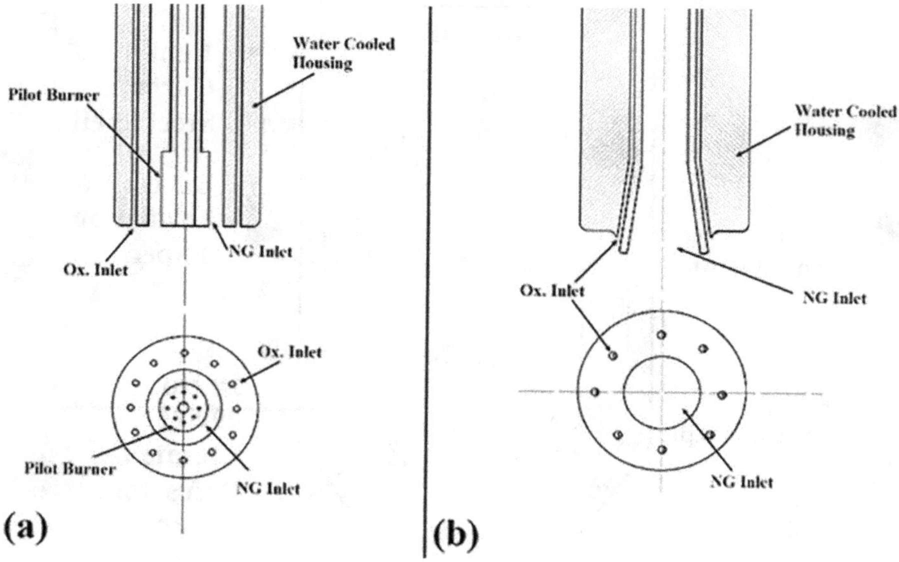

FIGURE 9.4 Schematic diagrams for (a) Preheat burner. (b) Main burner.

The pilot burner had a fiberglass flame detector that detected the pilot flame and started the flow of natural gas and oxygen through the preheat burner. The pilot burner was kept on all the time while there was a flow of natural gas and oxygen to ensure that there was an ignition source inside the reactor. The maximum flow rates for the preheat burner were 1,837 standard liters per minute (SLPM; at 0°C and 1 atm) of oxygen and 875 SLPM of natural gas. The pilot burner operated at flow rates of 9.4 SLPM of natural gas and 38 SLPM of oxygen. The burner nozzle was made of Inconel alloy to withstand the high temperature of the flame. A zirconia refractory block was installed at the tip of the burner to prevent the burner from excessive heating from the reactor.

The main burner, shown in Figure 9.5, fed natural gas and industrial oxygen to produce a high temperature reducing gas mixture. It was uniquely designed to make a swirling gas flow in the reactor to increase the residence time of the particles. The unique design of the burner also shortens the flame length and ensures a larger uniform temperature zone in the reactor. The maximum flow rates that the main burner can operate at are 954 SLPM of oxygen and 1,116 SLPM of natural gas. The burner nozzle was made of Inconel alloy and a zirconia refractory block surrounded the burner tip. This burner was designed to have a flame configuration of O-F-O as determined from the results of the laboratory flash reactor.

A plasma torch was procured and planned to be used as a supplementary heating source for the reactor. This plasma torch would reduce the gas flow rates giving more options to control the experiment conditions. However, this torch was not tested in the current work. Each burner was installed within a water-cooled shell to avoid the excessive heating of the burner parts as well as the reactor roof where all the burners were located.

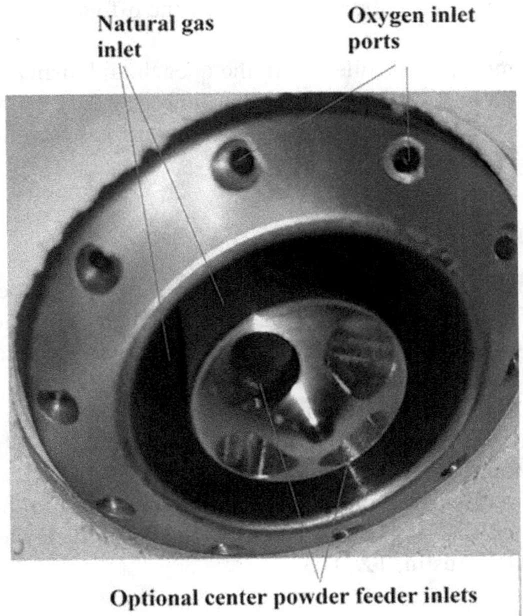

FIGURE 9.5 Main burner in the mini-pilot flash reactor.

9.2.3 Quench Tank

The quench tank was a stainless-steel chamber that allowed the reactor gas to cool down before flowing through the off-gas pipe. The design of the quench tank allowed the particles to settle on the walls of the chamber, and the gas temperature was lowered to less than 600°C. Nitrogen flowed through the inlets as shown in Figure 9.6 at a 45° angle downward to cool the gas and push the particles down. The maximum flow rate of the nitrogen gas was 13,479 SLPM, which was controlled automatically

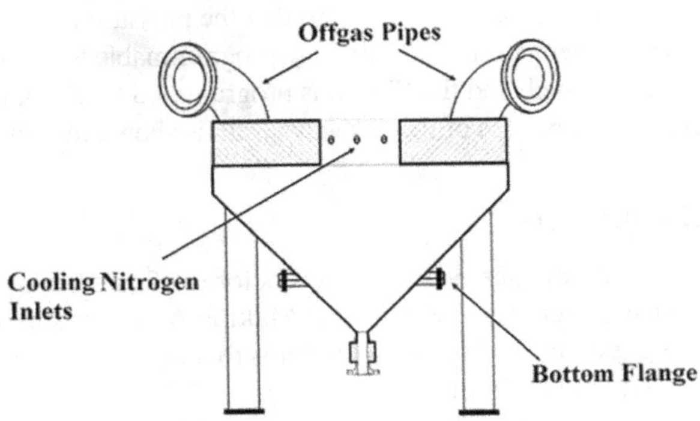

FIGURE 9.6 Schematic diagram for the quench tank.

by the gas valve train to keep the temperature of the off-gas in the quench tank less than 600°C.

The reduced samples were collected in the quench tank using pre-installed magnets, which captured part of the particles that flowed with the off-gas. The powder settled on the walls of the quench tank as well. The off-gas then flowed through the off-gas pipes to the flare stack.

9.2.4 Flare Stack

The flare stack was a long chimney extending to the exterior of the building to burn all the H_2 and CO left in the off-gas during the experiment. The flare stack consisted of a pilot burner, an air blower, and two flame detectors. The pilot burner kept an ignition source all the time during the operation of the flare stack. Two flame detectors were used in the flare stack. An infrared flame detector for the invisible hydrogen flame and an ultraviolet flame detector for a hydrocarbon flame. The air blower supplied an excess amount of air to burn the off-gas completely. The amount of air was controlled automatically based on the measured temperature at the top of the flare stack. The flare stack was supplied by Abutec Company, which was acquired subsequently by AEREON, Austin, TX, U.S.

9.2.5 Gas Valve Train

The gas valve train contained all the gas flow controllers and controlled the flow rates of the gases. The valve train system consisted of three main gas lines, that is, for natural gas, oxygen, and nitrogen. Each gas line contained flowmeters, mass flow controllers, manual valves, pressure gauges, check valves, and safety solenoid valves. The mass flow controllers, supplied by Alicat Scientific, Tucson, AZ, were used for controlling the flow rates of oxygen and natural gas with high accuracy (± 1% of the full scale).

The flow rates of nitrogen were controlled by high-accuracy flowmeters. The solenoid valves were automatically controlled to shut down the natural gas and oxygen flows in case of emergency. Each line contained two solenoid valves and two manual valves for redundancy. Pressure transmitters were installed on each line to measure the pressure at different positions of the lines to ensure that the pressure is within the designed values. All these components were controlled by programmable logic controller (PLC). The valve train was installed, and the PLC was programmed by PTI Combustion company, Tucker, GA, U.S. A picture of the gas valve train is shown in Figure 9.7.

9.2.6 Off-Gas Analyzer

The composition of the off-gas was analyzed before it flowed into the flare stack, using a NOVA gas analyzer Model 875A Steel Making Analyzer supplied by Tenova, ON, Canada. This gas analyzer used three systems to measure the volume percentage of H_2, CO, CO_2, CH_4, and O_2 in the off-gas.

An electrochemical sensor was used to measure O_2 content. This gas analyzer had a resolution of ±0.1%. The gas analyzer was calibrated before each run with a

FIGURE 9.7 The gas valve train system.

calibration gas mixture that contained a fixed composition of H_2, CO, CO_2, and CH_4. The composition of the off-gas was monitored all the time. During the preheat stage, the off-gas was analyzed to ensure that the flame was produced by complete combustion of the natural gas, and no CO, H_2, or CH_4 was present. During the experiment, the off-gas analysis was used to control the experiment conditions.

9.2.7 Water-Cooling System

All the burners were water-cooled during the operation of the reactor. The cooling system was connected to two main cooling sources: circulating water system and city water pipe in case of emergency. The temperature of the water coming to and from the burners was measured all the time using T-type thermocouple. The circulating water system consisted of 18 chillers which circulated 20 gal/min water to the system and back to the chiller where the water was cooled. Safety solenoid valves and manual valves were installed on the water delivery pipes. De-ionized water was used for the chillers to avoid any scale formation inside the water delivery piping and the burners.

In case of emergency, the safety valves shut off the water supply from the chillers and instantaneously turn on the water supply from the main city water pipe to the reactor at a flow rate of 20 gal/min. The emergency situation was defined as when there was a sudden electric shutdown or in the case of a sudden temperature increase in the roof components.

9.2.8 Concentrate Feeding System

The magnetite concentrate was fed into the reactor at a rate of 1–7 kg/h using a HA5171P-D pneumatic powder feeder supplied by HAI, Placentia, CA, U.S. Nitrogen gas at a flow rate of 11 SLPM was used as the carrier gas. The powder feeder had weighing cells that recorded the average feeding rate during the operation. The particles were fed through feeding inlets on the sides of the main burner. This was determined based on the results obtained from the laboratory flash reactor.

9.2.9 Gas Leak Detectors

Industrial gas leak detectors supplied by Honeywell, Lincolnshire, IL, U.S. were placed in different places around the reactor and inside the building to monitor any H_2, CO, and CH_4 gas leak as well as to measure the O_2 level in the building. All the detectors were connected to the main PLC to shut down all the gases in case of a major leak that required evacuation of the building and purging of the vessel.

9.2.10 Human–Machine Interface

The HMI consisted of the main PLC and a computer. The main PLC was connected to all the different parts of the system and to the computer so that the operator could monitor all the different parts and run the reactor. All the safety and emergency procedures relied on the PLC. The main PLC was supplied by ACS company, Boise, ID, U.S.

During the preheating stage, full combustion of natural gas with industrial oxygen is employed to heat the reactor at a ramp rate of 90–95°C/h. The operating temperature range was 1,200°C–1,550°C. During the experimental mode (in which particles were fed), natural gas and oxygen were fed at a predetermined ratio through the main burner to start the partial combustion to generate reducing gases, H_2 and CO, as well as provide the heat needed to maintain the process temperature. A screenshot of the computer monitor that displayed and controlled all the parameters of the operation is shown in Figure 9.8.

9.3 OPERATION OF THE MINI-PILOT FLASH REACTOR

The solid fed to this reactor was as-received, unscreened magnetite concentrate with a particle size range of less than 90 μm. For some experiments, the concentrate was sieved to particle size range 32–90 μm. All the components of the reactor were installed, and a leak test was performed by capping the off-gas pipes and pressurizing the vessel to 2.0 atm for 45 minutes to make sure that there were no leaks from any components. The system was preheated to the target temperature with a heating rate of 90–95 C/h, which was the maximum heating rate that could be used to avoid any damage to the refractories. The heating cycle was automatically controlled by the HMI, and the flow rates of natural gas and oxygen were varied based on the measured temperature of the reactor vessel.

Once the system reached the target temperature, natural gas and oxygen were switched to the main burner, and the main burner operated at full combustion while the preheat burner stopped flowing all gases. However, the pilot burner was kept ignited all the time with small flows of the fuel and oxygen. The system sent signals to the flare stack to turn on and the flare stack ignited its own flame for burning

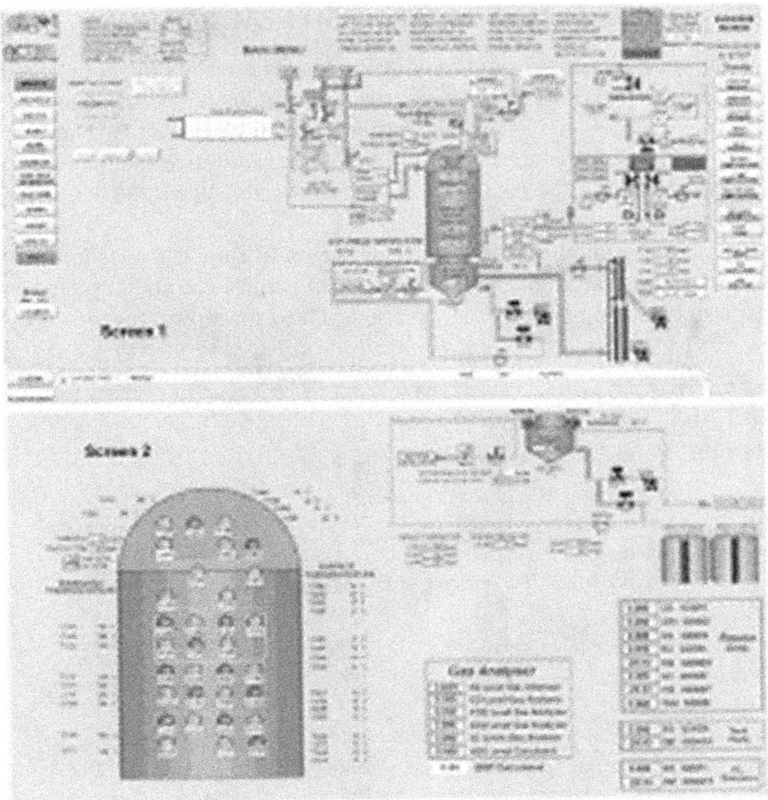

FIGURE 9.8 Screenshot of the computer monitor.

the off-gas. Once the system received a signal that the flare stack was operating, the operator changed the flow rates of the natural gas and oxygen through the main burner to the target experimental values.

The off-gas composition was measured during the run, and the hydrogen-reducing power (EDF) was calculated on the HMI. The flow rates of oxygen and natural gas were adjusted to achieve the target EDF. After feeding a predetermined amount of the magnetite concentrate, the system was shut down by switching all the gas streams to nitrogen. The flare stack was kept running for 5 minutes longer to burn any residual combustible gases flowing out of the reactor. The reactor was left to cool to a temperature below 400°C in the vessel and 25°C in the quench tank. This typically took 4 days. After the quench tank reached 25°C, the flange of the tank was opened, and the reduced sample was collected. The collected sample was analyzed by ICP to determine its reduction degree. The results of these runs were used to verify the CFD simulation model.

9.4 RESULTS FROM MINI-PILOT FLASH REACTOR RUNS

Table 9.1 shows the results from the MPFR runs. The results of these MPFR runs are expected to help in designing the industrial reactor, including the identification of the difficulties and how to improve the various aspects of operation. They were also used to develop a CFD model that would be used to optimize the operating conditions and reactor sizes to be used in an industrial reactor, as will be described in the subsequent chapter.

The swirl in the main burner caused problems in sample collection as will be described in the next section. Lower flow rates of natural gas and oxygen were used in the experiments to moderate the swirl effect. These low flow rates caused the inner

TABLE 9.1
Results of the Mini-Pilot Flash Reactor Runs

Inner Wall Temperature (°C)	Magnetite Concentrate Feeding Rate (kg/h)	Gas Flow Rate[a] Main Burner NG (SLPM)[b]	O_2 (SLPM)	H2 EDF[c]	Nominal Residence Time (s)	RD[d] (%)
1,200–1,130	5.0	404	321	0.76	12.5	65
1,290–1,220	1.8	410	293	0.84	12.0	79
1,290–1,210	2.9	410	293	0.96	12.0	82
1,290–1,230	2.5	358	270	1.00	13.3	83
1,290–1,240	3.5	512	327	1.07	10.2	76
1,330–1,230	4.7	330	200	1.36	15.3	89
1,330–1,230	4.5	330	200	1.44	15.3	87
1,330–1,230	5.2	500	290	3.00	10.6	80
1,330–1,230	4.3	500	290	3.00	10.6	82
1,355–1,260	5.5	235	190	0.03	18.3	7
1,350–1,300	4.0	255	209	0.15	17.0	49
1,350–1,270	4.5	275	212	0.20	16.2	31
1,340–1,280	5.0	280	209	0.21	16.2	37
1,350–1,290	4.6	280	230	0.50	15.6	80
1,400–1,300	6.3	300	240	0.82	14.4	88
1400–1,300	5.0	330	200	1.51	14.6	100
1,415–1,350	4.5	220	191	0.07	18.0	18
1,410–1,360	4.0	240	195	0.33	17.1	32
1,410–1,330	5.0	295	221	0.50	14.7	66
1,410–1,330	6.0	300	210	0.70	14.9	74
1,410–1,320	5.0	300	210	0.82	14.9	82

[a] The flow rates of NG and O_2 in the pilot burner were 9.6 and 37.6 SLPM, respectively. The flow rate of N_2 in the powder feeder was 10.7 SLPM.
[b] SLPM stands for standard liters per minute.
[c] EDF stands for excess driving force of reactant gas over the equilibrium value.
[d] RD stands for reduction degree.

wall temperature of the reactor to drop with time. Different experimental runs were designed and made in this reactor to yield a wide range of reduction degrees, deliberately at less than complete reduction, to better examine the effects of the operating conditions and validate the CFD model in these different conditions.

In this reactor, no soot was formed during the experiments. LECO analysis was performed on some of the samples collected from the experiments. Only one sample contained a small amount of carbon, less than 0.24% C, while all other samples contained no detectable amount of carbon. The LECO analysis was performed by RJ LEE Group (Monroeville, PA, USA).

The experiments in the mini-pilot reactor were performed at different temperatures and reducing powers of the gas with the aim of obtaining enough data for designing the industrial flash reactor. A nominal residence time for the particles was calculated, for a general reference, based on the total volumetric flow rate of the gases at the experiment temperature and the total volume of the reactor vessel which was 1.01 m^3. When the velocities of the gases and the particles at the injection points are rigorously applied, the actual residence time would be shorter. More accurate residence time distribution was obtained by the CFD simulation of the operation, as discussed in the subsequent chapter.

The results showed good reproducibility within $\pm 5\%$ of the average reduction degree by repeating the same experiment at least three times. This represents a very high degree of reproducibility, considering the complexity of the operation and design of this large unit.

9.5 DIFFICULTIES DURING OPERATION

The MPFR was the first of its kind, and the design and operation of such a large reactor posed a number of challenges. It took much effort to fix many of these difficulties. Others remained parts of the system design that must be avoided while designing the industrial reactor. This would provide a valuable finding from this type of testing program. In the following sections, we will discuss the major difficulties associated with operating the MPFR and how they were overcome.

9.5.1 Water-Cooling System

The water-cooling chillers encountered a pressurizing problem during the operation that caused them to over-pressurize and then crack causing water leaking. This caused running the system with low cooling efficiency resulting in increasing the temperature of the roof components to exceed the manufacturer specification (90°C). Therefore, the city water had to be used as the source for water cooling till fixing that problem. The chillers were operated using DI water while the city water contains salts, and these salts cause scale formation inside the water delivery pipes and the burners when used for a long time. The leakage was fixed by adding reinforcements for the chillers and replacing some of the flexible tubing connected to the chillers with reinforced tubes which avoided tubes being pinched and blocked.

9.5.2 Embedded Thermocouples in Reactor Vessel

The MPFR reactor was equipped with 12 embedded C-type thermocouples for measuring the temperature inside the reactor vessel at different positions and the roof. The complete run on the MPFR includes preheating the reactor to the target temperature, running the experiments at the target temperature for 2–3 hours then cooling the reactor to room temperature. After each complete run, three to four thermocouples routinely failed although C-type thermocouples are used in industry for similar applications, and they can read high temperatures up to 2,320°C. However, the C-type thermocouple wires are made of tungsten – 5% rhenium vs. tungsten – 26% rhenium. Tungsten is brittle at high temperatures, and due to the cyclic heating rate of the reactor from high to low temperature the wires of the thermocouples failed at a faster rate compared with industrial applications where the reactors are held at the target temperature for days.

All the embedded C-type thermocouples in the reactor had to be replaced by B-type thermocouples. The B-type thermocouple wires are made of platinum – 30% rhodium vs. platinum – 6% rhodium. Although the maximum temperature reading of B-type thermocouple is lower (1,860°C) than that of the C-type (2,230°C), the platinum wires of the B-type thermocouples enabled them to work at the cyclic heating rates in the reactor without failure. The programing of the HMI was changed to accommodate the new thermocouples.

9.5.3 Quench Tank Cracking

The reactor vessel consists of a carbon steel shell and layers of insulation while the quench tank is made up of stainless steel. The connection between the vessel and the quench tank is a stainless-steel cylinder, which is connected to the vessel with a flange and welded to the tank. In this region, the high-temperature gases coming from the vessel are quenched with a high flow of N_2 gas. However, this part is not well insulated and due to the rapid cooling of the hot gases, a crack was formed in this part, which caused a gas leak-out of the reactor. This crack was welded, and then a moldable ceramic fiber insulation, supplied by Cotronics Corp., NY, U.S., was used to form an insulation layer placed in this part, which avoided cracking and overheating of this part.

9.5.4 Sample Collection

The reactor's inner diameter was 80 cm while the opening where the gases and particles flowed through to the quench tank was only 20 cm. This opening was narrower than the reactor's inner diameter due to the limitation of overall reactor size to be built inside the facility at the University of Utah. That reduction in diameter required designing a conical part as shown in Figure 9.3 for reducing the size from 80 to 20 cm within a short height. Unfortunately, that conical shape resulted in particle accumulation on its surface during the run, which adhered to the alumina walls and stayed there forming a solid accretion layer. This accretion is hard to remove as shown in Figure 9.9. This accumulation resulted in a limited amount of sample

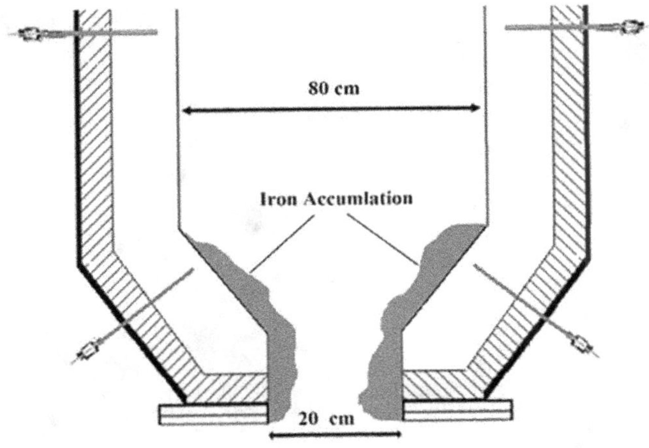

FIGURE 9.9 Product particles accumulated inside the vessel.

collected from the experiment compared with the amount fed. As more particles accumulated on it, the rougher the cone surface became, which in turn trapped more particles in this part.

9.6 CONCLUDING REMARKS

An MPFR was installed and operated on the University of Utah campus. This was the first ironmaking flash reactor in which the heating and reducing gas are produced by partial combustion of fuel gas. This reactor represented a huge step toward the development and commercialization of the flash ironmaking technology.

The reactor was heated completely by natural gas/oxygen flame and operated in the temperature range of 1,200°C–1,600°C. The concentrate feeding rate in this reactor was 1–7 kg/h. The reactor had a thick insulation layer to minimize heat loss and was lined with a high-quality refractory layer against corrosion and also for insulation. The operation of this reactor was completely automated with high safety precautions implemented.

Good results were obtained from the experimental runs in which a reproducibility of ±5% of the average reduction degree was achieved. Full reduction was reached at an average temperature of 1,350°C with H_2 EDF of 1.5. The results obtained from this reactor proved the concept of the flash ironmaking process at a large scale. These results together with other aspects of the overall project such as the CFD simulation would be used for the design of an industrial-scale pilot plant. This work further identified potential technical hurdles and associated safety problems and solutions that would be implemented in the design and operation of an industrial flash ironmaking reactor.

10 Development of Flash Ironmaking Technology – Computational Fluid Dynamics Design of Flash Ironmaking Reactors

A three-dimensional computational fluid dynamics (CFD) model was developed to simulate the gas-particle flow pattern, heat transfer, and chemical reactions inside a laboratory-scale flash. The obtained rate expressions discussed earlier were used in the CFD simulation. The temperature profiles and the reduction degrees obtained from the simulation results satisfactorily agreed with the experimental measurements.

10.1 COMPUTATIONAL FLUID DYNAMICS MODELING OF THE UTAH LABORATORY FLASH REACTOR

Along with the experimental data, CFD models are useful tools in process design and optimization of an industrial reactor. The CFD model can also be used to investigate the various transport phenomena taking place in the new process.

A three-dimensional CFD model described in Section 7.6 was modified to simulate the fluid flow, heat transfer, and chemical reactions involved in the laboratory flash reactor. User-defined functions (UDFs) for chemical reaction kinetics were implemented. Combustion mechanism that consisted of seven chemical reactions was incorporated via a separate chemical-kinetic mechanism data file [Kee et al., 1996; Eklund et al., 1990]. The CFD model was validated by using the experimental reduction degrees and measured temperature profiles.

The laboratory flash ironmaking reactor was described in Section 8.1.2. A non-premixed burner nozzle made of Inconel with four feeding crescent-shaped inlets for the fuel (H_2) and four cylindrical inlets for oxygen was used as the burner. Magnetite concentrate particles were fed through the two side-injection ports on the upper flange of the reactor at a constant rate with N_2 as the carrier gas. The side injection ports were installed on either side of the burner with an angle of 22.5° from the vertical axis of the reactor tube. The specific positions of the injection ports are shown in Figure 10.1. The quench chamber and the collection bin were not included in the simulation due to the negligible reactions in these regions where the temperature was lower than 500 K (227°C). Concentrate particles were fed continuously at 100–140 g/h for about an hour in each experiment after the reactor temperature

DOI: 10.1201/9781003342199-10

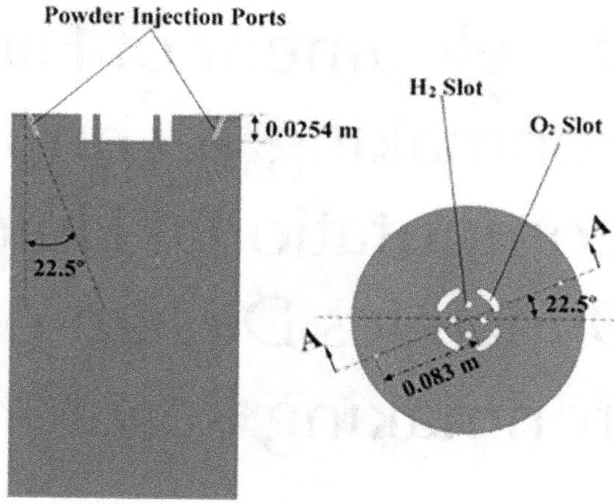

FIGURE 10.1 The positions of injection ports. (a) Injection ports plane (A-A section). (b) Top view of the reactor.

stabilizes for an hour. Therefore, the operation can be assumed to be at steady state. Only steady-state conditions were simulated in this work.

Three-dimensional mesh was generated using ICEM-CFD ANSYS with a total number of 442,364 hexahedral dominating hybrid cells, as shown in Figure 10.2.

10.1.1 Fluid Flow

The following steady-state continuity equation and the Reynolds averaged Navier–Stokes equation were solved:

$$\frac{\partial}{\partial x_i}(\rho u_i) = S_p \tag{10.1}$$

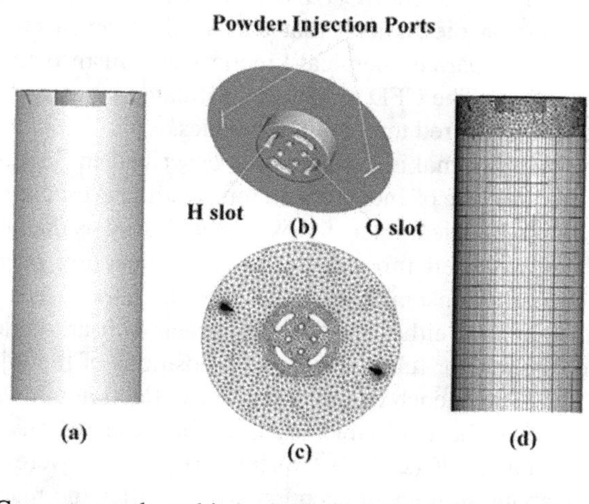

FIGURE 10.2 Geometry and meshing of the Utah Flash Reactor (UFR) and burner. (a) Reactor interior. (b) Burner. (c) Burner Mesh. (d) Hybrid Mesh for the entire reactor.

$$\frac{\partial}{\partial x_j}(\rho u_i u_j) = -\frac{\partial p}{\partial x_i} + \frac{\partial}{\partial x_j}\left[\mu\left(\frac{\partial u_i}{\partial x_j} + \frac{\partial u_j}{\partial x_i} - \frac{2}{3}\delta_{ij}\frac{\partial u_l}{\partial x_l}\right)\right] + \frac{\partial}{\partial x_j}\left(-\rho\overline{u_i' u_j'}\right) + \rho g_i + F_{p,i} \tag{10.2}$$

where ρ is the gas mixture density, which is calculated based on each of the constitutive component (Y_i represents the mass fraction of species i), as follows:

$$\rho = \frac{P}{RT\sum_i Y_i / M_{w,i}} \tag{10.3}$$

The source terms in Eqs. (10.1) and (10.2) arise from interactions between gas species and solid. The source term S_p represents the net rate of mass addition to the gas phase per unit volume due to the gas–solid reaction. $F_{p,i}$ represents the volumetric momentum exchange rate between the continuum phase (gas) and discrete phase (solid particles) due to the drag force exerted on the particle, which will be discussed in the next section.

The flow of the gas stream in the reactor was turbulent, especially in the upper region of the reactor. A turbulence model is needed for the calculation of the Reynolds stress term in Eq. (10.2); $-\rho\overline{u_i'u_j'}$. One of the weakness of the standard k-ε model lies in the model equation for the dissipation rate (ε), which causes erroneous prediction of the spread of the axisymmetric round-jet. The realizable k-ε model proposed by Shih et al. [1995] was intended to address these deficiencies of traditional k-ε models by adopting a new model equation for dissipation (ε) based on the dynamic mean-square vorticity fluctuation. These authors demonstrated that this method yields a better prediction of the dissipation of axisymmetric jets. Thus, the realizable k-ε model was chosen in this study to calculate the turbulent viscosity required for computing the Reynolds stress term in Eq. (10.2).

$$\frac{\partial}{\partial x_i}(\rho k u_i) = \frac{\partial}{\partial x_i}\left[\left(\mu + \frac{\mu_t}{\sigma_k}\right)\frac{\partial k}{\partial x_i}\right] + G_k + G_b - \rho\varepsilon - Y_M + S_k \tag{10.4}$$

$$\frac{\partial}{\partial x_i}(\rho\varepsilon u_i) = \frac{\partial}{\partial x_i}\left[\left(\mu + \frac{\mu_t}{\sigma_\varepsilon}\right)\frac{\partial \varepsilon}{\partial x_i}\right] + \rho C_1 S\varepsilon - \rho C_2\frac{\varepsilon^2}{k + \sqrt{\nu\varepsilon}} + C_{1\varepsilon}\frac{\varepsilon}{k}C_{3\varepsilon}G_b + S_k \tag{10.5}$$

The k and ε obtained from the above equations are used to calculate the turbulent viscosity according to the following equation:

$$\mu_t = \rho C_\mu k^2/\varepsilon \tag{10.6}$$

and the Reynolds stress term $-\rho\overline{u_i'u_j'}$ in Eq. (10.2) is calculated as

$$-\rho\overline{u_i' u_j'} = \mu_t\left(\frac{\partial u_i}{\partial x_j} + \frac{\partial u_j}{\partial x_i}\right) - \frac{2}{3}\left(\rho k + \mu_t\frac{\partial u_l}{\partial x_l}\right)\delta_{ij} \tag{10.7}$$

10.1.2 Heat Transfer

Heat transfer is of particular importance in the flash ironmaking process as the temperature distribution inside the reactor greatly affects the heating of the iron concentrate particles, which in return affects the kinetics of the particle reaction. The governing equation for the energy balance in the gas phase is given as

$$\frac{\partial}{\partial x_i}(\rho u_i h_g) = \frac{\partial}{\partial x_i}\left(\frac{k_{eff}\,\partial T}{\partial x_i}\right) + Q_r + S_g \tag{10.8}$$

where $h_g = \sum_i Y_i \int_{T_{ref}}^{T} c_{p,i} dT$ represents the specific enthalpy of the gas mixture.

The two source terms on the right-hand side (RHS) of Eq. (10.8) represent the net volumetric heat transfer by gas-phase radiation and heat addition/loss to the gas phase due to the reduction reaction of iron concentrate particles by H_2 per unit volume, respectively. A correct prediction of the volumetric radiation source term in Eq. (10.8) is critical in obtaining an accurate temperature distribution due to the fact that the radiation plays a significant role in heating particles. The radiative transfer equation is as follows, in which for simplicity the gas mixture was assumed to be gray and, thus, the spectral dependency term is neglected:

$$\frac{dI}{ds} = \kappa \frac{n^2 \sigma T^4}{\pi} - \kappa I - (\kappa_p + \sigma_p) I + E_p + \frac{\sigma_p}{4\pi}\int_0^{4\pi} I(\vec{s}_i)\Phi(\vec{s}_i, \vec{s})\, d\Omega_i \tag{10.9}$$

The most absorbing species in the relevant gas mixture is H_2O and thus the contributions from other gas species were neglected. The weighted-sum-of-gray-gases model [Smith et al., 1982; Coppalle and Vervisch, 1983] (WSGGM) was used to evaluate the absorptivity (κ) of the gas mixture, which is a reasonable compromise between the oversimplified gray gas model and a complete model that considers particular absorption bands for nongray gases. The scattering coefficient of the gas mixture was assumed zero.

The existence of the particles in the gas stream adds two extra terms to Eq. (10.9) representing the attenuation and emitting effects of particles, namely the third term and the fourth term on the RHS, respectively. The equivalent particulate absorption and scattering coefficients were calculated by averaging the values for all the particles contained within the control volume of interest when tracking the discrete phase, as follows:

$$\kappa_p = \frac{1}{V}\sum_{i=1}^{N} A_{p,i}\varepsilon_{p,i},\ \sigma_p = \frac{1}{V}\sum_{i=1}^{N}(1-\sigma_{p,i})(1-\varepsilon_{p,i})A_{p,i},\ \text{and}$$

$$E_p = \frac{1}{V}\sum_{i=1}^{N} A_{p,i}\frac{\varepsilon_{p,i}\sigma T_{p,i}^4}{\pi} \tag{10.10}$$

where N is the total number of particles existing in the control volume, $A_{p,i}$ is the surface area of particle i, and V is the volume of the control volume of interest. The radiative properties of single particles will be discussed in Section 10.1.7.

10.1.3 Species Transport

As multiple species were involved during the experiment, each component of the gas mixture has to be solved individually. The mass conservation equation for each of the constitutive components in terms of mass fraction is given by

$$\frac{\partial}{\partial x_j}\left(\rho Y_i u_j\right) = -\frac{\partial J_j}{\partial x_j} + R_i + S_{p,i} \tag{10.11}$$

where Y_i is the mass fraction of species i, R_i is the net rate of production of the species i by chemical reaction (detailed chemical reactions will be covered later) and S_i is the net rate of mass addition to the species i from the particle phase.

A simplified Fick's-law correlation was used for the mass diffusion flux $\boldsymbol{J}_i$, which is expressed as

$$J_i = -\left(-\rho D_{i,m} + \frac{\mu_t}{Sc_t}\right)\nabla Y_i - D_{T,i}\frac{\nabla T}{T} \tag{10.12}$$

The coefficient of the first term in Eq. (10.12) combines the effects of molecular diffusion and turbulent diffusion. The turbulent Schmidt number Sc_t was chosen to be 0.7 in this work. The second term accounts for diffusion due to thermal gradient.

10.1.4 Particle Tracking

Particle Tracking Method and Equations were the same as those described in Section 7.6.

10.1.5 Boundary Conditions

All the boundary conditions were chosen to match the experimental conditions. A mass flow rate was imposed at the H_2 and O_2 inlets. The operating pressure was kept at 86.1 kPa (the barometric pressure at Salt Lake City). A nonslip condition was applied to the reactor wall for the gas flow inside the reactor. At the outlet, the flow was assumed to be fully developed where the gradients for all variables in the exit direction were zero.

The temperatures of the gases at the inlet were set at room temperature 298 K (25°C). The wall temperature along the longitudinal direction of the reactor was measured and used as a first-type boundary condition. Two typical wall temperature profiles under different experimental conditions are shown in Figure 10.3. Despite the significantly different amounts of fuel and oxygen in the two cases, the two temperature profiles were essentially the same. Thus, it was assumed that the wall temperature profiles under other experimental conditions remained the same, as the amounts of fuel and oxygen fed in most of the other cases were similar to these two cases. For simplicity, the radiative properties of all the inner walls were assumed to be gray and diffuse. The values of 0.8 and 0.4 [Bergman et al., 2011] were chosen for the emissivity of the heavily oxidized stainless-steel tube wall of the reactor and polished stainless-steel surface of the burner, respectively.

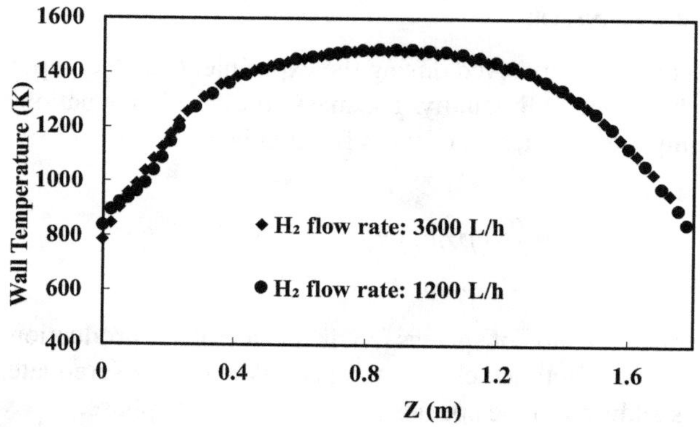

FIGURE 10.3 Measured wall temperature profiles of the laboratory flash reactor. (a) H_2 flow rate 3,600 L/h, O_2 flow rate 579 L/h. (b) H_2 flow rate: 1,200 L/h, O_2 flow rate: 178 L/h (Flow rates are calculated at 298 K (25°C) and 86.1 kPa).

10.1.6 Numerical Details

The gas-phase-governing equations were discretized and solved using the commercial CFD software package ANSYS FLUENT 15.0. Three-dimensional mesh was generated using ICEM-CFD ANSYS with a total number of 442,364 hexahedral dominating hybrid cells. Tetrahedral mesh was used only in the top part of the reactor to capture the complex geometric configuration of the burner and powder-feeding ports, which are displayed in Figure 10.2. Mesh independency was confirmed by halving and doubling the number of cells without changing the computational results. Total particle streams of 10,050 were released from the injection ports to establish a statistical representation of the spread of the particles due to turbulence. The calculation was carried out by a steady-state, pressure-based solver. The SIMPLE scheme [Ferziger, 2002] was chosen for handling the pressure-velocity coupling. A second-order upwind scheme was chosen for momentum, species transport and energy equation discretization for the convection term. Two-way coupling calculations of the gas and particle phases were performed.

The overall computational diagram is shown in Figure 10.4. The calculation started with a guessed solution of the gas-phase variables. Then, the governing equations of the gas phase were solved using the SIMPLE algorithm [Ferziger, 2002] by ten iterations. Based on this intermediate gas-phase solution, the particle streams were released from the injection ports. The particle velocities were determined by numerically integrating the equation of particle motion. As the particle velocities were computed, the particle temperature and mass were also updated. With the new particle-phase information, the interaction source terms over the computational domain, namely the momentum and mass exchanges, between the two phases were updated. After these updates, check whether the convenience criteria were met. If not, the calculation was shifted back to the gas-phase iterations and repeated the above process until the convergence criteria were met to obtain a converged solution.

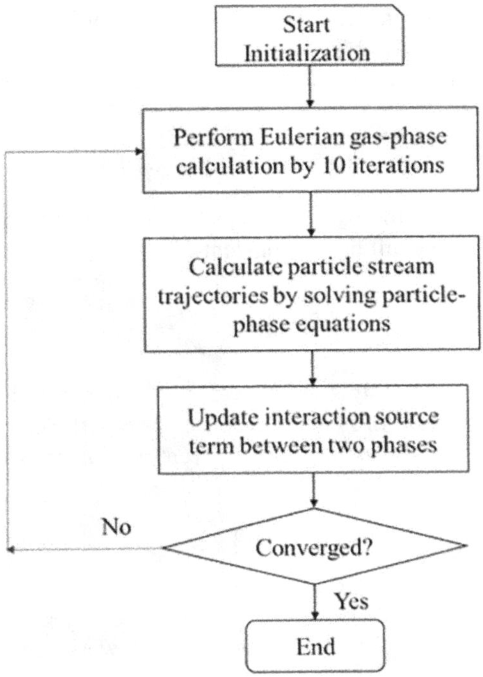

FIGURE 10.4 Overall block diagram for the computation. [Fan et al., 2016b].

10.1.7 Laboratory Flash Reactor Runs – with Hydrogen

The CFD model was first applied to the runs with hydrogen. Partial combustion of H_2 by O_2 injected through a non-premixed burner was simulated.

The H_2 partial combustion mechanism used in this study consists of seven chemical reactions involving six species, which are listed in Table 10.1 [Sohn and Perez-Fontes, 2016; Olivas-Martinez, 2013].

TABLE 10.1
H_2-O_2 Partial Combustion Mechanism

Reaction	A ($cm^3 \cdot mol^{-1} \cdot s^{-1}$)	γ	E_a ($kJ \cdot mol^{-1}$)
1 H2+O2=OH+OH	0.170×10^{14}	0.0	2.015×10^5
2 H+O2=OH+O	0.142×10^{15}	0.0	6.862×10^4
3 OH+H2=H2O+H	0.316×10^8	1.8	1.268×10^4
4 O+H2=OH+H	0.207×10^{15}	0.0	5.753×10^4
5 OH+OH=H2O+O	0.550×10^{14}	0.0	2.929×10^3
6 H+OH=H2O+M	0.221×10^{23}	−2.0	0.000
7 H+H=H2+M	0.653×10^{18}	−1.0	0.000

To take the turbulence–chemistry interaction into consideration, the eddy dissipation concept (EDC) [Gran and Magnussen, 1996] approach was adopted. The forward reaction rate constant for each of the elementary reactions is given by

$$k_{f,i} = AT^{\gamma} \exp\left(-\frac{E_{a,i}}{RT}\right) \tag{10.13}$$

For the reverse reactions, the backward reaction rate constants were linked to the forward reactions by the equilibrium constant:

$$k_{b,i} = \frac{k_{f,i}}{K_{c,i}} \tag{10.14}$$

The rate expression used in this study was a global nucleation and growth rate expression for the reduction of magnetite concentrate particles by H_2 was described in detail in Chapter 7. The general form is given by [Fan et al., 2016a; Chen et al., 2015a; 2015b]:

$$\frac{dX}{dt} = n \cdot k_0 \exp\left(-\frac{E}{RT}\right)\left[p_{H_2}^m - \left(\frac{p_{H_2O}}{K_e}\right)^m\right] d_p^s \cdot (1-X)\left[-\text{Ln}(1-X)\right]^{1-1/n} \tag{10.15}$$

The kinetics parameters in Eq. (10.15) used in this work are listed in Table 10.2 [Fan et al., 2016a].

By substituting the definition of reduction degree X into Eq. (10.15), the term dm_p/dt (which is also the mass balance equation of the particle) is obtained as

$$\frac{dm_p}{dt} = -\omega_O^0 m_p^0 \cdot n \cdot k_0 \exp\left(-\frac{E}{RT}\right)\left[p_{H_2}^m - \left(\frac{p_{H_2O}}{K_e}\right)^m\right] d_p^s \cdot (1-X)\left[-\text{Ln}(1-X)\right]^{1-1/n} \tag{10.16}$$

The particle heat transfer coefficient was evaluated using the following correlation of Ranz and Marshall [1952]:

$$Nu = \frac{hd_p}{k_g} = 2.0 + 0.6\, Re_d^{1/2} Pr^{1/3} \tag{10.17}$$

TABLE 10.2
Kinetics Parameters

Kinetics Parameter	Value
N	1
M	1
S	0
k_o ($atm^{-1} \cdot s^{-1}$)	1.23×10^7
E ($kJ \cdot mol^{-1}$)	196

A value of 0.8 was chosen for both the particle emissivity ε_p and the particle scattering coefficient σ_p, recommended by Hahn and Sohn [1990a; 1990b].

The particle specific heat was approximated as a mass fraction average of those for iron and magnetite during the reduction process. The physical properties of the particle are presented elsewhere [Fan, 2019].

$$c_{p,d} = X \cdot c_{p,Fe} + (1 - X) \cdot c_{p,Fe_3O_4} \tag{10.18}$$

The sizes of the particles remain essentially unchanged during an experiment as SEM micrographs revealed (not shown here). The solid particle density equals the instantaneous particle mass divided by its volume as

$$\rho_p = \rho_{p,0}\left(1 - \omega_O^0 X\right) \tag{10.19}$$

10.1.7.1 Combustion Mechanism Validation

The combustion mechanism listed in Table 10.1 was validated elsewhere [Sohn and Perez-Fontes, 2016]. The predicted temperature and concentrations of major species (H_2, H_2O, N_2, O_2) as well as intermediate radical species by CFD simulation incorporating the combustion mechanism in Table 10.1 were compared with the experimental data reported in the literature [Barlow and Carter, 1994;1996.] for non-premixed hydrogen jet flames. Sufficiently good agreement was achieved between the predicted and the experimental values [Sohn and Perez-Fontes, 2016].

10.1.7.2 Temperature Validation

Temperature along the centerline of the reactor was measured using a long K-type thermocouple during the experiment. The comparison of the computed temperature profiles and corresponding experimental measurement is given in Figure 10.5. It is evident that the gas temperature first spiked due to the flame and then dropped to

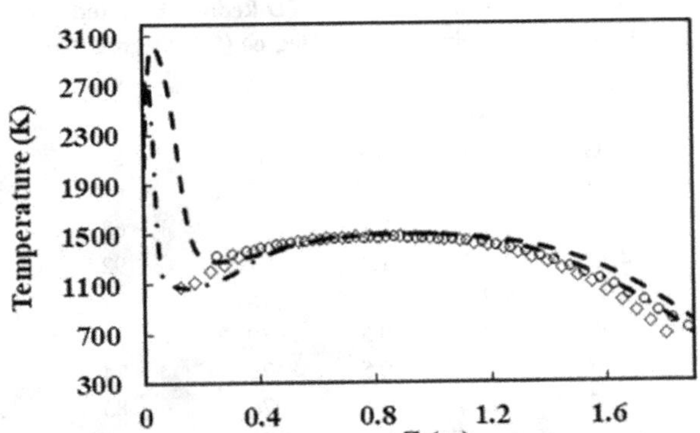

FIGURE 10.5 Gas-phase temperature along the centerline of the reactor. O, ---: experimental and calculated temperature profiles for H_2 flow rate 3,600 L/h, O_2 flow rate 579 L/h; □, –·–,: experimental and calculated temperature profiles for H_2 flow rate 1,200 L/h, O_2 flow rate: 178 L/h [Fan et al., 2016b].

the isothermal temperature of the reactor. Due to experimental difficulties, the temperature in the flame region was not measured. The computed and experimental gas temperatures agree well with each other overall except for some deviations at the end of the reactor. The reason for the latter is that this part of the reactor approached the exit of the reactor tube in which the tube was not surrounded by the insulation material and was thus exposed to the ambient air directly. Despite the disagreement, it does not cause significant errors in the calculated reduction degree of the magnetite concentrate particles as the chemical reaction at a temperature lower than 1,273 K (1,000°C) is relatively slow according to Eq. (10.15), and the reduction degree achieved within the short residence time of particle while descending in this region was negligible.

10.1.7.3 Reduction Degree

Table 10.3 shows the test conditions, experimental and calculated degrees of reduction for the experiments selected for simulation from those listed in Table 8.2. A comparison of the calculated reduction degrees (averaged over all particle streams) and experimental values is shown in Figure 10.6. The reduction degrees obtained from CFD simulations satisfactorily agree with the experimental values, except for some scatter in the region of low reduction degree. Good prediction for higher reduction degrees is important in the future design and optimization of larger flash reactors, as this novel process is aimed at producing iron at a high reduction degree (≥ 90%).

TABLE 10.3
Experimental and Computational Fluid Dynamics (CFD) Run Conditions and Results (All Solid Feeding Modes = SS; All N_2 Flow Rates = 2.8 L/min[a])

Flame Config.	H_2 Flow Rate (L/min)[a]	O_2 Flow Rate (L/min)[a]	Conc. Feed Rate (g/min)	EDF	CFD Redn. Degree (%)	Exp. Redn. Degree (%)	Max. Flame Temp. (K)	Avg. Res. Time (s)
H-O-H	15.3	2.16	1.9	0.5	84	82	2,722	5.4
H-O-H		2.72	2.1	0.06	63	57	2,805	5.6
H-O-H	20	2.96	1.8	0.5	80	84	2,851	5
H-O-H		3.22	1.9	0.3	78	80	2,879	5.1
H-O-H		3.70	2.0	0.1	51	63	2,901	5.2
H-O-H	40	4.32	2.2	1.4	91	92	2,966	3.6
H-O-H		6.40	2.1	0.5	75	77	3,017	3.7
O-H-O		6.40	2.3	0.5	80	87	2,790	4.2
H-O-H	60	9.65	1.9	0.5	72	74	3,030	2.9
O-H-O		9.65	2.1	0.5	76	83	2,898	3.3
H-O-H		10.1	1.67	0.3	70	69	3,034	2.9
H-O-H		10.65	2.3	0.3	62	72	3,041	2.9
H-O-H		11.2	2.0	0.2	54	65	3,047	3

[a] Flow rates are at 298 K (25°C) and 0.85 atm at Salt Lake City (1 atm = 101.3 kPa).

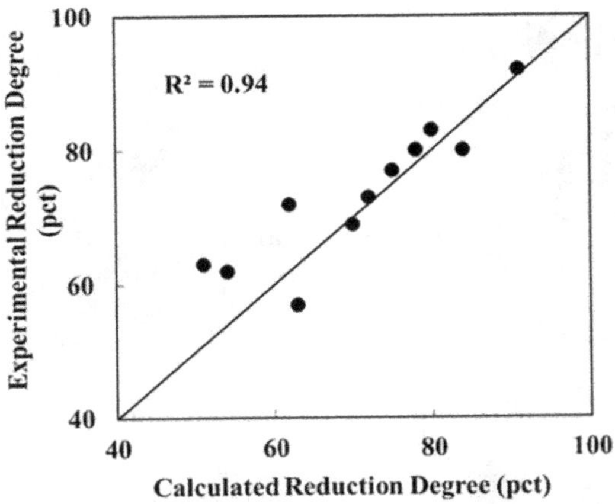

FIGURE 10.6 Comparison between the calculated reduction degrees vs. experimental results. [Fan et al., 2016b].

10.1.7.4 Velocity Field

A typical velocity field inside the reactor with H_2 flow rate of 3,600 L/h is shown in Figure 10.7. It can be seen that a recirculation zone was formed in the top part of the reactor as a result of the entrainment of high-velocity particle-laden jets coming out of the injection tubes. After the recirculation region, the gas stream travels downward along the longitudinal axis of the reactor tube. The recirculation zone played an important role in dispersing the concentrate particles and in increasing the particle residence time in the top part of the reactor, which can be seen from the particle trajectories in Figure 10.8. The effect the recirculation zone had on the temperature profile can be seen in Figure 10.9. High-temperature bumps were formed in the recirculation region as gas streams at higher temperatures originating from the combustion zone recirculated upward. In this context, the recirculation promotes the uniformity of temperature in the top part of the reactor.

10.1.7.5 Temperature Distribution

The flame configuration inside the reactor was changed in some experiments by swapping the H_2 and O_2 feeding ports in the burner nozzle. In the flame configuration with oxygen surrounding hydrogen (designated as O-H-O), O_2 was fed through the crescent-shaped inlets with H_2 feed streams confined to the center. The reverse flame configuration (designated as H-O-H) had the two gas streams reversed. The values of maximum flame temperature were lower in the O-H-O flame configuration compared with the H-O-H configuration in which the same amounts of O_2 were consumed for combustion. A comparison of the temperature distributions with different fuel and oxygen feeding configurations is displayed in Figure 10.9. It was found that the temperature in the top side parts of reactor was elevated in the O-H-O flame configuration. The reason for this is that with the O-H-O flame configuration, the high-temperature flame region was closer to the particle-laden jets and more

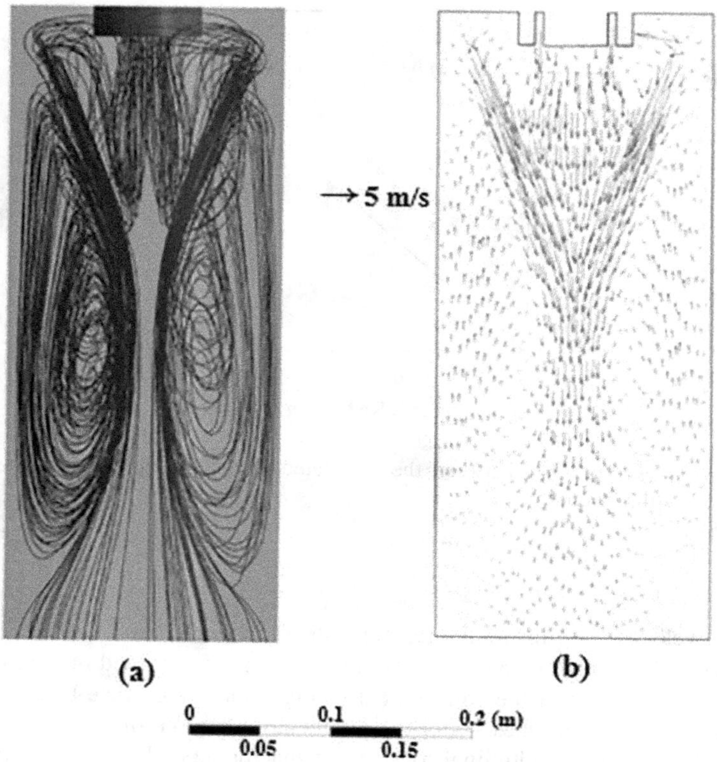

FIGURE 10.7 Velocity fields. (a) Streamlines. (b) Velocity vector for H_2 flow rate 3,600 L/h, O_2 flow rate 579 L/h with H-O-H configuration (A-A section shown in Figure 6.2. 40% of the total reactor length shown here). The maximum velocity is 36 m/s. (Adapted from Fan et al., [2016b]).

high-temperature gas streams were entrained into the recirculation zone, and hence, a better temperature homogeneity was achieved. The consequence of this also leads to an increase in the average particle reduction degrees in the O-H-O flame configuration runs, in which the same amounts of O_2 and solid feeding rate were used, maintaining the same excess driving force (EDF) as in the H-O-H flame configuration.

10.1.7.6 Species Distribution

The species distribution inside the reactor directly affects the local EDF and thus the local chemical reaction rate. The local EDF is an indicator of how much excess H_2 exists locally compared with the gas mixture in equilibrium with iron and wüstite, and is mathematically definedby the following equation:

$$EDF_{local} = \frac{\left(\frac{p_{H_2}}{p_{H_2O}}\right)_{local} - \left(\frac{p_{H_2}}{p_{H_2O}}\right)_{eq}}{\left(\frac{p_{H_2}}{p_{H_2O}}\right)_{eq}} = \frac{\left(\frac{p_{H_2}}{p_{H_2O}}\right)_{local} - \frac{1}{K_e}}{\frac{1}{K_e}} = K_e\left(\frac{p_{H_2}}{p_{H_2O}}\right)_{local} - 1 \quad (10.20)$$

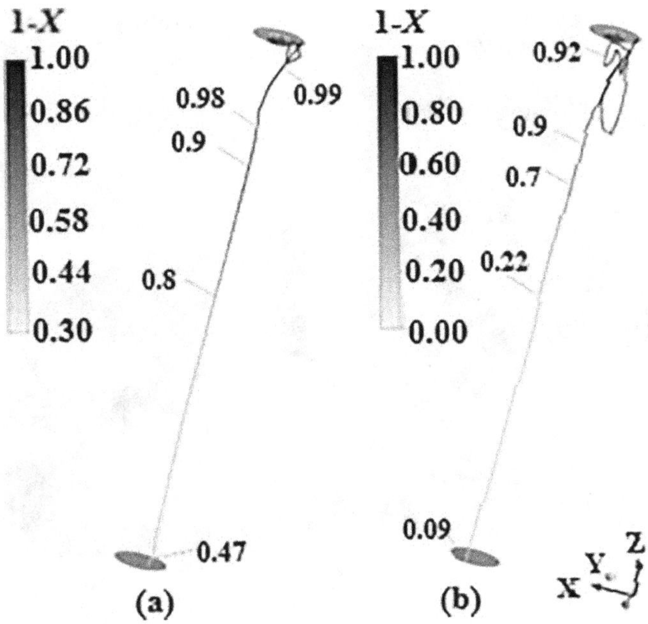

FIGURE 10.8 Representative particle trajectories for H_2 flow rate 3,600 L/h, O_2 flow rate 579 L/h with H-O-H flame configuration. The final reduction degrees in (a) and (b) are 0.53 and 0.81, respectively. (Adapted from Fan et al., [2016b]).

Figure 10.10 shows the typical species distributions and the corresponding local EDF distribution. The oxidation reactions are seen to take place very fast. It is also seen that H_2 and H_2O were uniformly distributed in most parts of the reactor, which is also observed under other experimental conditions. The local EDF shown in Figure 10.10(c) indicates that the local EDF in the isothermal region was around 0.6.

10.1.7.7 Particle Residence Time

The particle residence time is one of the most important experimental parameters in the flash ironmaking process. The particle residence time in this work is controlled by varying the gas flow rate. A total number of 10,050 particle streams were released from the two-side injection ports in all simulation runs to establish a statistical representation of the particle spreading due to turbulence. The average particle residence time in each of the simulation runs is calculated by averaging all the particle streams. The results are given in Fan [2019]. It is noted that τ_{iso} represents the average particle residence time where the particle temperature reaches the isothermal temperature [1,448 ± 25 K (1,175 ± 25°C)] of the reactor. The reduction reaction of the magnetite concentrate particles by H_2 takes place mostly during τ_{iso}. Trajectory (a) in Figure 10.8 did not pass through the flame zone and thus the particle did not react much before it entered the isothermal region, while trajectory (b) crossed the flame zone and about 15% reduction degree was achieved before the particle entered the isothermal zone.

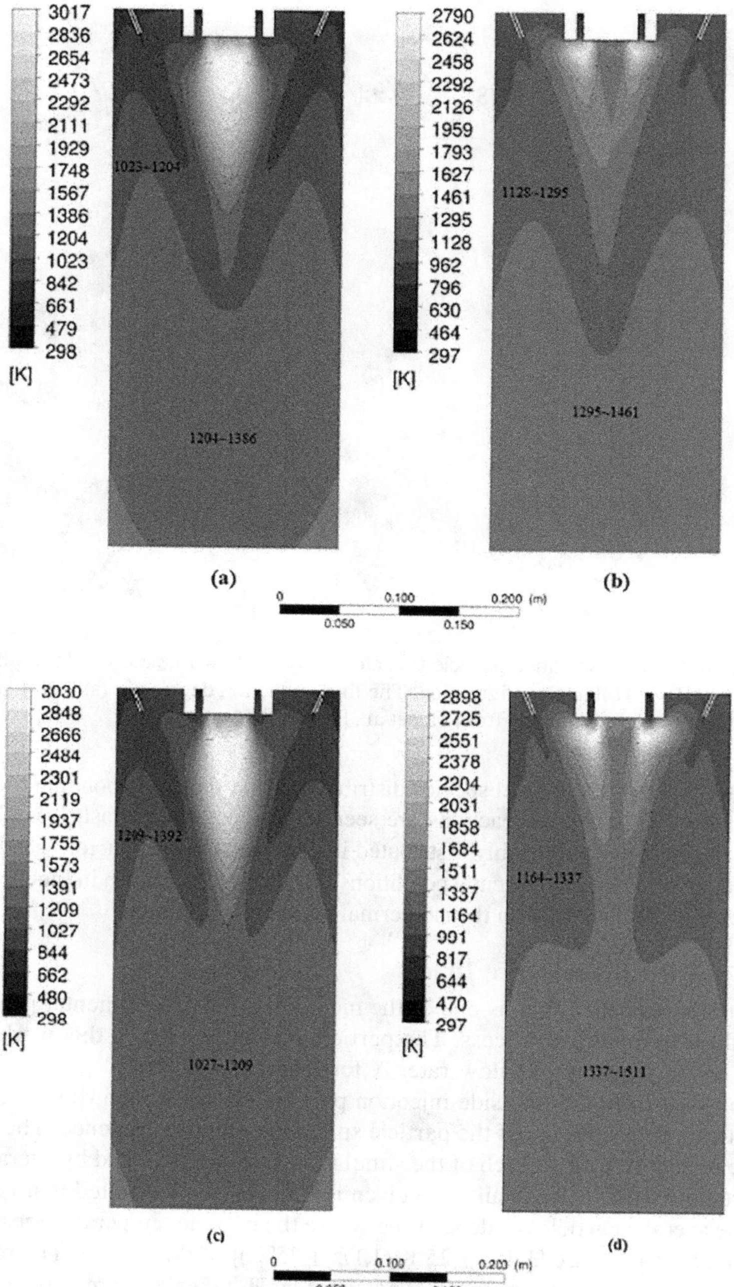

FIGURE 10.9 Temperature distribution for H_2 flow rate 2,400 L/h, O_2 flow rate 384 L/h (a) H-O-H flame configuration and (b) O-H-O flame configuration. Temperature distribution for H_2 flow rate 3,600 L/h, O_2 flow rate 579 L/h (c) H-O-H flame configuration, and (d) O-H-O flame configuration. (Adapted from Fan et al. [2016b]).

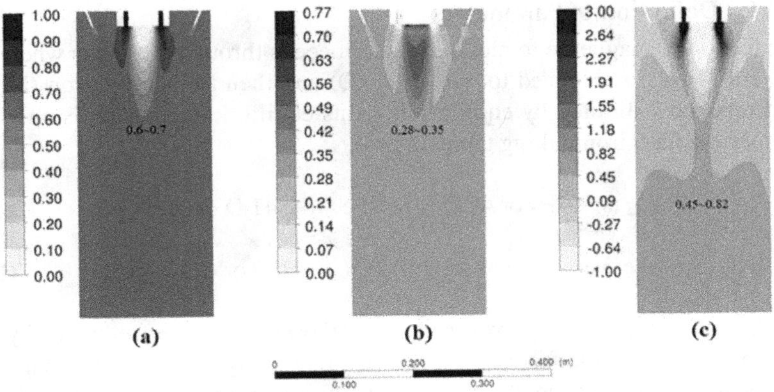

FIGURE 10.10 Species distribution: (a) H_2O mole fraction, (b) H_2 mole fraction (c) EDF distribution for H_2 flow rate 3,600 L/h, O_2 flow rate 579 L/h with H-O-H flame configuration (A-A section, Figure 6.2). (Adapted from Fan et al., [2016b]).

10.1.7.8 Concluding Remarks on Runs with Hydrogen

A three-dimensional CFD model was developed to simulate the gas particle two-phase flow, heat transfer, and chemical reaction in a laboratory flash reactor. The computed values of reduction degree and centerline temperature were compared with the experimentally measured results and satisfactory agreements were obtained. A recirculation zone was formed in the top part of the reactor. After the recirculation region, the gas stream travels downward along longitudinal axis of the reactor tube. The recirculation zone played an important role in dispersing the concentrate particles and increasing the particle residence time. The values of maximum flame temperature were lower in the O-H-O flame configuration compared with the H-O-H configuration, although the same amounts of O_2 were consumed in the two cases. Better temperature homogeneity was achieved in the O-H-O flame configuration. The non-mixed gaseous combustion reached equilibrium composition very fast and within a small region of the reactor tube. The values of the average particle residence time in the O-H-O burner configuration were larger than in the H-O-H flame configuration, other conditions being the same.

10.1.8 Laboratory Flash Reactor Runs – with Methane

The CFD model was also applied to the runs with methane. In the same laboratory flash reactor, described above, methane was partially oxidized by oxygen to form a reducing H_2+CO gas mixtures. The test variables in this work included the particle residence time, the extent of excess reducing gases, particle feeding mode, and flame configuration. More than 80% reduction was achieved in this reactor despite its low operating temperature (1,170°C). The same three-dimensional CFD model used to simulate the operation with hydrogen was applied in this case. The partial combustion of methane by oxygen was also simulated in this study.

10.1.8.1 Definition of Parameters

The reduction of magnetite to metallic iron proceeds through two steps where magnetite (Fe_3O_4) is first reduced to wüstite (FeO) and then to metallic iron (Fe). The second reaction is limited by equilibrium as its equilibrium constant is only about unity near the flash ironmaking temperatures.

$$Fe_3O_4\ (s) + 4H_2 \text{ or } 4CO\ (g) = 3Fe\ (s) + 4H_2O \text{ or } 4CO_2\ (g) \tag{10.21}$$

$$FeO\ (s) + H_2 \text{ or } CO\ (g) = Fe\ (s) + H_2O \text{ or } CO_2\ (g) \tag{10.22}$$

The reduction of magnetite by hydrogen is endothermic, while the reduction by CO is exothermic. Based on the fact that the reaction rate of H_2 is much faster than that of CO in the temperature range, the excess reducing gas was defined in terms of excess H_2 and the overall heat effect should be close to that of H_2, which is endothermic.

HSC 5.11 thermodynamics software [Roine, 2002] was used in the design of experimental conditions. The designed H_2 EDF values, defined by Eq. (10.20), were first selected. The values of H_2 EDF tested in this work were 0.5 and 1.0. The concentrate feeding rate was 2.2 ± 0.2 g/min based on the feeder limitation and accuracy. Due to the variation in the magnetite feeding rate during the experiments, the EDF actually achieved varied within ± 0.1 of the target values. The flow rates of input gases (H_2, CH_4, O_2, and N_2) to be used under given magnetite concentrate feeding rates were then determined such that the off-gas composition met the selected H_2 EDF value. The "Equilibrium Composition" module of HSC was used to determine those flow rates iteratively. The calculation was performed for complete reduction of the concentrate particles at 1,448 K (1,175°C) under a total pressure of 0.86 atm (the atmospheric pressure at Salt Lake City, Utah). A summary of the experimental conditions obtained from this method is shown in Table 10.4. The flow rates of the input gases obtained remained unchanged during the entire duration of the experiment. The temperature in the reactor was controlled by adjusting the power to the electric heating elements.

10.1.8.2 Governing Equations

The governing equations for the particle phase were a series of ordinary differential equations, which described the particle movement, particle temperature change and oxygen removal. Here, only the oxygen removal equation is listed, as given by the following equation:

$$\left.\frac{dX}{dt}\right|_j = k_j \cdot \left[p_j^{m_j} - \left(\frac{p_{jO}}{K_j} \right)^{m_j} \right] \cdot n_j\ (1-X)\left[-\text{Ln}(1-X)\right]^{1-1/n_j} \cdot d_p^{-s_j}; j = H_2 \text{ or } CO \tag{10.23}$$

where k_j is the reaction rate constant for reducing gas j, $k_j = k_{o,j} \exp\left(-\frac{E_j}{RT}\right)$; K_j is the equilibrium constant for the reduction of FeO by gas j; n_j is the Avrami parameter; m_j is the reaction order with respect to gas j.

The rate expression used for the reduction of magnetite concentrate particles by single gas H_2 or CO was the global nucleation and growth rate expression. The kinetic

TABLE 10.4
Summary of the Experimental Conditions

Exp. No#	Gas Flow Rates (L/min)				Conc. Feeding Rate (g/min)	Feeding Mode	Flame Config.	H_2 EDF	Nominal Residence Time (s)	τ (s)	Exp. RD (%)	Cal. RD (%)
	CH_4	H_2	O_2	N_2								
1	5.0	2.0	4.0	2.8	2.0	SS	O-F-O	1.1	10.5	7.9	83	95
2	5.0	2.0	4.0	2.8	2.4	SS	F-O-F	1.1	10.5	7.1	76	90
3	5.0	2.0	4.5	2.8	2.0	SS	F-O-F	0.5	10.5	6.8	64	75
4	5.0	2.0	4.5	2.8	1.8	SS	O-F-O	0.5	10.5	7.5	72	85
5	5.0	2.0	4.0	2.0	1.9	B	O-F-O	1.1	10.5	8.2	83	97
6	5.0	2.0	4.0	2.0	2.1	B	O-F-O	1.0	10.5	8.2	82	96
7	5.0	2.0	4.5	2.0	2.2	B	O-F-O	0.5	10.5	7.7	65	86
8	5.0	0.13	4.2	2.0	2.1	B	O-F-O	0.5	11	8.3	59	89
9	5.0	0.13	3.75	2.0	1.9	B	O-F-O	1.0	11	8.4	78	99
10	5.0	0.13	3.75	2.8	2.2	SS	O-F-O	1.0	11	8.4	81	98
11	5.0	0.13	3.75	2.0	2.2	B	F-O-F	1.0	11	8.0	33	92
12	5.0	0.13	3.75	2.0	2.4	B	O-F-O	1.0	11	8.4	81	99
13	5.0	0.13	4.2	2.8	2.5	SS	O-F-O	0.5	11	8.0	68	87
14	10.0	2.0	8.2	2.8	2.0	SS	O-F-O	1.1	7.6	5.6	46	83

TABLE 10.5
Summary of the Experimental Condition

Reducing Gas	Temperature Range (K)	$k_{o,j}$	E_j (kJ/mol)	m_j	N_j	s_j
H_2	1,423–1,623	1.23×10^7	196	1	1	0
CO	1,423–1,623	1.07×10^{14}	451	1	0.5	0

parameters in Eq. (10.23) are listed in Table 10.5 which were used in the simulation. Since the amount of O_2 fed was less than the stoichiometric amount required for full combustion, a mixture of H_2+CO was generated in the partial combustion. In the temperature range investigated, most of the reduction work was done by H_2. It was also found that a synergistic effect occurred for the reduction of magnetite concentrate particles by H_2+CO mixtures, as mentioned in Sections 7.5.5, 7.5.6 and 7.6.6.

10.1.8.3 Combustion Mechanism

The partial combustion of methane was described in this study by four chemical reactions involving six species that are listed in Table 10.6 [Jones and Lindstedt, 1988]. The EDC model [Gran and Magnussen, 1996] was adopted to account for the turbulence–chemistry interaction. This model takes account of the effects of turbulent mixing as well as detailed chemical mechanisms in turbulent combustion [ANSYS Inc., 2019].

The chemical reactions are represented in the following general form:

$$a\ \mathrm{A(g)} + b\ \mathrm{B(g)} = c\ \mathrm{C(g)} + d\ \mathrm{D(g)} \tag{10.24}$$

The forward reaction rate constant for each of the elementary reaction is given by:

$$k_{f,i} = AT^{\gamma}\exp\left(\frac{-E_{a,i}}{RT}\right) \tag{10.25}$$

In the CH_4-O_2 partial combustion mechanism, reactions 2 and 4 in Table 10.6 are treated as reversible for which the corresponding backward reaction rate constants are calculated by

$$k_{b,i} = \frac{k_{f,i}}{K_{c,i}} \tag{10.26}$$

TABLE 10.6
CH_4-O_2 Partial Combustion Mechanism

Reaction	A	γ	E_a (J·mol⁻¹)	$K_{c,i}$ at 3,073 K
1 CH4+0.5O2=CO+2H2	4.40×10^{14} ($m^{2.25}\cdot mol^{-0.75}\cdot s^{-1}$)	0.0	1.26×10^5	2.8×10^{10}
2 CH4+H2O=CO+3H2	3.00×10^{11} ($m^3\cdot mol^{-1}\ s^{-1}$)	0.0	1.26×10^5	1.6×10^9
3 H2+0.5O2=H2O	2.50×10^{19} ($m^{2.25}\ mol^{-0.75}\ s^{-1}\ K$)	−1.0	1.67×10^5	18
4 H2O+CO=CO2+H2	2.75×10^{12} ($m^3\cdot mol^{-1}\cdot s^{-1}$)	0.0	8.37×10^4	1.3×10^{-1}

where $K_{c,i}$ is the equilibrium constant of reaction i based on molar concentrations, rather than on activities. The net rates are then calculated as the difference between the forward reaction rate and backward reaction rate

$$R_i = k_{f,i} C_A^a C_B^b - k_{b,i} C_C^c C_D^d \tag{10.27}$$

Reactions 1 and 3 in Table 10.6 are treated as irreversible reactions, and thus only forward reactions are considered.

The forward reaction rate for $CH_4 + 0.5O_2 = CO + 2H_2$ is calculated by

$$R_{i,1} = k_{f,1} C_{CH4}^{0.5} C_{O2}^{1.25} \tag{10.28}$$

The forward reaction rate for $H_2 + 0.5O_2 = H_2O$ is calculated by

$$R_{i,3} = k_{f,3} C_{H2}^{0.25} C_{O2}^{1.5} \tag{10.29}$$

The boundary conditions and numerical techniques used in this work were the same as in Section 10.1.5. Mass flow rate boundary conditions were imposed at the gas inlets. The operating pressure was kept at 86.1 kPa (the barometric pressure at Salt Lake City). The wall temperature along the axial direction of the reactor was measured and used as fixed-value thermal boundary condition. Other details of the boundary conditions can be found in Fan et al., [2016b].

Commercial software package ANSYS Fluent was used to solve the governing equations mentioned above. The partial combustion mechanism above was prepared into a CHEMKIN [Kee et al., 2000.] format and was imported into the model. The chemical reaction kinetics for the concentrate particles were implemented using the UDFs. We should keep in mind that the flame region is only a small portion of the reactor. The combustion mechanism list above was mainly responsible for the quick oxidation of the methane that generates H_2 and CO in this region, which were subsequently used in the reduction of the concentrate particles. Outside the flame region, the gas composition was determined entirely by the water-gas shift reaction.

10.1.8.4 Experimental Results

The effect of the flame configuration on particle reduction was investigated in this work. In the F-O-F flame configuration experiments, the particles fed through the center of the burner tended to melt and became molten droplets when they went through the flame region (> 2,000°C), which was observed in the previous H_2 reduction experiments [Fan et al., 2016b] as shown in Figure 8.8. This melting lowered the reduction degree in burner feeding compared with side feeding under the same flow rates and EDF, which was verified in the experiments. The reason is that the active solid surface area for chemical reaction was decreased after melting compared with side feeding in which case the particles remained solid and retained their irregular shape and reactivity, as shown in Figures 10.11 and 10.12. Therefore, side feeding was mostly used in the F-O-F flame configuration experiments in this work.

The values of H_2 EDF tested in this work were limited to 0.5 and 1.0, which are within the range of values expected in an industrial flash ironmaking reactor. As expected, an increase in the EDF value under the same nominal residence time and temperature led to higher reduction degrees, which is seen from Figure 10.13. It was also found that higher reduction degrees were achieved under the O-F-O flame configuration compared with that obtained under the F-O-F flame configuration. This has to do with the temperature distribution inside the reactor, which will be further discussed in the CFD results. Figure 10.11 shows the SEM micrographs for the samples collected under the O-F-O flame configuration with burner feeding at EDF = 1. No melting of the particles was observed. The particles were angular, irregularly shaped and porous. The SEM micrograph for the sample collected under the F-O-F flame configuration with two-side feeding mode at EDF = 1 shown in Figure 10.12 also shows the particles to be angular, irregularly shaped and porous.

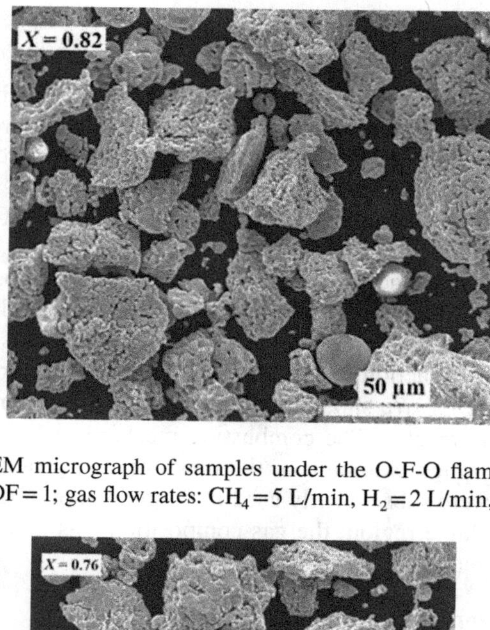

FIGURE 10.11 SEM micrograph of samples under the O-F-O flame configuration with burner feeding at EDF = 1; gas flow rates: CH_4 = 5 L/min, H_2 = 2 L/min, O_2 = 4 L/min.

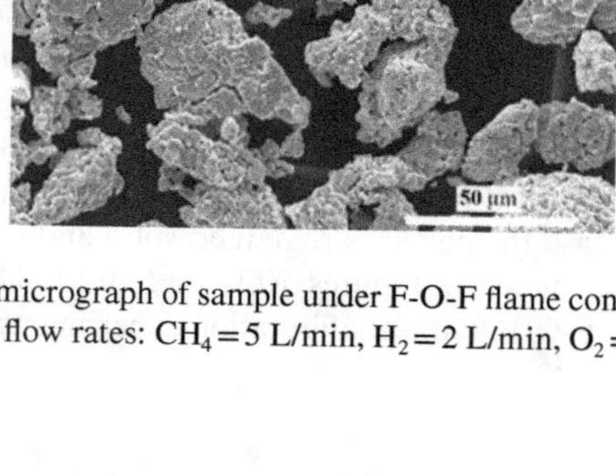

FIGURE 10.12 SEM micrograph of sample under F-O-F flame configuration with two-side feeding at EDF = 1; gas flow rates: CH_4 = 5 L/min, H_2 = 2 L/min, O_2 = 4 L/min.

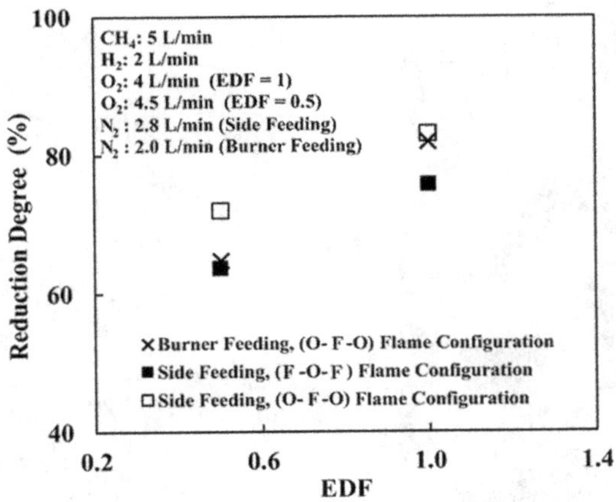

FIGURE 10.13 Effect of EDF on reduction degree under different feeding modes and flame configurations. [All flow rates are at 298 K and 86.1 kPa, the atmospheric pressure at Salt Lake City].

The reduction degrees from all the experiments were listed in Table 10.4. It is noted that the reduction degrees achieved with methane were lower than that obtained using H_2 under the same H_2 EDF and shorter nominal residence time [Fan et al., 2016b]. The reason will be discussed subsequently.

10.1.8.5 CFD Simulation Results

A typical velocity field inside the reactor is shown in Figure 10.14. In the top part of the reactor, a recirculation zone was formed as a result of the entrainment of the high-velocity particle-laden jets coming out of the injection tubes. This kind of recirculation zone was also observed in the simulation of H_2 reduction experiments [Fan et al., 2016b]. After this recirculation region, the gas stream started to travel downward along the axis of the reactor tube. The recirculation zone played an important role in the concentrate particles dispersion and in increasing the particle residence time. The effect the recirculation zone had on the temperature profile can be seen in Figure 10.15. High-temperature bumps were formed in the recirculation region as gas streams at higher temperatures originated from the combustion zone recirculated upward. The recirculation enhances the temperature homogeneity in the top part of the reactor.

A comparison of the temperature distributions with different fuel and oxygen feeding configurations is displayed in Figure 10.15. The F-O-F feeding configuration led to higher flame temperatures. For the O-F-O feeding configuration, the burner feeding mode resulted in a slight decrease in the flame temperature. The O-F-O flame configuration also led to elevated temperature in the top part of the reactor as the high-temperature flame was more evenly distributed in the redial direction compared with that under the F-O-F flame configuration. The consequence of this also leads to an increase in the average particle reduction degrees in the O-F-O flame configuration runs for otherwise the same operating conditions. Figure 10.15(c) also explains the melting phenomenon encountered in Figure 8.8 where the concentrate particles were fed with the burner feeding mode under the F-O-F flame configuration.

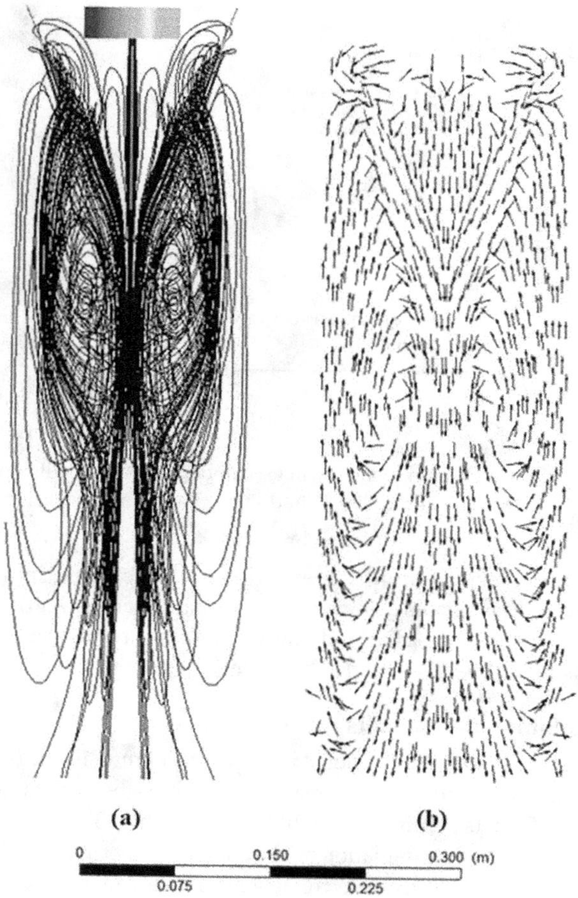

FIGURE 10.14 (a) Streamlines and (b) Velocity vector for $CH_4=5$ L/min, $H_2=2$ L/min and $O_2=4$ L/min with O-F-O configuration (A-A section shown in Figure 4. 40% of the total reactor length shown here). The maximum velocity is 24 m/s. [All flow rates are at 298 K and 86.1 kPa, the atmospheric pressure at Salt Lake City].

The particles in this case went through the whole flame region experiencing complete melting. In contrast, most of the particles fed in Figure 10.15(a) did not melt although the same burner feeding mode was used. This is because in the O-F-O flame configuration, the high-temperature flame region was shifted toward the edge of burner leaving the center region below the melting point of the concentrate. The reason for this shift of high-temperature region is that the fuel was fed at a higher volumetric flow rates and the gases generated from the oxidation of methane generated additional volume, which pushed the gas outward. Under the same O-F-O flame configuration with the same fuel, oxygen, and concentrate particles flow rates, similar reduction degrees were achieved through the burner feeding mode and two-side feeding mode, which can be seem from Figures 10.15(a) and 10.14(b) as the two have similar temperature distributions in the top part of the reactor.

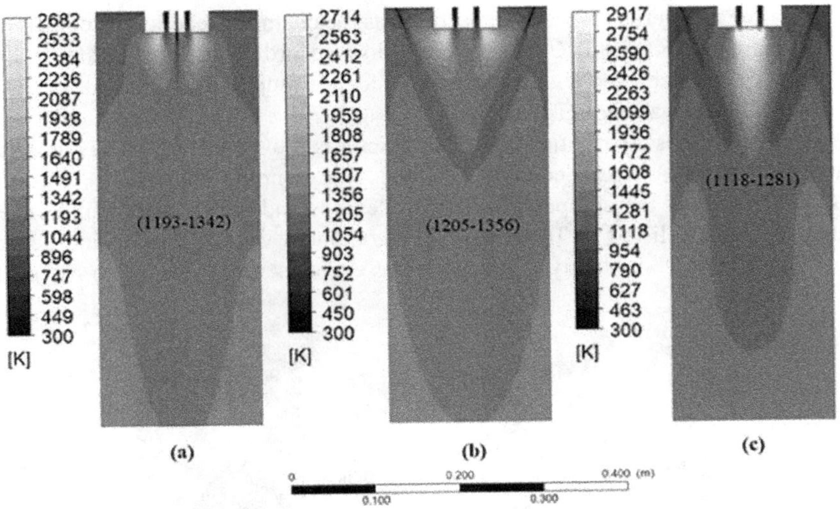

FIGURE 10.15 Temperature distribution for CH_4=5 L/min, H_2=2 L/min and O_2=4 L/min (a) O-F-O flame configuration and burner feeding mode; (b) O-F-O flame configuration and two-side feeding mode; (c) F-O-F flame configuration and two-side feeding mode.

Temperature along the centerline of the reactor was measured using a long K-type thermocouple during the experiment. The comparison of the computed temperature profiles and corresponding experimental measurement is given in Figure 10.16. It is evident that the computed and experimental gas temperatures agree well with each other overall except some deviations at the end of the reactor. The reason for the latter is that this part of the reactor close to the exit of the reactor tube was not surrounded by the insulation material and was thus exposed to the ambient air directly. Despite the disagreement, it does not cause significant errors in the calculated reduction

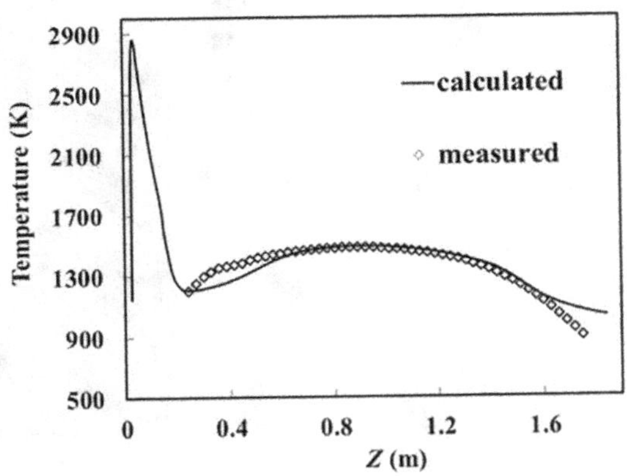

FIGURE 10.16 Comparison of calculated gas-phase temperature and the measured values along the centerline of the reactor for CH_4=5 L/min, H_2=2 L/min and O_2=4.5 L/min with F-O-F flame configuration and two-side feeding mode.

degree of the magnetite concentrate particles as the chemical reaction at temperature lower than 1,273 K (1,000°C) is relatively slow according to Eq. (10.23), and the reduction degree achieved within the short residence time of particle while descending in this region was negligible.

A typical species distribution inside the reactor is shown in Figure 10.17. It is seen that the conversion of methane into CO and H_2 was complete and fast.

Comparisons of the computed reduction degrees and the corresponding experimental values are listed in Table 10.4. The computed values of reduction degrees were found to be consistently higher than the experimental values. The reason for

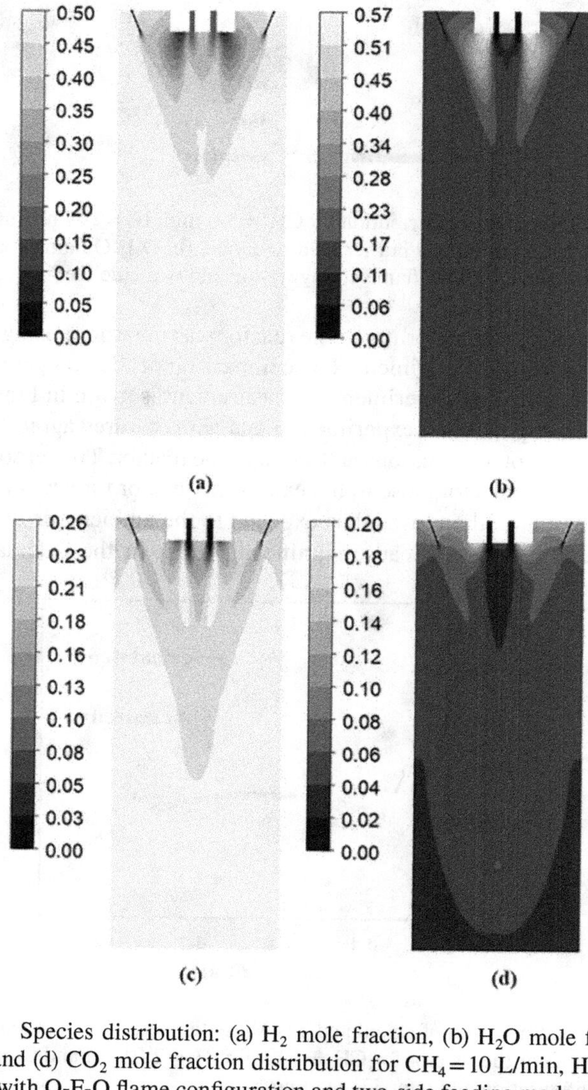

FIGURE 10.17 Species distribution: (a) H_2 mole fraction, (b) H_2O mole fraction, (c) CO mole fraction, and (d) CO_2 mole fraction distribution for $CH_4=10$ L/min, $H_2=2$ L/min and $O_2=8.2$ L/min with O-F-O flame configuration and two-side feeding mode.

this is that during the experiments, soot was formed at the tip of the burner and on the inner wall of the reactor tube. The soot formation was due to the large metal surface of the reactor tube which served as a catalyst for CH_4 cracking. But the combustion mechanism listed in Table 10.6 did not take into consideration of the soot formation.

Therefore, the actual amount of CO generated from the partial combustion was less than the amount calculated in the CFD simulation. Furthermore, the amount of H_2O generated was greater than the theoretical value as the oxygen that would have oxidized with the soot instead oxidized hydrogen, which led to a lower EDF. The reduction of magnetite concentrate particles was negatively affected under a lower EDF of H_2 compared with the CFD simulation.

Particle residence time is one of the most important experimental parameters in the flash ironmaking process. In the design of experimental conditions, nominal residence time was used. The particle residence time was calculated by tracking particle streams. A total number of 5,000 particle streams were released from the two side-injection ports in all simulation runs to establish a statistical representation of the particle spreading due to turbulence. The average particle residence time in each of the simulation runs is calculated by averaging all the particle streams. The calculated results (designated as τ) listed in Table 10.4 represent the average particle residence time where the particle temperatures are at least equal to the target temperature [1,448 K (1,175°C)] of the experiment.

10.1.9 Concluding Remarks

The reduction of magnetite concentrate particles was investigated in a laboratory-scale flash reactor by partially oxidizing methane with oxygen to generate a mixture of H_2+CO as the reducing gases. The experimental results indicated that more than 80% reduction degrees can be achieved in this reactor despite its low maximum temperature possible. Different factors such as residence time, excess reducing gas, feeding mode and flame configuration that affect the reduction of the concentrate particles were tested.

A three-dimension CFD model was developed to simulate the gas particle two-phase flow, heat transfer and chemical reaction in a laboratory flash reactor. The calculated centerline temperatures satisfactorily agreed with the measured values. However, the computed values of reduction degrees were consistently higher than the experimental values due to the soot formation. The soot formation, which was not accounted for in the CFD simulation, lowered the experimental values of reduction degrees. The soot formation did not occur when the burner and the wall were lined with refractory material in another larger bench reactor, and thus is not expected to be a problem in an industrial flash reactor. Thus, the problem of soot formation was not further investigated in this work.

The particles melted when passing through the flame region in the burner feeding mode with F-O-F flame configuration, which negatively affected the particle reductions as the active solid surface area decreased upon melting compared with particles remaining solid in the two-side feeding mode. The O-F-O flame configuration yielded higher reduction degrees than the F-O-F flame configuration (other conditions being the same) as better temperature homogeneity in the top part of the reactor and longer average particle residence time in the O-F-O were achieved.

10.2 COMPUTATIONAL FLUID DYNAMICS MODELING OF THE PILOT-PLANT-SCALE FLASH REACTOR

CFD modeling was performed to simulate the mini-pilot flash reactor (MPFR), described in Chapter 9, using a program named ANSYS Fluent®17.1. The simulation results of the reduction degrees of concentrate particles and the composition of the off-gas from the reactor experimental data were found to have a satisfactory agreement.

The developed model of the MPFR was used to study the effect of the oxygen/natural gas ratio, the total input gas flow rate, and the concentrate powder-feeding locations on the reduction degrees of the concentrate particles. The optimum operating conditions of the MPFR required to achieve high reduction degrees of iron oxide concentrate were suggested.

An industrial reactor was designed to produce 0.3–1 million tonnes/year of metallic iron of > 90% metallization using the flash ironmaking technology (FIT).

10.2.1 Introduction

A simplified schematic of the MPFR used for the purpose of CFD modeling is shown in Figure 9.3 with dimensions of 0.8 m inner diameter, 2.1 m length, and lined with three layers of refractory, insulation, and carbon steel. The mini-pilot-scale study aimed to test the effects of system variables like feeding rates, oxygen to natural gas ratio, and temperature. The solid feed to the MPFR was the magnetite concentrate with a mass flow rate of 1–7 kg/h. The burner of the MPFR was designed to make a swirling flow in the reactor. Swirl was introduced to increase the residence time of the magnetite concentrate particles, thus increasing the reduction degree. The unique design of the burner shortened the flame zone and ensured a larger uniform temperature zone in the reactor.

The CFD platform used in this work was essentially the same as that used for the CFD modeling of the laboratory flash reactor described in Section 10.1. The model used in the simulation program was validated by comparison of the results with experimental data in terms of the outlet gas composition and the product reduction degree. Satisfactory validation achieved for the model compared with the experimental runs thus decreased the number of costly experiments using the MPFR. Reliable simulations reduce the time required for the experimental investigation.

We calculated the following using the CFD model:

1. The percentage of heat loss through the walls with respect to the heat generated by the combustion of fuel.
2. The particle flow pattern and distribution inside the reactor.
3. The gas flow pattern inside the reactor.
4. Contours of the gas temperature and H_2, CO. H_2O, and CO_2 concentrations.

The flow rates of natural gas and oxygen in the MPFR affected the operating temperature of the reactor. The oxygen/natural gas ratio changed the mole percentages of hydrogen, carbon monoxide, carbon dioxide, and water vapor, thus changed the value of the EDF defined below. The EDF at certain operating temperatures was the driving force for the reduction of magnetite concentrate to produce metallic iron.

10.2.2 CFD Simulation

Natural gas is partially oxidized with oxygen in a uniquely designed burner, producing a short flame that provides heat (1,423–1,873 K) and a reducing gas mixture (hydrogen and carbon monoxide). A CFD model was developed to simulate the MPFR runs, in which the kinetics of magnetite concentrate reduction, separately determined in a drop-tube reactor, was incorporated. The reduction degrees and composition of the off-gas obtained from the CFD model show reasonable agreements with experimental results.

Six typical runs of the MPFR listed in Table 10.7 were simulated.

The mini-pilot reactor vessel is shown schematically in Figure 9.3. The burner used in the MPFR is shown in Figure 9.5. The reactor vessel is made up of the layers of refractory, insulation, and carbon steel. The inclined angle of the oxygen inlet ports (total of ten ports) causes a swirl flow in the reactor, which increases the particles residence time. This particular design of the burner shortens the flame length and ensures a larger uniform temperature zone in the whole reactor.

The following assumptions were made in the simulation:

1- Steady-state conditions: The gas phase responds to changes much more quickly compared with condensed phases such as walls, as the density of the gas is about three orders of magnitude lower. Thus, the accumulation

TABLE 10.7
The Experimental Conditions of the MPFR Runs that Were Simulated

Parameters	Run 1	Run 2	Run 3	Run 4	Run 5	Run 6
Concentrate feeding rate, kg h^{-1}	2.5	4.3	5.0	5.0	4.6	4.0
Particle size range, μm	32–90	less than 90	less than 90	32–90	less than 90	less than 90
Mass average particle size, μm (used for simulation)	45	32	32	45	32	32
Natural gas flow rate[a] m^3h^{-1}	25.16	30.56	20.36	24.80	17.36	15.86
Natural gas input temperature, K	300	300	300	300	300	300
O_2 flow rate[a] m^3h^{-1}	19.85	19.67	14.27	21.53	16.35	14.81
O_2 input temperature, K	300	300	300	300	300	300
Total inlet gas flow rate[a] m^3 h^{-1}	45.01	50.23	34.63	46.33	33.71	30.67
O_2 to natural gas mole ratio	0.79	0.64	0.70	0.87	0.94	0.93
Inner wall temperature[b] K	1,483–1,563	1,503–1,603	1,573–1,673	1,403–1,473	1,563–1,623	1,573–1,623
Inner wall temperature, K (used for simulation)	1,526	1,548	1,626	1,440	1,594	1,599

[a] Flow rates are calculated at 298 K and 0.85 atm, the barometric pressure at Salt Lake City(1atm=101.32 kPa).
[b] Wall temperatures were measured during feeding of the concentrate in the experiment.

term for the gas-phase energy balance is much smaller than the accumulation term for condensed phases. Thus, the steady-state condition is assumed for the gas-particle phase.

2- Inter-particle collisions between magnetite concentrate particles in the reactor are neglected as the volume of the solid particles occupying the reactor is of three orders of magnetite lower than the volume of the gas.

10.2.2.1 Incorporation of Natural Gas Combustion

The reaction kinetics for the partial combustion of natural gas [with the actual composition of 96 mol% CH_4, 2 mol% C_2H_6, and 2 mol% nitrogen] was incorporated in the simulation. Natural gas was considered as 100.6 mol% CH_4 (1 mol% of C_2H_6 equivalent to 2.6 mol% of CH_4 in heat production and 2 mol% of CH_4 in hydrogen and carbon monoxide production, both for generating a representative hot gas mixture from the partial combustion; thus considering these two factors, 1mol% of C_2H_6 was treated as being equivalent to 2.3 mol% of CH_4) and 2 mol% nitrogen to avoid the complexity of including ethane. The kinetics of natural gas oxidation used in this work were discussed in Section 10.1.8.

During a run the wall temperature somewhat varied with time. Since our model was a steady-state model, the wall boundary condition was represented by an average temperature that gives a rate constant which represents the average rate constant over the period of change, computed using a separate MATLAB® program. This averaging method was thought to be somewhat better than the arithmetic average of the varying temperature. These inner wall temperatures used in the simulation are shown in Table 10.7.

The gas-phase-governing equations were discretized and solved using the commercial CFD software package ANSYS FLUENT 17.1. Three-dimensional mesh was generated using ICEM-CFD ANSYS with a total number of 398,380 hexahedral cells. Mesh optimization was confirmed by doubling the number of cells without changing the computational results. Figure 10.18 shows the meshing of the MPFR. Total particle streams of 480 were released from the injection ports to establish a statistical representation of the spread of the particles due to turbulence. Turbulence effect on the particle dispersion was considered by using the stochastic tracking model. In the stochastic tracking approach, ANSYS Fluent predicts the turbulent dispersion of particles by integrating the trajectory equations for individual particles, using the instantaneous fluid velocity, $\overline{u} + u'(t)$, along with the particle path during the integration. The path in this manner is computed for a sufficient number of representative particles (termed the "number of tries"). A stochastic method (random walk model) is used to determine the instantaneous gas velocity. In the discrete random walk (DRW) model, the fluctuating velocity components are discrete piecewise constant functions of time. Their random value is kept constant over an interval of time given by the characteristic lifetime of eddies. The particle trajectories and velocities were determined by numerically integrating the equation of particle motion. As the particle trajectory was computed, the particle temperature and mass were obtained. The calculation was carried out by a steady-state pressure-based solver and the convergence was achieved by having a residual less than 10^{-3} in the momentum equation, less than 10^{-6} in the energy balance equation, and less than 0.01% variation in

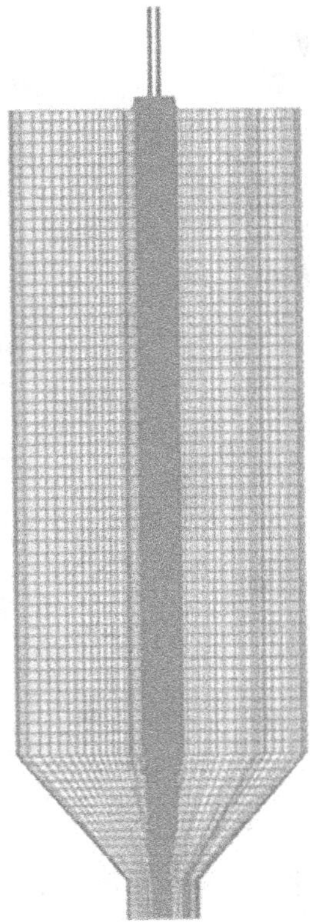

FIGURE 10.18 The meshing of the mini-pilot flash reactor.

the mass-weighted average temperature at the exit of the MPFR for more than 500 consecutive iterations. A second-order upwind scheme was chosen for momentum, species transport and energy equation discretization for the convection term.

10.2.3 Results and Discussion

The simulated temperature distribution of the six runs show largely uniform temperature distributions and short flame lengths thanks to the particular design of the burner. Higher temperatures compared with other runs were noticed for Runs 5 and 6 in which the higher mole ratio of oxygen to natural gas was at 0.94 and 0.93, respectively. Lowest temperature is noticed for Run 2 where the mole ratio of oxygen to natural gas was the lowest at 0.64.

The model was able to predict the contents of the reducing gases H_2 and CO at the exit with accuracies greater than 93% compared with experimental data in most of the runs.

The simulated particle streams in Figure. 10.19 show a swirl flow because of the particular design of the burner.

The experimental reduction degrees and the calculated values are listed in Table 10.8.

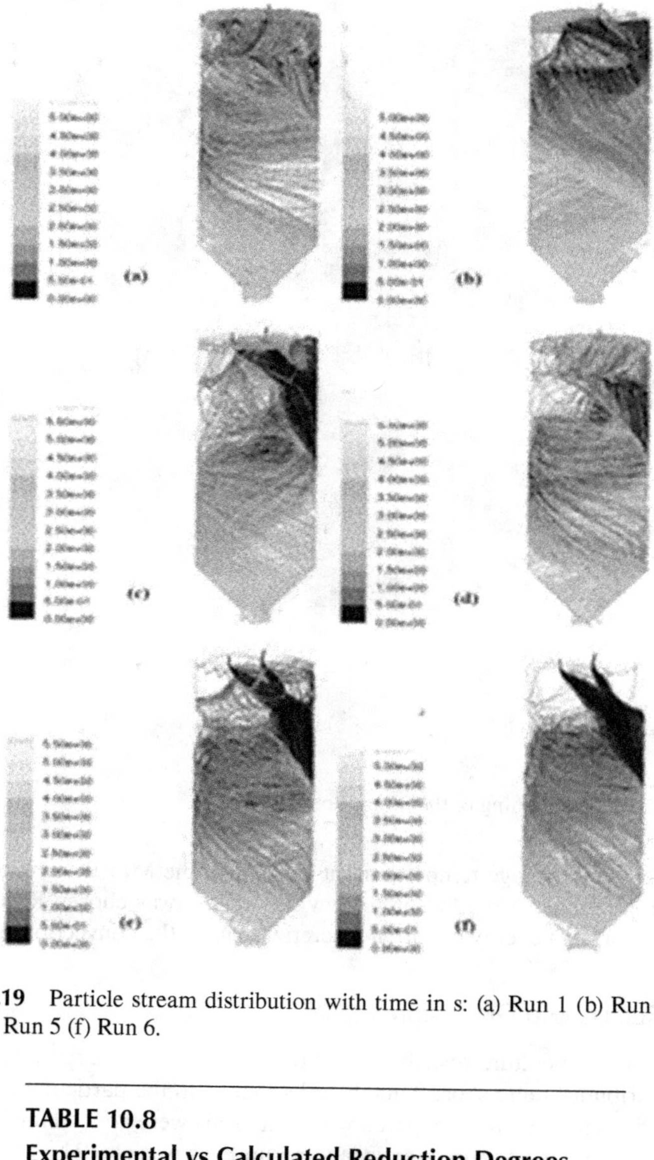

FIGURE 10.19 Particle stream distribution with time in s: (a) Run 1 (b) Run 2 (c) Run 3 (d) Run 4 (e) Run 5 (f) Run 6.

TABLE 10.8
Experimental vs Calculated Reduction Degrees

Run	Experimental (%)	Simulation (%)
1	94.0	99.8
2	80.0	84.5
3	94.5	99.6
4	74.0	99.8
5	72.5	99.5
6	50.0	85.0

The reduction degrees agree well for the first three runs. The agreement is not as good for the last three runs. The reason for this is likely to be because of the neglect of particle interactions for Runs 4–6. The temperature of the particle-gas stream in the main reaction zone was largely uniform, and this value is used to represent the reactor temperature. These runs had higher ratios of oxygen to natural gas and thus higher temperatures than the other runs, above 1,577°C, which is higher than the melting point of iron at 1,538°C. Particle agglomerate together more readily at these high temperatures, as shown previously during flash smelting of copper [Kimura et al., 1986; Kemori et al., 1988; Themelis et al., 1988]. This might have caused lower reduction rates in the actual cases than in the simulation. Particle coalescence slows down the reaction and cause lower reduction degree in experimental compared with the simulation in those runs. Although the gas temperature in Run 1 is high at above 1,850 K, the mass feeding rate of concentrate powder in this run is nearly half the values in Runs 4–6,. The low mass feeding rate greatly decreases the probability of droplets to coalesce in Run 1. This points to the need for improving the CFD model to account for particle coalescence at high temperatures.

The temperature of the particle-gas strean in Run 1 was also above 1,577°C, but the solid feed rate in this run was only about one-half of the values in Runs 4–6. The lower solid feed rate in Run 1 together with the fact that portions of the particles usually get stuck on the wall significantly lowered the possibility of particles in the gas stream to agglomerate in Run 1. The scanning electron microscope (SEM) micrographs of the samples from Runs 1–3 in Figure 10.19 indicate that particle did not fuse in these runs, yielding the satisfactory agreements between the experimental and CFD results. The micrographs of the sample collected from Run 4 shown in Figure 10.20 indicates a larger amount of big particles, pointing to greater extent of particle agglomeration.

10.2.4 Concluding Remarks

A MPFR for flash ironmaking has been simulated using the CFD method. The model proposed in this work described the fluid flow and heat transfer within the MPFR. The model also predicted the reduction degrees with good agreement for those experiments where the gas temperature was below the melting/fusion point of iron (1,811 K). The off-gas composition obtained by incorporating the UDF for reduction kinetics of H_2 and CO gases with iron concentrate, previously determined in a drop-tube reactor at our research lab, agreed with the off-gas composition readings from the experiments. The model can be used in the design and optimization of much larger reactors when the gas operating temperature is within the range of 1,423–1,811 K. The model had lower degree of agreement for reduction degree in the experiments where the gas temperature was above iron melting, which caused the particles to melt and coalescence. Particle/droplet coalescence caused the assumption of no interactions between particles less accurate. This points out to the need to incorporate the particle/droplet coalescence into the model to improve its performance at operating temperatures above 1,811 K.

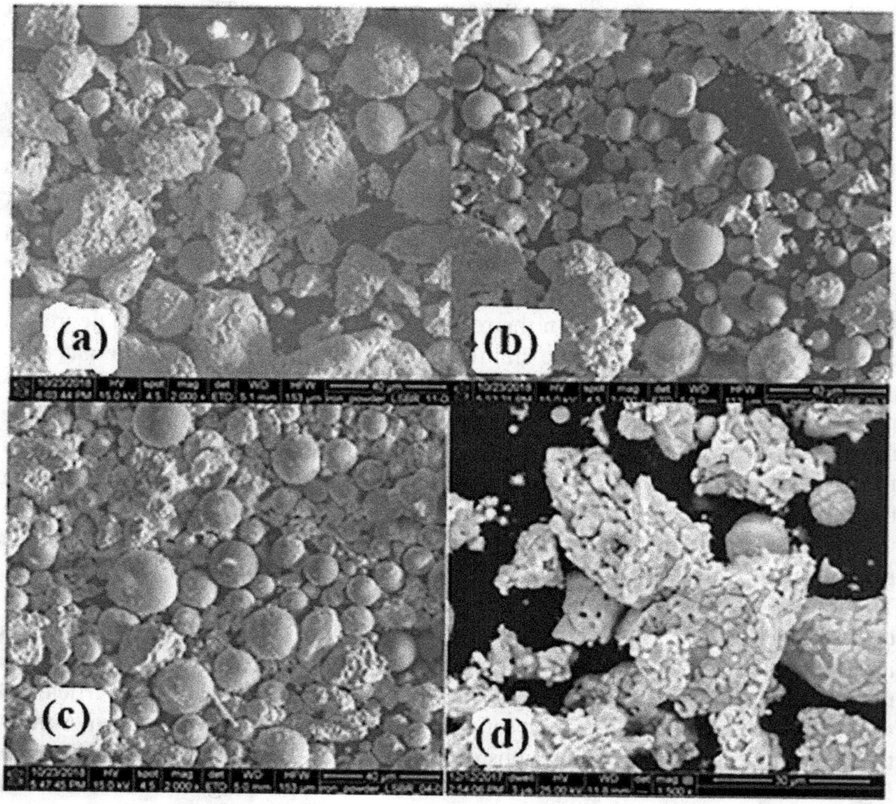

FIGURE 10.20 SEM micrographs of samples from (a) Run 1, (b) Run 2, (c) Run 3, and (d) Run 4. (Adapted from Abdelghany et al., [2020]).

10.3 OPTIMIZATION OF MINI-PILOT FLASH REACTOR OPERATING CONDITIONS WITH CFD

The operating conditions of the MPFR were optimized by using the CFD model described in the previous section, after adding more comprehensive boundary conditions. The effects of operating conditions including O_2/natural gas ratio and gas flow rates on the product gas composition, temperature, heat loss to the walls of the MPFR, and reduction degree of the magnetite concentrate were tested. An optimized value of oxygen/natural gas ratio and total inlet gas flow rate are suggested.

For the purpose of incorporating more realistic boundary conditions for a continuous operation, the thickness and thermal properties of the wall layers are shown in Table 10.9. The reactor was fed natural gas (equivalent to 100.6% CH_4 and 2% N_2) with oxygen to provide heat through a particular burner shown schematically in Figure 10.21. It is noted that the oxygen inlet ports (a total of ten ports) has an inclined angle which is causing a swirl flow in the reactor, which increased the residence time of particles. This particular design of the burner shortened the flame length and

TABLE 10.9
Thicknesses, and Thermal Properties of Refractory, Insulation, and Carbon Steel Layers in the Mini-Pilot Reactor (MPFR)

Property	Refractory Layer	Insulation Layer	Carbon Steel Shell
Top part thickness, mm	321.5	79.4	19.1
Side part thickness, mm	177.8	73	9.5
Bottom part thickness, mm	193.5	73	9.5
Density, kg/m^3	2,891	1,081	7,850
Specific heat, J/g/K	$0.0003 \times T + 0.3621$	0.714	0.47
Thermal conductivity, W·m·K	$10^{-6} \times T^2 - 0.0032 \times T + 4.5396$	$3 \times 10^{-8} \times T^2 + 4 \times 10^{-5} \times T + 0.1797$	52

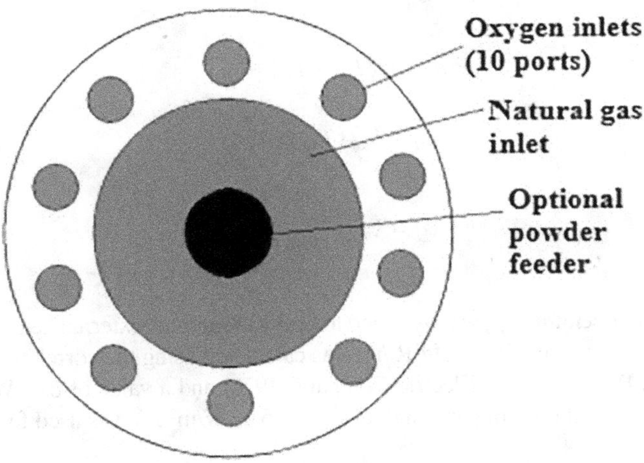

FIGURE 10.21 Schematic of the particular burner used in the MPFR.

ensured a larger uniform temperature zone in the whole reactor. The ratio of oxygen and natural gas was adjusted to provide the hydrogen and carbon monoxide reducing gases, and the minimum ratio to provide sufficient reducing gases was found to be 0.7. The changes in the oxygen/natural gas ratio and the total inlet gas flow rates were tested by the CFD model in seven simulation runs listed in Table 10.10.

10.3.1 Realistic Boundary Conditions

More comprehensive boundary conditions were used in the CFD model, rather than the constantwall temperature used in the previous section. The CFD model of the MPFR was adjusted to include the effect of the layers of refractory, insulation, and carbon steel surrounding the fluid zone and the free convection on the outer wall of the reactor in addition to the internal heat transfer. The following equations were used to calculate the heat flux to the wall:

TABLE 10.10
The Simulation Runs of the MPFR

Run No	Natural Gas in (SLPM[a])	Oxygen in (SLPM[a])	Oxygen /Natural Gas Mole Ratio	Conc. Feed Rate (kg/h)	Nominal Resident Time (s)
1	1,065	745	0.7	5.0	4.0
2	598	438	0.7	5.0	7.0
3	411	315	0.7	5.0	10.0
4	1,005	805	0.8	5.0	4.0
5	575	460	0.8	5.0	7.0
6	403	322	0.8	5.0	10.0
7	352	305	1.0	5.0	10.0

[a] SLPM: Standard liters per minute.

$$q = h_f\left(T_{w_i} - T_f\right) + q_{rad} \tag{10.30}$$

$$q = h_{ext}\left(T_{ext} - T_{w_o}\right) \tag{10.31}$$

$$q = k_{th_w}\frac{\left(T_{w_o} - T_{w_i}\right)}{\Delta x} \tag{10.32}$$

The ambient temperature T_{ext} was assumed to be 300 K and the external heat transfer coefficient on the outer wall of the MPFR h_{ext} was calculated using the correlation in General Electric Data Book [General Electric Company, 1970] and a value of 6.4 (W/m^2 K) was used for the side and bottom parts and a value of 6.0 (W/m^2 K) was used for the top part of the MPFR.

A number of simulation runs were performed on this more comprehensive model to study the effect of the following on the metallization degree of the magnetite concentrate in the MPFR.

10.3.2 Effect of the Inlet Oxygen to Natural Gas Ratio with the Same Total Gas Flow Rate

Natural gas is partially combusted by industrial oxygen to produce the reducing gases (H_2+CO) in the temperature range of 1,150°C–1,600°C. The flow rates of natural gas and oxygen in the MPFR affect the operating temperature of the reactor and the mole percentages of hydrogen, carbon monoxide, carbon dioxide, and water vapor, thus affecting the chemical driving force of the gas. For the total gas flow rates of 726, 1,036, and 1,810 SLPM, an oxygen/natural gas ratio of 0.7 resulted in a metallization degree of 86.4 % $\pm$ 8.4%. Run 7 with a ratio of oxygen/natural gas of 1.0 gave an operating temperature of the MPFR above the melting point of iron. This was found to cause problems in this model as particles/droplets will tend to coalesce

TABLE 10.11
Computational Fluid Dynamics Simulation Results

Run No	Oxygen /Natural Gas Mole Ratio	Total Inlet Gas Flow Rate in SLPM[a]	Temp. (K) Half-Length of MPFR	Temp. (K) Outlet of MPFR	Metallization (%)
1	0.7	1,810	1,519	1,484	94.8
2	0.7	1,036	1,491	1,466	84.7
3	0.7	726	1,462	1,438	78
4	0.8	1,810	1,798	1,777	99.9
5	0.8	1,036	1,665	1,647	99.9
6	0.8	726	1,569	1,548	95.5
7	1.0	726	1,893	1,879	99.3

[a] SLPM: Standard liters per minute.

together and the results of this model will not be accurate as we assumed no particle interactions. An oxygen/natural gas ratio of 0.8 achieved a high metallization degree of 97.7% $\pm$2.2%. These effects have been summarized in Table 10.11.

10.3.3 Effect of Total Gas Flow Rate with Constant Oxygen/Natural Gas Ratio

The total gas flow rate at a constant oxygen/natural gas ratio affects the operating temperature and the particle residence time. Although the latter decreases at a higher gas flow rate, the resulting higher operating temperature affects the reduction degree more strongly. This explains the results shown for Runs 4–6 in Table 10.11.

10.3.4 Comparison of the Simulated and the Equilibrium Gas Compositions

The product gas compositions for H_2, CO, H_2O, and CO_2 from simulation at the reactor exit were compared with the equilibrium gas composition using HSC program and high accuracy were achieved in all runs as shown in Table 10.12. It should be noted that the effect of reduction of magnetite concentrate was included in the calculation.

10.3.5 Profiles of Metallization Degree

Figure 10.22 shows the mass average iron mass fraction profile where the horizontal axes of the plots represents the axial direction (Z-direction) in the reactor starting from the inlet on the reactor top and ending with 2.1 m at the outlet of the reactor. It should be noted that iron mass fraction represents the fraction of metallic iron mass with respect to the whole particle mass including mass of iron oxide while the metallization fraction represents the fraction of iron in the metallized form with respect to the total mass of iron in the particle. The values of the iron mass fraction calculated

TABLE 10.12
Comparison of Simulated Gas Mole % at Outlet vs. Equilibrium Values

Run No	Sim. / HSC	H_2	CO	H_2O	CO_2
1	Sim.	54.0	30.4	12.3	2.8
	HSC	54.0	30.6	12.6	2.8
2	Sim.	52.6	30.0	14.0	3.3
	HSC	51.6	29.9	15.0	3.5
3	Sim.	50.4	29.4	16.2	4.0
	HSC	49.2	29.1	17.5	4.2
4	Sim.	48.5	30.2	18.3	3.1
	HSC	47.5	30.1	19.2	3.2
5	Sim.	48.1	29.7	18.6	3.6
	HSC	47.2	29.6	19.4	3.7
6	Sim.	47.9	29.2	18.8	4.1
	HSC	47.0	29.1	19.7	4.3
7	Sim.	37.9	28.1	28.7	5.2
	HSC	35.7	31.0	27.6	5.7

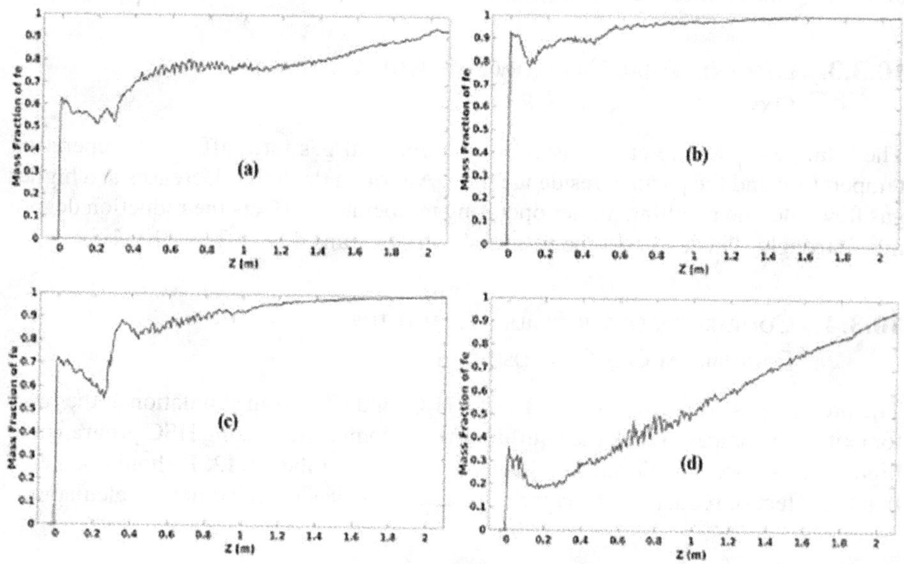

FIGURE 10.22 Profile of mass-averaged iron mass fraction along the length of MPFR: (a) Run 1, (b) Run 4, (c) Run 5, and (d) Run 6.

at the exit of the reactor in Figure 10.22 were used to calculate the metallization degrees previously shown in Table 10.11.

$$Metallization\ fraction = \frac{W_{FeM}}{W_{FeT}} \tag{10.33}$$

where W_{Fe_T} and W_{Fe_M} represents the total weight of iron and metallic iron, respectively, in the reduced sample.

The larger total flow rate made the particles occupy more area in the MPFR as we can see from the number density distribution in Figure 3.8 as Run 4 made particles occupy larger area compared with Runs 5 and 6. The Runs 1 and 4 made particles occupy larger area compared with Run 5 and 6 because of the swirl effect and the larger flow rates. The more area is occupied by particles the higher metallization degree achieved as the concentrate particles become more uniformly exposed to reducing gases.

The profiles of mass average iron mass fraction along the MPFR shown in Figure 10.22 has irregular variations because of the eddies inside the reactor cause higher mass-averaged particles to be present at areas closer to the top part of the reactor. Farther down in the Z-direction, the less eddies will be present and we can see the mass average iron mass fraction increasing with fewer irregular variations. It is noticed from Figure 10.22 that Run 4 with a total gas flow rate of 1,810 SLPM and 0.8 oxygen/natural gas ratio will have the highest iron mass fraction in the shortest length of the reactor. That means we can achieve the 99.9% metallization using operating conditions of Run 4 in a shorter reactor, almost two third of the current reactor as we achieved the 99.9% metallization degree after 1.4 m of the 2.1 m length of the MPFR.

10.3.6 Heat Loss to the Surroundings

Calculating the heat loss from the generated heat of the inlet gases to the walls of the reactor helps us to evaluate the percentage of energy loss in the reactor and gives us a better view on how energy efficient is our reactor with certain operating conditions. We calculated the amount of heat generated from the reaction of natural gas with oxygen and the heat lost to the surroundings of the MPFR through the walls. We calculated the percentage of heat loss in each of the seven runs and summarized those numbers in Table 10.13.

We notice from Table 10.11 that Runs 3, 6, and 7 has the highest percentage of heat loss because of the longer nominal residence time (10 seconds) in those runs. Also, the runs with the lowest nominal residence time have the lowest percentage of heat loss. Finally, Runs 1 and 4 have shown the lowest percentage of heat loss compared

TABLE 10.13
Heat Generated from the Combustion of Natural Gas, Heat Loss Through the Walls of the Reactor, and Percentage Heat Loss

Run No	Heat Generated (W)	Heat Loss (W)	Percentage Heat Loss
1	61,670	22,080	35.8
2	36,450	20,250	55.5
3	25,480	19,400	76.1
4	87,870	29,180	33.2
5	47,370	24,960	52.7
6	29,270	22,270	76.1
7	43,880	31,140	71.0

with all other runs due to lowest nominal residence time of gas in the MPFR. Using operating conditions of Runs 1 and 4, means that we get the lowest percentage of heat loss to the walls of the reactor with respect to the generated heat from gases reaction.

Based on the results, we can easily recommend the operating conditions of Run 4 to be the optimum operating conditions because of the higher metallization degree and lowest percentage of heat loss through the walls. We can even achieve the metallization degree in Run 4 using shorter reactor

10.4 COMPUTATIONAL FLUID DYNAMICS MODELING – DESIGN OF INTERMEDIATE-SIZE FLASH IRONMAKING REACTORS

As an intermediate step to the full industrial-scale flash ironmaking reactor, the design of a medium-sized flash reactor was investigated in this work. The verified rate expressions and CFD models were used for the design of a reactor with a capacity of 100,000 tonnes/year of metallic iron. The CFD simulation provided information such as temperature and species distribution, gas and particle flow patterns that are essential for the proper design of reactor design.

10.4.1 Introduction

Before going to full industrial-scale flash ironmaking reactor that can produce 1–3 million tonnes/year of metallic iron (comparable with a modern blast furnace), a reactor with a capacity of 100,000 tonnes/year of iron was designed to further test the feasibility of FIT in this range of production rate. For the proper design and scale-up of such reactors, it is essential to have information on the temperature and species distribution, gas and particle flow patterns. This information is difficult or even impossible to obtain from experiments. With CFD modeling, it is possible to gain such insights into these critical parameters that are essential in reactor design.

The product of the flash ironmaking process can be either in the solid state or in the molten state depending on the operating temperatures. In this chapter, two types of pilot-scale reactors will be designed. The first type is to produce metallic iron in solid state. The typical operating temperature in this case is around 1,300°C. The solid-state product collected could be charged into an electric arc furnace in the steelmaking process. The second type is to produce iron in the molten state, which is typically operated at a temperature of around 1,600°C, and can lead to direct steelmaking combined with flash reduction or charged into a basic oxygen furnace or an electric arc furnace without further treatment.

10.4.2 Geometries and Dimensions

Sketches of possible configurations of flash ironmaking reactors are shown in Figure 10.23. Depending on the operating conditions, the main body of the reactor is either made up of a cylindrical part and a conical part or a cylindrical shaft only. Under solid-state operating conditions, a conical part near the exit of the reactor is needed for solid particle collection. If iron is produced in the molten state, a bath settler is needed below the shaft. Our focus is mainly on the shaft part of the

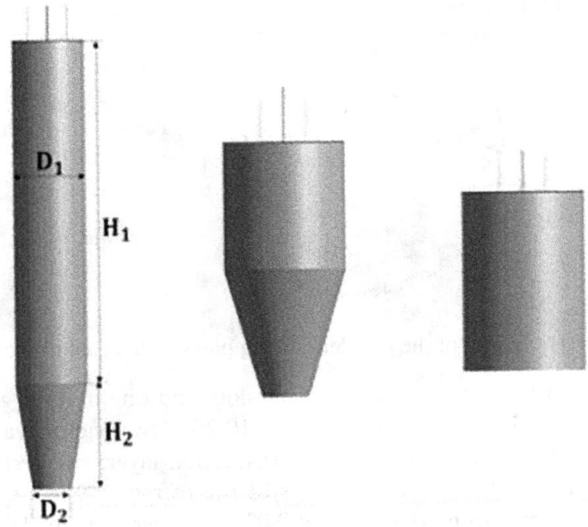

FIGURE 10.23 Sketches of possible configurations of flash ironmaking reactors.

reactor in this work as the reduction of concentrate particles mostly happens during their travel in the shaft.

With the same reactor volume, the design with a large height to diameter ratio leads to a long and thin reactor, while a small height to diameter ratio leads to a short and fat one as shown in Figure 10.23. In this study, two typical diameters, 4m and 6m, were tested. The diameter of the long and thin reactor was set to be 4m. A diameter of 6 m was used for the short and fat reactor. The number of burners to be used is also an important factor in reactor design. Reactors with one burner and four burners were tested in this work. Before deciding on the number of burners to be used, the optimal value for the diameter was first determined under the one-burner design. The dimensions of the reactors simulated are listed in Table 10.14. The concentrate is fed through four feeding ports installed on the roof of the reactor. The four powder-feeding ports are distributed evenly (90° apart), as shown in Figure 10.24. The distance between each feeding port and the centerline of the reactor is equal to half of the radius.

TABLE 10.14
Dimensions of Reactors with One-Burner

D_1 (m)	D_2 (m)	H_1 (m)	H_2 (m)	Preheat Temp. (°C)	Designed Product Temp. (°C)	Run No
4.0	2.0	12.0	6.0	600	1,300	1
4.0	2.0	10.0	6.0	1,000	1,300	2
6.0	2.0	6.0	6.3	600	1,300	3
6.0	2.0	6.0	5.0	1,000	1,300	4
4.0	–	13.0	–	1,000	1,600	5
6.0	–	9.0	–	1,000	1,600	6

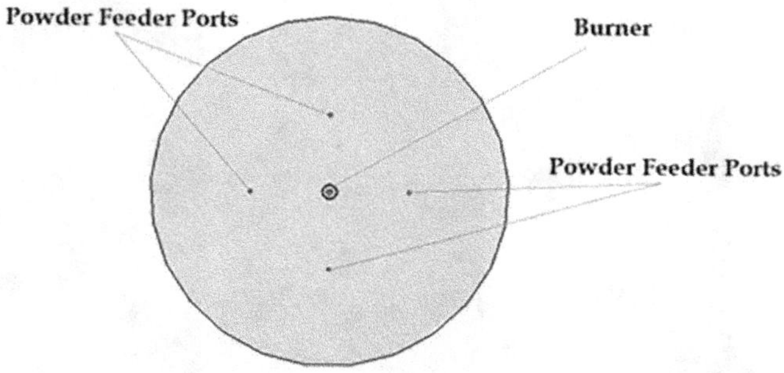

FIGURE 10.24 Distribution of the powder-feeding ports on the roof of the reactor.

A non-premixed burner with two oxygen slots and one natural gas slot is used in the simulation, which is shown in Figure 10.25. The reactor wall consisted of three layers, namely, a refractory layer, an insulation layer, and a steel shell layer, as shown in Figure 10.26. The thicknesses of the refractory, insulation, and steel shell layers are kept at 0.15 m, 0.08 m, and 0.0254 m, respectively. Wall materials at those thicknesses were proved to be efficient in a pilot flash ironmaking reactor constructed on the campus of the University of Utah that was designed to operate from 1,200°C to 1,600°C. The properties of the wall materials are listed in Table 10.15.

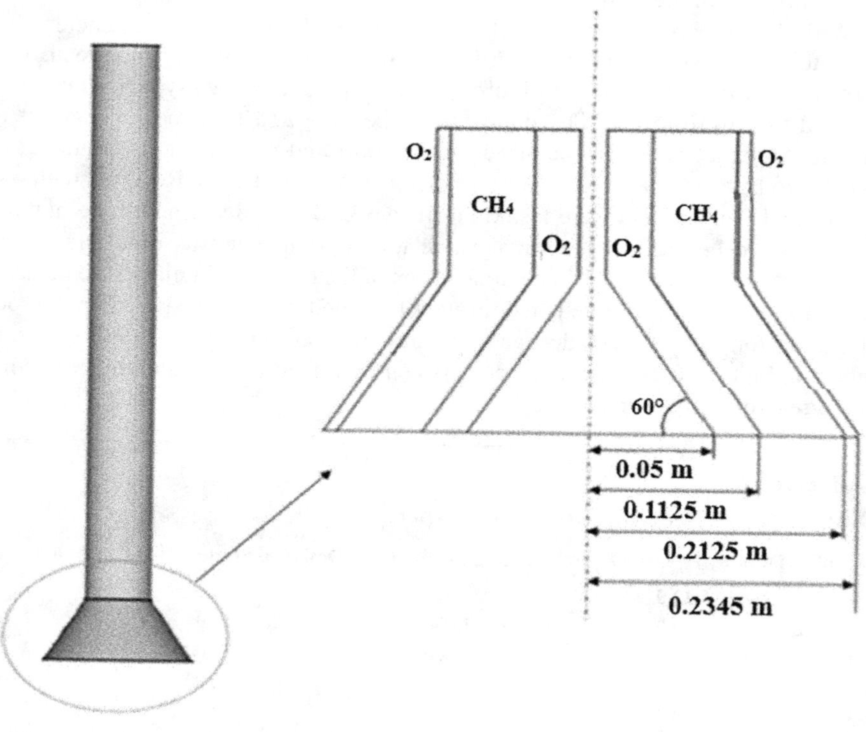

FIGURE 10.25 Burner configuration.

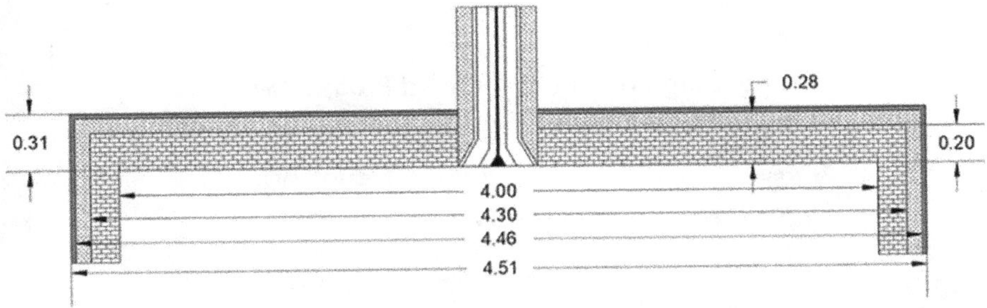

FIGURE 10.26 Reactor wall structure (unit m).

TABLE 10.15
Wall Material Properties

	Thermal Conductivity ($W{\cdot}m^{-1}{\cdot}K^{-1}$)	Density ($kg{\cdot}m^{-3}$)	Specific Heat ($J{\cdot}kg^{-1}{\cdot}K^{-1}$)
Refractory	$10^{-6}\,T^2 - 0.0032\,T + 4.5396$	2,890	$0.2965T + 362$
Insulation	$3\times10^{-8}\,T^2 + 4\times10^{-5}\,T + 0.1797$	1,081	714
Steel shell	50	7,850	470

10.4.3 Operating Conditions

The fuel entering the reactor consisted of fresh natural gas and recycled H_2, which were partially oxidized by oxygen to generate the heat needed for the reaction as well as the reducing gases CO and H_2. The recycled H_2 is recovered from the off-gas which always contains a significant amount of hydrogen because the reduction of FeO by hydrogen is limited by equilibrium. The actual composition of natural gas was 96% CH_4, 2% C_2H_6, and 2% nitrogen by volume. It makes the CFD simulation program, including the combustion calculation, rather complicated to add the oxidation of C_2H_6, even if its amount is small. Considering that the amount is small and also that its oxidation produces the same gases as CH_4, the amount of C_2H_6 was converted to the equivalent amount of CH_4. In terms of heat production, 1 mol% of C_2H_6 is equivalent to 2.6 mol% of CH_4 and in terms of hydrogen and carbon monoxide production, 1 mol% of C_2H_6 is equivalent to 2 mol% of CH_4. Thus, to correct for the presence of the small amount of C_2H_6, 1 mol% of C_2H_6 was treated as being equivalent to 2.3 mol% of CH_4. As a result, natural gas was considered as 98.1% CH_4 and 1.9% N_2. The input gases were preheated to a specified temperature before charging into the reactor to reduce the amount of input gases needed. In this work, two preheat temperatures (600°C and 1,000°C) were investigated. The operating conditions are summarized in Tables 10.16–10.18.

The concentrate feeding rate was calculated based on 340 normal operating days in a year, 70 wt.% total iron content in the concentrate and product metallization of 95%.

As seen from Tables 10.16 and 10.17, when the preheat temperature was lower, the flow rates of fuel and oxygen had to be increased to maintain the temperature in the reactor at 1,300°C.

TABLE 10.16
Operating Conditions for Solid Product with Input Gases Preheated to 600°C

Feed	Flow Rate (kg/s)	Preheat Temp. (°C)
Natural gas	1.15	600
Recycled H_2	0.43	600
Oxygen	2.19	600
N_2 (carrier gas)	0.07	25
Concentrate	5.20	25

TABLE 10.17
Operating Conditions for Solid Product with Input Gases Preheated to 1,000°C

	Flow Rate (kg/s)	Preheat Temp. (°C)
Natural gas	0.91	1,000
Recycled H_2	0.36	1,000
Oxygen	1.54	1,000
N_2 (carrier gas)	0.07	25
Concentrate	5.20	25

TABLE 10.18
Operating Conditions for Molten Product with Input Gases Preheated to 1,000°C

	Flow Rate (kg/s)	Preheat Temp. (°C)
Natural Gas	0.91	1,000
Recycled H_2	0.36	1,000
Oxygen	1.54	1,000
N_2 (carrier gas)	0.07	25
Concentrate	5.20	25

10.4.4 Meshing and Mathematical Model

The pilot reactors were designed to have either a single burner in the center or four symmetrically distributed burners with four powder-feeding ports evenly distributed and have the same radial position equal to half the radius of the reactor. The symmetry of the reactor was used to decrease the computational cost by taking a quarter of the reactor as a representation of the whole reactor. This decreases the time required for simulation as the number of mesh cells decreases accordingly.

The typical mesh for the pilot reactor is shown in Figure 10.27. The mesh consisted of hexahedral cells only.

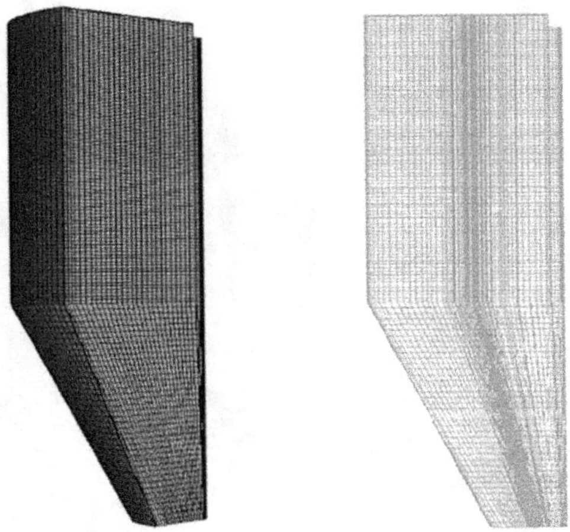

FIGURE 10.27 Typical mesh for the pilot reactor, total number of cells 250,996: (a) 3D view. (b) Cross-section view.

Steady-state conditions were simulated in this work. The Euler-Lagrange approach was used to model the two-phase flow, in which the gas phase was treated as a continuum in the Eulerian frame of reference while the solid phase was tracked in the Lagrangian mode. A two-way coupling approach between the gas phase and solid phase was used in the simulation. The governing equations for the gas and solid phases have already been covered in the previous chapters, which will not be repeated here. The same CH_4-O_2 combustion mechanism as in the last chapter was used. In a large-scale pilot reactor, efficient cooling method, such as copper stove cooling technology, is usually necessary to cool the outer surface of the reactor to an acceptable temperature. The incorporation of such cooling system significant complicated the model. For simplicity, the outer surface of the reactor in this work was set to be 60°C for all the calculations.

The mass-weighted average metallization degrees at the exit of reactor in all the designs listed in Table 10.14 were aimed to be 95%.

10.4.5 One-Burner Design

An example of the velocity vector field in the plane that passes through the center of two powder-feeding ports is shown in Figure 10.28. It is seen from the figure that due to the high-velocity jets erupting from the burner nozzles, recirculation zones formed in regions close to the reactor inner wall in the top part of the reactor. In design No. 1, the particles entering the reactor may be pushed to the reactor wall by the hot, high-velocity gas coming out of the flame region [Fan, 2019]. The effect of this is twofold. On the one hand, as the particles were close to the high-temperature flame region, the particles may fuse and melt, which would negatively affect the reduction of those particles as observed in the laboratory-scale flash reactor. On the other hand, the particles being pushed toward the wall may cause the sticking problem. The accumulation of

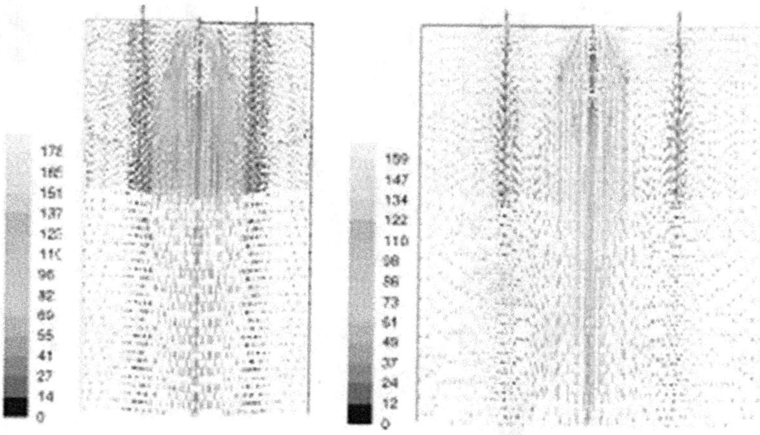

FIGURE 10.28 Velocity vector field in the plane that passes through the center of two powder-feeding ports: (a) Run No. 1. (b) Run No. 3 under operating conditions listed in Table 10.16 (unit in m/s).

such particles would severely affect the smooth operation of the reactor. In designs No. 3, No. 4, and No. 6, the concentrate particles were less affected due to the larger diameter [Fan, 2019]. Design No. 3 gave a better flow field in this case.

A typical particle distribution is shown in Figure 10.29. This and other particle distributions indicated [Fan, 2019] that when the reactor diameter was 4 m, the particle concentration near the wall was higher than that when the reactor diameter was increased to 6 m. Although high concentration near the wall was not completely eliminated with a larger diameter design in Runs No. 3 and No. 4, the particles were more evenly distributed in the top part of the reactor. In Run No. 6, the effect of larger diameter was very obvious. The particle distribution near the wall region was significantly reduced.

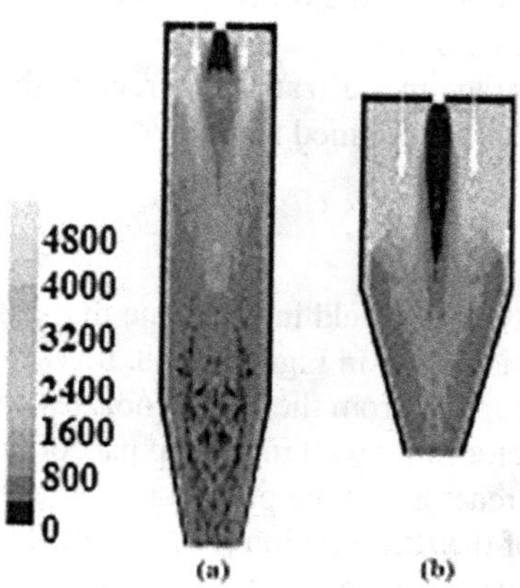

FIGURE 10.29 Particle number density (particles/cm^3) in the plane that passes through the center of two powder-feeding ports: (a) Run No. 1. (b) Run No. 3 under operating conditions listed in Table 10.16.

TABLE 10.19
Product Temperature in Each Run

Run No	Product Temp. (°C)	Reduction Degree (%)
1	1,302	96
2	1,285	94
3	1,303	96
4	1,287	94
5	1,605	97
6	1,590	96

The temperature distributions in the same plane above indicated [Fan, 2019] that in the design with a diameter of 4 m, the particles were close to the flame region in the top part of the reactor, especially in design No. 1, which may cause the particles to melt as mentioned above. For the reactor with a diameter of 6 m, the particle-laden streams were farther away from the flame region. The flame lengths in the design with 6 m diameter were longer than that in the design with 4 m diameter. The average product temperature and reduction degree at the exit of the reactor in each run are listed in Table 10.19.

Heat loss is another criterion to look at in the reactor design. The heat loss through the walls in each run was calculated and listed in Table 10.20. To help better evaluate the energy efficiency of each reactor, the percentage of the energy loss (percentage of heat loss from the heat generated from combustion plus the amount of sensible energy of the input gases) is also calculated. The numbers indicated that reactors with a diameter of 6 m had a smaller value of heat loss than reactors with a diameter of 4 m, as expected, but this result gives a numerical indication of how the heat loss compares between the two cases. The reason for this is that reactors with a diameter of 6 m had a smaller surface area per volume compared with reactors with a diameter of 4 m.

The species distributions in designs No. 5 and No. 6 indicated [Fan, 2019] that the main component gases outside the flame region reached equilibrium quickly and were uniformly distributed. The mole fractions of H_2, H_2O, and CO at the exit of two reactors (outside the flame region) were the same at 0.48, 0.32, and 0.13, respectively.

TABLE 10.20
Heat Loss and Heat Generated from Partial Combustion

Design No	Heat Loss (MW)	Heat Generated (MW)	Sensible Heat of Input Gases from Preheating (MW)	Percentage (%)
1	0.72	19.26	6.82	2.76
2	0.55	12.31	9.92	2.47
3	0.62	19.26	6.82	2.38
4	0.45	12.31	9.92	2.02
5	0.68	20.58	13.60	1.99
6	0.61	20.58	13.60	1.79

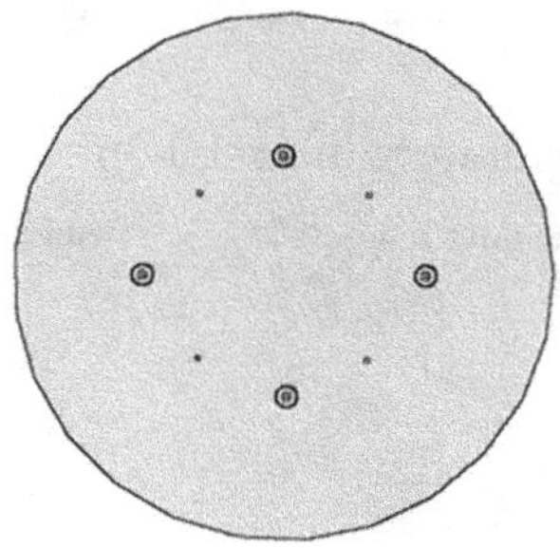

FIGURE 10.30 Distribution of the burners on the roof of the reactor. The large openings are burners and the small openings are powder-feeding ports.

10.4.6 Four-Burner Design

The flash ironmaking reactor with the one burner had an uneven distribution of gaseous species and temperature as well as high particle concentrations near the wall in the top part of the reactor, as seen from Figures 10.29 due to the strong recirculation flow. In this section, the four-burner design is discussed. The distribution of the four burners on the roof of the reactor is shown in Figure 10.30.

The four burners were evenly distributed 90° apart on the roof. The distance between the burner and the centerline of the reactor was equal to half of the radius. The powder-feeding ports were symmetrically placed in-between two burners. The distance between the powder-feeding port and the centerline of the reactor was also equal to half of the radius. The burners used in this case were different from the ones used in the one-burner design. The radial velocity was eliminated by replacing the conical burner tip with a straight concentric design, as shown in Figure 10.31. The natural gas stream was in the middle and was surrounded by two oxygen streams.

The same dimension as design No. 3 in Table 10.14 was used in this simulation. The reactor diameter was chosen as 6 m. The same three layers of walls were also

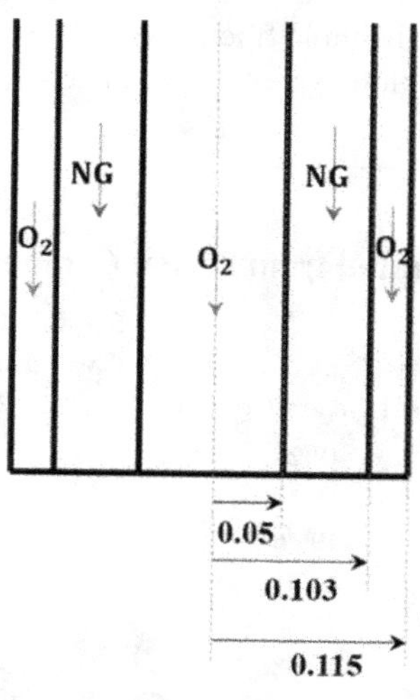

FIGURE 10.31 Burner configuration and dimensions (unit in m).

used in this design. The operating conditions listed in Table 10.16 were used as the reactor was designed to produce solid iron particles.

The vector field in the plane that crosses the center of two powder-feeding ports indicated [Fan, 2019] that the radial-velocity component near the burner was greatly reduced compared with the one in the one-burner design. The number density of the concentrate particles close to the wall was also greatly reduced, making particles sticking to the wall much less likely [Fan, 2019].

Compared with the one-burner design, the particle stream regions in the four-burner design were less exposed to the high temperature of the flame [Fan, 2019]. The consequence of this is that melting of the concentrate particles was less likely to happen so that the reduction of the concentration particles was not affected. Better temperature homogeneity was also seen in this reactor as four burners were used. As a result, the energy generated from the partial combustion was more uniformly distributed inside the reactor. The average product temperature at the exit of the reactor was 1,278°C rendering the mass average reduction degree of the product to be 93%. The heat loss of this reactor was 0.65 MW, which is somewhat greater than the heat loss in design No. 3. This is expected as the high-temperature region is closer to the wall than in the case of one burner.

No noticeable change in the CO and CO_2 mole fractions outside the flame region was seen as the reduction of magnetite concentrate particles was done by H_2. The mole fractions at the exit of the reactor for H_2, H_2O, CO, and CO_2 are 0.47, 0.31, 0.13, and 0.025, respectively.

10.4.7 Summary

Flash ironmaking reactors of different geometrical dimensions with a capacity of producing 100,000 tonnes/year of metallic iron were designed. The metallization degrees of product from these reactors were sufficiently high for use in the subsequent steelmaking step. In the one-burner design, reactors with a diameter of 6 m gave better particle and temperature distributions than reactors with a diameter of 4 m. Better energy efficiency in terms of heat loss was also seen for reactors with a diameter of 6 m. A high particle number density near the wall was less likely in design No. 6. A reactor with a diameter of 6 m and four burners was also simulated. The larger burner number led to a better particle distribution. The particle distribution in this reactor showed a lower probability of particle sticking compared with design No. 3 with a single burner design. All these results may be expected qualitatively, but the CFD simulations present the possibility of yielding quantitative effects of design variation.

10.5 Computational Fluid Dynamics Modeling – Design of Full-Scale Industrial Flash Ironmaking Reactors

10.5.1 Introduction

An industrial ironmaking plant should have a capacity to produce at least 0.3–1.0 million tonnes/year of iron with > 90% metallization to be competitive with modern blast furnaces, which typically produce 0.3–3.0 million tonnes/year of iron. The same model that was previously described was used for designing two industrial reactors capable of production in this range. A multi-burner configuration was shown in the previous section to have a number of advantages, but for the larger reactors the computational time and difficulties were rather prohibitive for this work. Thus, the industrial reactors were designed to have a single burner in the center of the reactor with four feeding ports. Furthermore, the simulated reactor was designed to produce solid iron particles as opposed to molten iron, as the near-term application of the novel FIT is to produce direct reduced iron (DRI) rather than to operate for direct steelmaking. It is hoped that what we learn from this work will provide helpful information and insight into designing future industrial flash ironmaking reactors.

The industrial production is the ultimate goal of any new technology. Our novel FIT has shown good potential for industrial production. The operating temperature of our novel technology is 1,423–1,873 K which is close to that of the blast furnace. We determined that producing 0.3–1.0 million tonnes/year of metallic iron using the FIT was compatible with industrial-scale production as the modern blast furnaces can produce from 0.3–3.0 million tonnes/year of metallic iron.

10.5.2 Model

The same model that has been previously described in Section 10.1 was used for designing two industrial reactors. One minor change from the referred model was that the standard k-ε model used in the industrial reactor instead of the realizable k-ε that was used in the MPFR. The reason for that is we used constant angle feeding for the inlet gases instead of swirl flow in the industrial reactor as for large reactors, swirl does not last through the whole length of the reactor even if it was applied.

10.5.3 Dimensions and Operating Conditions

Table 10.21 shows the operating conditions of the two industrial reactors. The smaller reactor produces 0.3 million tonnes of iron per year. while the larger reactor produces 1.0 million tonnes of iron per year.

The schematic representation of the industrial reactor and its burner are shown in Figures 10.32 and 10.33, respectively, while the dimensions for the two reactors are shown in Table 10.22. It is noted that the radial location of the powder feeders was half the radius of the reactor. Furthermore, the volumetric flow rates of oxygen in inlets one and two were equal.

TABLE 10.21
Operating Conditions of the Two Industrial Reactors

Parameter	Reactor 1	Reactor 2
Target production of iron in million tonnes/year.	0.3	1.0
Feed of the magnetite concentrate in million tonnes/year.	0.415	1.38
Natural gas feeding rate in m^3/s	17.5	45.8
Oxygen feeding rate in m^3/s	13.2	34.7
Expected excess of hydrogen at full reduction (EDF)	0.3	0.3

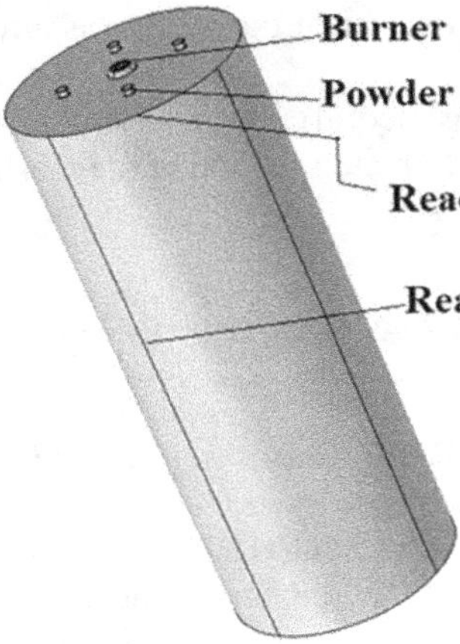

FIGURE 10.32 Schematic representation of the industrial reactor.

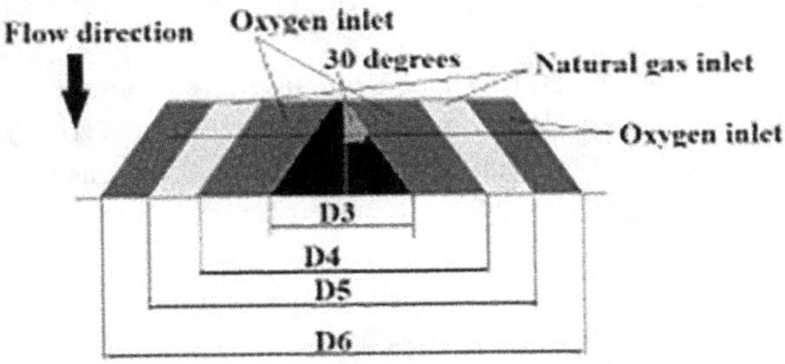

FIGURE 10.33 Schematic representation of the burner.

TABLE 10.22
The Dimensions of the Industrial Reactors

Parameter	Definition	Reactor 1 (m)	Reactor 2 (m)
H	Height	35.0	35.0
D1	Inner diameter	7.0	12.0
D2	Diameter of powder feeder (4 feeders)	0.05	0.30
D3	Inner diameter of the oxygen inlet 1	0.02	0.02
D4	Outer diameter of the oxygen inlet 1	0.26	0.8
D5	Outer diameter of the natural gas inlet	0.44	1.6
D6	Outer diameter of the oxygen inlet 2	0.51	1.8

10.5.4 Meshing

The industrial reactors were designed to have a single burner in the center of the reactor with four feeding ports evenly distributed and have the same radial position equal to half the radius of the reactor. The symmetry of the reactor was used to decrease the computational time by taking a quarter of the reactor as a representation of the entire reactor.

The mesh consisted of 264,000 hexahedral cells in the smaller reactor and 279,000 hexahedral cells in the larger reactor. The top section of the meshing for Reactor 2 is shown in Figure 10.34.

10.5.5 Mass-Weighted Average Gas Composition and Product Metallization at the Outlet

A velocity of 100 m/s was used for the inlet gases in Reactor 1, while for the larger Reactor 2 the area of the burner was increased to increase the use of generated heat from gas reactions. As a result, the inlet velocity of 37 m/s was used for the inlet gases in Reactor 2. The product from reactors at the exit point can tell us how efficient the design was. As shown in Table 10.23, the metallization percentage of the concentrate from the two reactors is nearly identical and both were above 90%. The higher temperature of the gas mixture in the second reactor indicates that the design did use the generated heat from natural gas combustion better than the first reactor.

The products from the reactor exit can tell us the efficiency of the design. As shown in Table 10.23, the metallization degrees of the products from the two reactors were nearly identical at > 90%. The higher temperature of the gas mixture in Reactor 2 indicates that heat loss was lower from it, as expected.

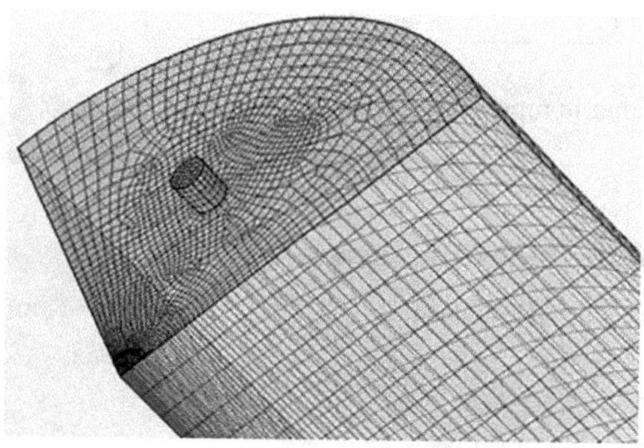

FIGURE 10.34 Meshing of the top section for a quarter of Reactor 2.

TABLE 10.23
Mass Weighted Average Gas Composition, EDF, and Metallization Degree of the Product at Reactor Exit

Reactor	T (K)	H_2	CO	CO_2	H_2O	EDF	Metallization (%)
1	1,519	40.2	26.3	5.9	24.1	0.34	91.2
2	1,578	39.9	26.6	5.7	24.8	0.29	91.4

10.5.6 Profiles of Metallization Degree

Particles need to be distributed through the volume of the reactor to increase their reduction. A higher density near the wall in Reactor 1, which is not favorable, was noticed [Fan, 2019]. A higher density near the walls means that particles are more likely to collide with the wall and stick to the wall. Reactor 2 showed a good distribution of particles and low probability of particles sticking on the wall. Sticking was not considered at the inner walls of the industrial reactors and total reflection of particles hitting the wall was assumed in the model.

The mass-weighted average mass fraction of metallic iron is plotted in Figures 10.35 and 10.36. Both Reactors 1 and 2 reached a 90% reduction at the length of 35 m, but Reactor 1 reached a 70% reduction faster. The irregular variations in the curves arose from the recirculating flows in the reactors as particles will stay longer on average in those zones. The particle temperature did not reach 1,811 K, which is the melting temperature of iron, as we designed the reactor to produce solid iron particles.

10.5.7 Heat Loss

Calculating the heat loss from the generated heat of the inlet gases to the walls of the reactor helps us to evaluate the percentage of energy loss in the reactor and gives us a better view on how energy efficient is our industrial reactor. The amount of heat generation and the percentage of heat loss in each of the two reactors are summarized in Table 10.24.

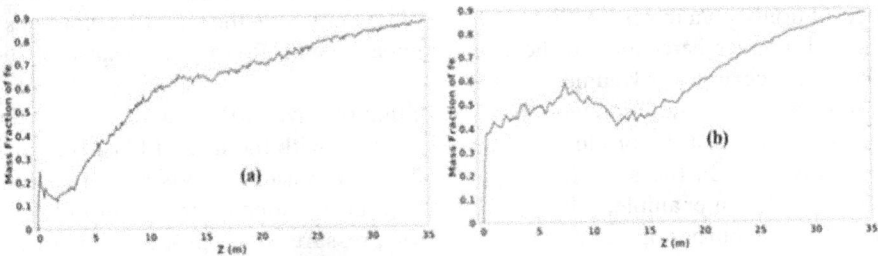

FIGURE 10.35 The mass-weighted average mass fraction of metallic iron in particles. (a) Reactor 1. (b) Reactor 2.

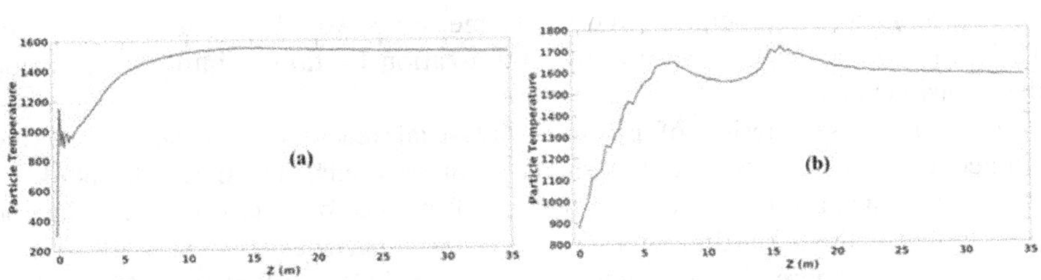

FIGURE 10.36 The mass-weighted average particle temperature in K. (a) Reactor 1. (b) Reactor 2.

TABLE 10.24
Heat Generated from the Combustion of Natural Gas, Heat Loss from the Walls, and Percentage Heat Loss

Reactor	Heat Generated (MW)	Heat Loss (MW)	Percent Heat Loss
1	89.8	1.8	2.0
2	327	3.8	1.2

The numbers indicate that the design of Reactor 2 with a smaller surface area per volume lost only about half of the percentage heat loss of Reactor 1.

10.5.8 Concluding Remarks

Two industrial reactors with different configurations were designed. The metallization degrees of product from these reactors were sufficiently high for use in the subsequent steelmaking step. The symmetry of each reactor was used to decrease the computational time by simulating one-quarter of the reactors and specifying the symmetry faces of that quarter. The burner in Reactor 2 was modified from the burner used in Reactor 1 by increasing the relative cross-sectional area of the burner to decrease the inlet gas velocity and generate a split in the flame. This modification in Reactor 2 caused higher outlet gas temperature, temperature distribution along the reactor, concentrate particle temperature, and slightly higher metallization degree of the product. Particle distribution showed that Reactor 2 had a better distribution with a lower likelihood of particles sticking on the wall. Reactor 2 also showed a lower percentage of heat loss compared with Reactor 1 because of the lower surface area per volume.

Overall, this simulation work has shown that the size of the reactor used in the novel FIT, even at the production rate comparable with the largest blast furnaces currently used in the steel industry, can be quite reasonable vis-à-vis the blast furnaces. As an example, a flash reactor of 12 m in diameter and 35 m in height with a single burner operating at atmospheric pressure would produce 1.0 million tonnes of iron per year. The height can be further reduced by using multiple burners or preheating the feed gas. Furthermore, the total volume of the reactor can be greatly reduced by operating the reactor under elevated pressures, from the points of residence time and reaction kinetics. Obviously, the cost of the reactor per unit volume and those of operation and safety measures would increase accordingly. Thus, the actual design will require optimization by taking into consideration these various factors.

The CFD-based design of potential industrial reactors for flash ironmaking pointed to a number of features that should be incorporated. The flow field should be designed in such a way that a larger portion of the reactor is used for the reduction reaction but at the same time excessive collision of particles with the wall must be avoided. Furthermore, a large diameter-to-height ratio that still allows a high reduction degree should be used from the viewpoint of decreased heat loss. This may require the incorporation of multiple burners and solid feeding ports.

11 Flash Ironmaking Flow Sheet Development and Process Analysis

11.1 HYDROGEN-BASED FLASH IRONMAKING TECHNOLOGY

11.1.1 Introduction

Based on the potential advantages discussed in the previous chapters, detailed process flow sheets for different versions of the flash ironmaking technology (FIT), depending on the fuel type, and economic analyses have been developed [Pinegar et al., 2011; 2012; 2013a; 2013b].

Ironmaking was simulated using two different process configurations, the one-step and one-step reduction processes. In the two-step process, the off-gas from the final reduction process, which contains a significant amount of hydrogen even at equilibrium, is used to pre-reduce the magnetite (or hematite) feed in a separate flash furnace. The simulation results show that the required fresh hydrogen would increase with higher excess driving force and operating temperature, but not significantly when hydrogen is preheated. Compared with the average blast furnace process, the flash ironmaking process would consume much less energy and emit little carbon dioxide [Sohn, 2007; Pinegar et al., 2011; 2012; 2013a], when hydrogen or natural gas for flash ironmaking and coal for the blast furnace are considered as the starting materials in the respective processes.

Flow sheets of the flash ironmaking process were constructed and its operation was simulated using the commercially available software METSIM [Bartlett et al., 2014]. A flow sheet built by connecting streams and unit operation modules calculates material and energy balances by simulating each unit operation in the order specified [Guo, 2003]. The program calculates or incorporates mass, energy, composition, temperature, volume, and pressure of each stream, heat loss, and other important parameters for the evaluation of processes by applying several simulation constraints.

(**Note**: In energy balance, a certain item may be included in the input category or in the output category, as long as the total input and output amounts agree. In the calculation of the "energy requirement" of a process, however, there have been different approaches as to which of the input items to select, which often causes confusion. Comparing different processes, the total difference is largely unaffected by the choice of placement of certain energy items, but the magnitude of the 'energy requirement' of each process depends on this choice. This affects the numerical value of the % difference. Thus, it is recommended to use the difference in energy consumption, rather than the % difference when comparing different processes. For a more detailed discussion on this subject, the reader is referred to Sohn and Martinez-Olivas [2014]).

DOI: 10.1201/9781003342199-11

11.1.2 Flow Sheet Development and Process Simulation

In the process METSIM model, Phase Splitter (SPP), Mixer (MIX), Heat Exchanger (HTX), Flash Separator (FLA), and Component Splitter (SPC) unit operations were used. The description of each unit operation is given in Table 11.1. Process flow sheets of the hydrogen-based one-step and two-step reduction processes are shown with reactor labels and stream numbers in Figures 11.1 and 11.2, respectively. The descriptions of unit operations and streams are shown in Tables 11.2 and 11.3 for the one-step process, and Tables 11.4 and 11.5 for the two-step process. The thick streamlines indicate that these streams are at high temperature and with large amounts of sensible heat.

TABLE 11.1
Descriptions of Unit Operations Used in Process Flow Sheets

Unit Operation Name	Code	Description and Use
Phase splitter	SPP	Splits streams by different phases
Stream mixer	MIX	Mixes streams into one stream
Heat exchanger	HTX	Transfers energy from one stream to another
Flash separator	FLA	To flash water to a gas and calculate an equilibrium state of outlet gas after vapor condensation
Component splitter	SPC	Splits certain components from other components

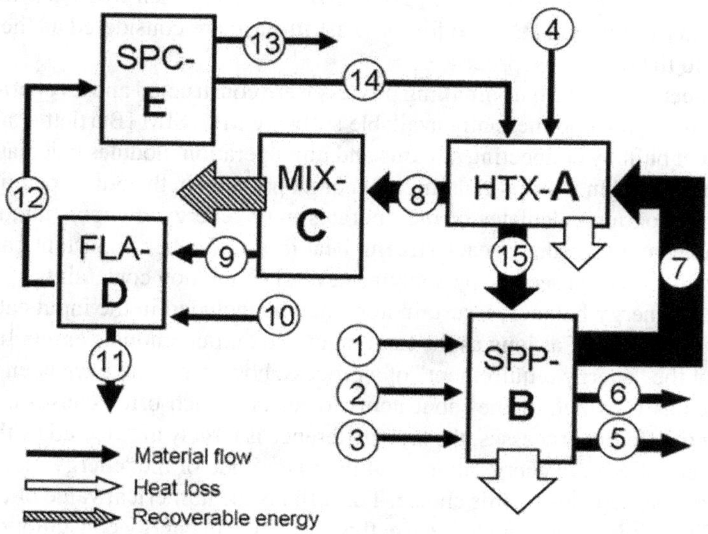

FIGURE 11.1 Block diagram of METSIM flow sheet for the one-step ironmaking process.

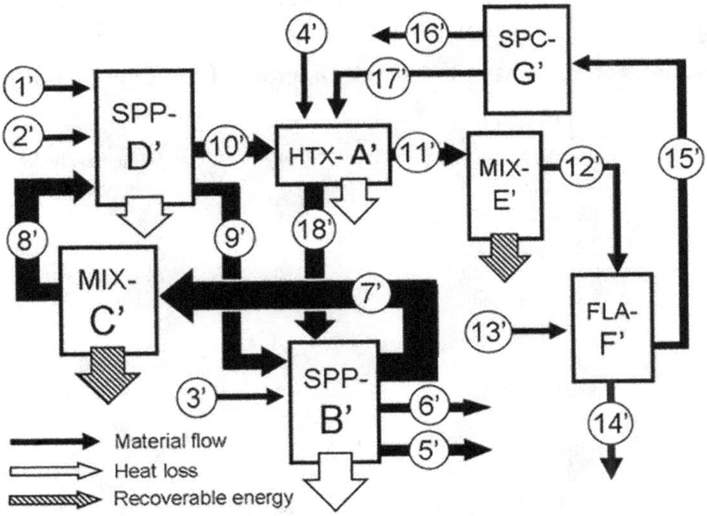

FIGURE 11.2 Block diagram of METSIM flow sheet for the two-step ironmaking process.

TABLE 11.2
Descriptions of Unit Operations in the Block Diagram of the One-Step Process

Unit Operation Label	Description
HTX-A	Heat exchanger for preheating hydrogen by the sensible heat of the reactor off-gas
SPP-B	Ironmaking reactor
MIX-C	Waste heat boiler modeled to recover energy¤(Modeled by MIX because there is no specified heating/cooling stream HTX requires)
FLA-D	Wet scrubber
SPC-E	Water vapor removal process

The temperature of all input streams was assumed to be 25°C. In the water removal process, no hydrogen loss was assumed, and all water vapor was removed. The annual iron production rate was set to be 1 million tonnes with 300 days of operation.

In the one-step process (Figure 11.1), ore, flux, oxygen, and preheated hydrogen streams (Streams 1–3 and 15, respectively) were fed into the ironmaking reactor SPP-B. The reactions taking place in SPP-B are shown in Table 11.6. Each reaction was assumed to proceed to completion, except for the last step of iron oxide reduction, that is, the reduction of wustite to iron, which was assumed to have a reduction extent of 0.99. This reaction is limited by equilibrium, and the hydrogen to water vapor ratio at this reduction extent was set to be 50% in excess of the equilibrium value.

TABLE 11.3
Descriptions of Streams in the Block Diagram of the One-Step Process

No.	Description	No.	Description
1	Iron ore (95.4% Fe_3O_4, 0.4% CaO, 3% SiO_2, 1% Al_2O_3 and 0.2% MgO)	9	Waste heat boiler outlet gas (90°C)
2	Flux (90.5% CaO, 2.5% SiO_2, and 7% MgO)	10	Cooling water (25°C)
3	Oxygen	11	Slurry (50°C)
4	Fresh hydrogen	12	Hydrogen with saturated water vapor (50°C)
5	Molten metal (1,500°C–1,600°C)	13	Removed water vapor (50°C)
6	Slag (1,500°C–1,600°C) (40.9% CaO, 33.6% SiO_2, 10.9% Al_2O_3, 5% MgO, and 9.6% FeO)	14	Recycled hydrogen (50°C)
7	Hot reactor off-gas (With more than 51.9% H_2, 1,500°C—1,600°C)	15	Preheated hydrogen (to 900°C)
8	Waste heat boiler inlet gas		

TABLE 11.4
Descriptions of Unit Operations in the Block Diagram of the Two-Step Process

Unit Operation Label	Description
HTX-A'	Heat exchanger for preheating hydrogen by the sensible heat of the prereduction reactor off-gas
SPP-B'	Ironmaking reactor
MIX-C'	Waste heat boiler using the iron-producing reactor off-gas (Modeled by MIX because there is no specified heating/cooling stream HTX requires)
SPP-D'	Prereduction reactor
MIX-E'	Waste heat boiler using the prereduction reactor off-gas (Modeled in the same manner as MIX-C')
FLA-F'	Wet scrubber
SPC-G'	Water vapor removal process

TABLE 11.5
Descriptions of Streams in the Block Diagram of the Two-Step Process

No.	Description	No.	Description
1'	Iron ore (95.4% Fe_3O_4, 0.4% CaO, 3% SiO_2, 1% Al_2O_3, and 0.2% MgO)	10'	Prereduction reactor off-gas (900°C)
2'	Flux (90.5% CaO, 2.5% SiO_2, and 7% MgO)	11'	Waste heat boiler inlet gas
3'	Oxygen	12'	Waste heat boiler outlet gas (90°C)
4'	Fresh hydrogen	13'	Cooling water (25°C)
5'	Molten metal (1,500°C–1,600°C)	14'	Slurry (50°C)
6'	Slag (1,500°C to 1,600°C) (40.9% CaO, 33.6% SiO_2, 10.9% Al_2O_3, 5% MgO, and 9.6% FeO)	15'	Hydrogen with saturated water vapor (50°C)
7'	Ironmaking reactor off-gas (With more than 51.9% H_2, 1,500°C–1,600°C)	16'	Removed water vapor (50°C)
8'	Inlet gas to prereduction reactor	17'	Recycled hydrogen (50°C)
9'	Reduced iron ore (wustite) and flux (900°C)	18'	Preheated hydrogen (to 750°C)

TABLE 11.6
Reactions in the One-Step and Two-Step Processes

Description	Reaction	Reaction Extent
Fuel combustion	(A) $2H_2+O_2=2H_2O$	1
Reduction of iron oxide by hydrogen	(B) $Fe_3O_4+H_2=3FeO+H_2O$	1
	(C) $FeO+H_2=Fe+H_2O$	0.99
Slagmaking	(D) $CaO(s)=CaO(l)$	1
	(E) $SiO_2(s)=SiO_2(l)$	1
	(F) $Al_2O_3(s)=Al_2O_3(l)$	1
	(G) $MgO(s)=MgO(l)$	1

The kinetics of the reduction reaction was described in Chapters 7–10. Molten metal, slag, and off-gas (Streams 5–7, respectively) were produced as outputs from SPP-B.

In the two-step process (Figure 11.2), fuel combustion (Reaction A), the reduction of wustite to iron (Reaction C), and slagmaking reactions (Reactions D–G) took place in the ironmaking reactor SPP-B' while the reduction of magnetite to wustite (Reaction B) took place in the prereduction reactor SPP-D'. In this configuration, the gas from SPP-B' (Stream 8') containing a sufficient content of hydrogen to reduce magnetite to wustite entered SPP-D', while the wustite (stream 9') from SPP-D' entered SPP-B'. Thus, oxygen and preheated hydrogen (Streams 3' and 18') were fed to SPP-B', and ore and flux (Streams 1' and 2') were fed into SPP-D'. The off-gas (Stream 7') from SPP-B' went through the first waste heat boiler MIX-C' before entering the prereduction reactor SPP-D'. MIX-C' was used as a waste heat boiler to recover the sensible heat of Stream 7' in such a way that the temperature of the stream into SPP-D' (Stream 8') was controlled to make the temperature of wustite (Stream 9') and off-gas (Stream 10') 900°C. Then, the off-gas (Stream 10') from the prereduction reactor went through the subsequent heat recovery steps.

The one-step and two-step processes have similar steps for heat recovery and hydrogen recycling around the iron reduction part. The off-gas from the iron ore reduction reactor (Stream 7 or 10') passed through the heat exchanger (HTX-A or HTX-A') to preheat hydrogen. The off-gas (Stream 8 or 11') then entered a waste heat boiler (MIX-C or MIX-E') before a scrubber (FLA-D or FLA-F'), and the off-gas (Stream 9 or 12') exited at 90°C. A mixer was used to calculate the energy recovered for steam generation from the off-gas.

In the scrubber, the water vapor content in the off-gas was reduced to the saturation level at 50°C with cooling water (Stream 10 or 13'), and all the condensed water was separated from the gas stream (Stream 12 or 15') into the slurry (Stream 11 or 14'). The water vapor in the off-gas (Stream 13 or 16') was separated with a water vapor removal system (SPC-E or SPC-G'), with all the remaining hydrogen (Stream 14 or 17') recycled. The latter was combined with fresh hydrogen (Stream 4 or 4') and preheated in the heat exchanger (HTX-A or HTX-A') before being fed to the ironmaking reactor (SPP-B or SPP-B').

The heat loss from the reactors was estimated based on an industrial copper flash smelting furnace, as it is expected that flash ironmaking would be carried out in a similar furnace [Pinegar et al., 2011]. The heat loss from heat exchangers was assumed to be 5% of total energy [J. A. T. Jones, VP Process Engineering and Technology, WorleyParsons Gas Cleaning, Irving, TX, Personal communication, August 2010].

11.1.3 Simulation Results

11.1.3.1 Material and Energy Balances

Tables 11.7 and 11.8 list the material and energy balances for the one-step and two-step processes producing 1 million tonnes of iron per year with the ironmaking reactor operating under atmospheric pressure and a temperature of 1,500°C at an excess driving force of 0.5 as the standard case. The hydrogen preheating temperature was set at 900°C for the one-step process, and 750°C for the two-step process. Material

TABLE 11.7
Material Balances of the Industrial-Scale One-Step and Two-Step Ironmaking Processes Producing 1 Million Tonnes of Iron per Year at an Operating Temperature of 1,500°C, the Excess Driving Force of 0.5 and a Hydrogen Preheating Temperature of 900°C in the One-Step Process and 750°C in the Two-Step Process, Compared with the Average Blast Furnace Process

		One-Step Process	Two-Step Process	BF Process [Choi, 2010]
Input (tonnes/tonne of iron)	Hydrogen	0.10[a]	0.08[a]	
	Oxygen	0.43	0.30	0.68
	Iron ore	1.46 (magnetite)	1.46 (magnetite)	1.43 (hematite)
	Flux	0.06	0.06	0.29
	Nitrogen			2.21
	Coke			0.46
	Total ironmaking	2.05	1.90	5.07
	Cooling water for scrubber	17.9	15.6	
Output (tonnes/tonne of iron)	Molten metal	1.0	1.0	1.05 (contains 4.5% C)
	Slag	0.14	0.14	0.21
	Condensed water vapor	0.71	0.63	
	Total discharged gas	0.2	0.13	3.81
	Carbon dioxide			1.60[b]
	Nitrogen			2.21
	Total ironmaking	2.05	1.90	5.07
	Cooling water in slurry	17.9	15.6	22.3[c]
	Carbon dioxide from $CaCO_3$ and $MgCO_3$ calcination	0.04	0.04	

[a] The actual amount may be slightly higher if heat loss from the hydrogen recycling step is considered.

[b] This number does not include the carbon dioxide emissions from ore/coke preparation.

[c] Water requirement for U.S. Steel's Fairfield No.8 Blast Furnace (production rate: 5,500 tonnes/day) [Camlic and Goodman, 1997; United States Steel Corporation, 1985] including water for top gear, stack plates, bosh channels, tuyere jacket, tuyeres and tuyere coolers, hearth and subhearth, and stove valves, as the reference years of water usage is older than that of the production rate, this number may be somewhat larger than the number for a modern plant.

TABLE 11.8
Energy Balances of the Industrial-Scale One-Step and Two-Step Ironmaking Processes Producing 1 Million Tonnes of Iron per Year with an Operating Temperature of 1,500°C and the Excess Driving Force of 0.5, and a Hydrogen Preheating Temperature of 900°C in the One-Step Process and 750°C in the Two-Step Process, Compared with the Average Blast Furnace Process

		One-Step Process	Two-Step Process	BF Process[a]
Input (GJ/tonne of iron)	Fuel combustion	8.28[b]	6.02[b]	8.33
	Heat recovery	−2.80	−0.99	−1.32[f]
	Subtotal	5.48	5.03	7.01
	Energy input for ore/coke preparation			5.68
	Total ironmaking[g]	5.48	5.03	12.7
	Energy input for $CaCO_3$ and $MgCO_3$ calcination	0.26[c]	0.26[c]	
Output (GJ/tonne of iron)	Reduction	0.91	0.91	2.09
	Slagmaking			−0.16
	Sensible heat of molten metal	1.27 (1,500°C)[d]	1.27 (1,500°C)[d]	1.35 (1,600°C)
	Sensible heat of slag	0.24 (1,500°C)[d]	0.24 (1,500°C)[d]	0.47 (1,600°C)
	Slurry ($H_2O(l)$)	1.93	1.69	
	Sensible heat of off-gas			0.26 (90°C)
	Removed water vapor	0.01	0.006	
	$CaCO_3$ decomposition			0.33
	Heat loss in the reactor	0.78	0.78	2.6
	Heat loss in the heat exchanger	0.34[e]	0.14[e]	0.07[f] (heat loss in heat recovery)
	Subtotal	5.48	5.03	7.01

(Continued)

TABLE 11.8 (*Continued*)
Energy Balances of the Industrial-Scale One-Step and Two-Step Ironmaking Processes Producing 1 Million Tonnes of Iron per Year with an Operating Temperature of 1,500°C and the Excess Driving Force of 0.5, and a Hydrogen Preheating Temperature of 900°C in the One-Step Process and 750°C in the Two-Step Process, Compared with the Average Blast Furnace Process

		One-Step Process	Two-Step Process	BF Process[a]
Output (GJ/tonne of iron)	Pelletizing			3.01[40]
	Sintering			0.65[40]
	Cokemaking			2.02[40]
	Total ironmaking[g]	5.48	5.03	12.7
	$CaCO_3$ and $MgCO_3$ decomposition at 1,000°C [Biswas, 1981]	0.26[c]	0.26[c]	

[a] Energy balance was calculated by METSIM based on the material balance.

[b] Fuel comtbustion energy generated in the process was calculated by subtracting the energy used for iron ore reduction from the balance of heat formation of all output components from the entire system and input components into it. The high heating value of natural gas was used for this calculation. If the low heating value is used, the total energy requirements for the one-step and two-step processes will decrease, respectively, to 3.74 GJ and 3.49 GJ/tonne of iron.

[c] The energy requirement for $CaCO_3$ and $MgCO_3$ calcination to generate CaO and MgO for the flux was calculated by METSIM. Calcination was assumed to be done at 1,000°C separately from the ironmaking process.

[d] If the temperature of iron and slag is replaced by 1,600°C (same as the blast furnace process), the total energy for the one-step and two-step processes increases to 5.86 and 5.32 GJ/tonne of iron, respectively, by the increased sensible heat of iron and slag.

[e] Heat losses in a heat exchanger were assumed to be 5% of the total sensible heat of input streams.

[f] Heat loss in the heat recovery step was estimated to be 5% of the energy difference between the hot off-gas (calculated to be 426°C) and cooled reactor off-gas (90°C), and the rest was estimated to be the recovered energy.

[g] Hydrogen and coking coal are considered as the starting raw materials. This energy requirement comparison does not include electrical power required for the plant operation.

and energy flow diagrams for the processes for the standard case with all the stream values and conditions are shown in Figures 11.3 and 11.4.

The amount of hydrogen required to produce a tonne of iron at an excess driving force of 0.5 as a function of operating temperature is shown in Figure 11.5 for the one-step and two-step processes. The results indicate that the fresh hydrogen requirement increases almost linearly as the ironmaking reactor operating temperature increases.

When hydrogen is considered as the starting fuel/reductant material, the energy requirements are 5.48 and 5.03 GJ/tonne of iron, respectively, for the one-step and two-step processes, compared with 12.7 GJ/tonne of iron for the blast furnace. These energy requirements were obtained using the high heating value (HHV) of hydrogen used as the fuel and reductant. Although 2.26 GJ more energy was generated in the one-step ironmaking reactor than in the two-step process due to higher hydrogen consumption, the difference in the total energy requirement was estimated to be just 0.45 GJ/tonne of iron. This is because the one-step process can use more sensible heat in the reactor off-gas than the two-step process. Compared with the average blast furnace process assuming that coking coal is the starting material, the flash ironmaking processes consume 7.2 and 7.7 GJ/tonne of iron less energy, respectively. If the low heating value (LHV) of hydrogen is used, the energy requirements would be 3.74 GJ/tonne of iron for the one-step process and 3.49 GJ/tonne of iron for the two-step process.

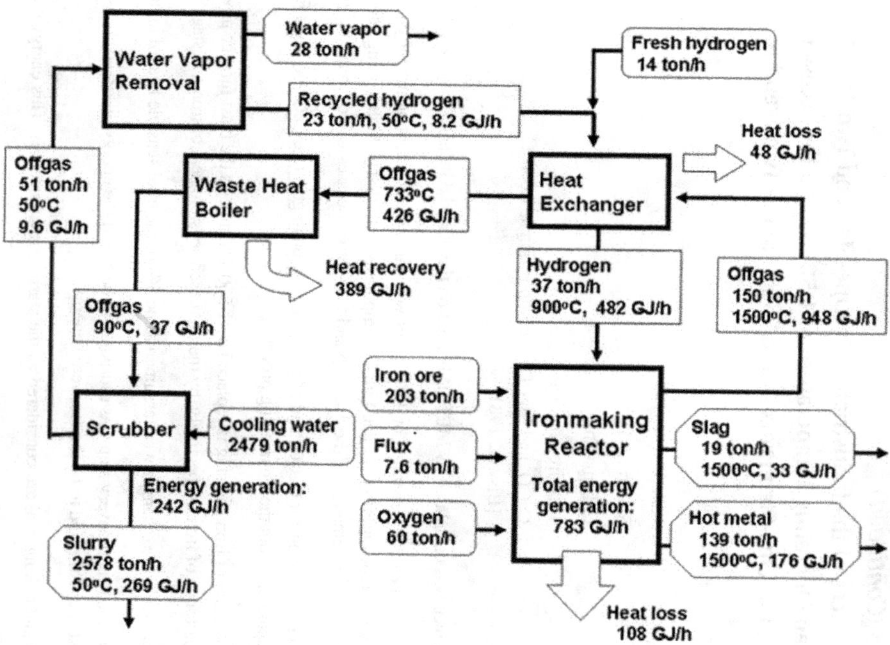

FIGURE 11.3 Material and energy flow diagram for the industrial-scale hydrogen-based one-step ironmaking process producing 1 million tonnes of iron per year at an operating temperature of 1,500°C, an excess driving force of 0.5 and the reactor feed gas temperature of 900°C (300 days of operation in 1 year).

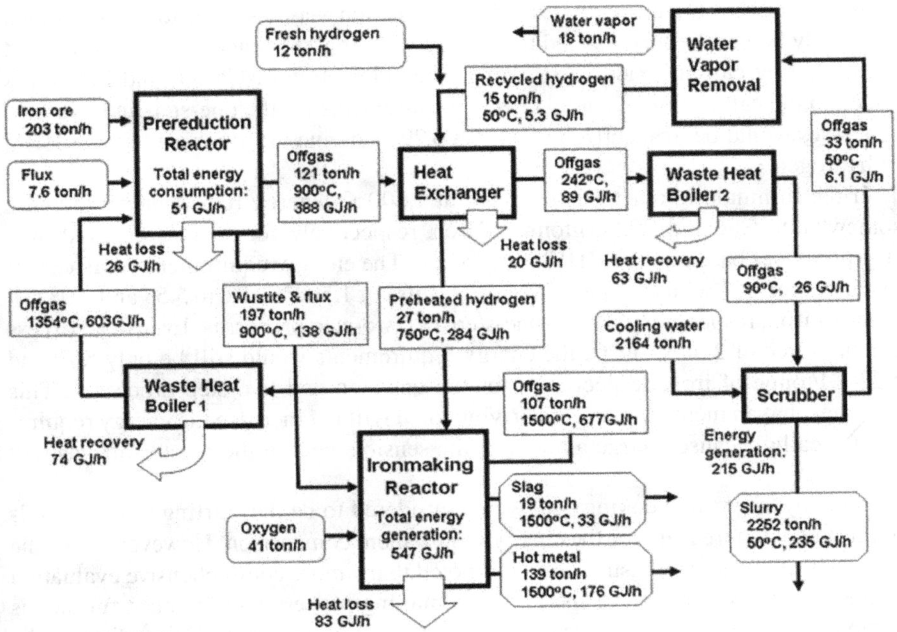

FIGURE 11.4 Material and energy flow diagram for the industrial-scale hydrogen-based two-step ironmaking process producing 1 million tonnes of iron per year at an operating temperature of 1,500°C, an excess driving force of 0.5 and the reactor feed gas temperature of 750°C (300 days of operation in 1 year).

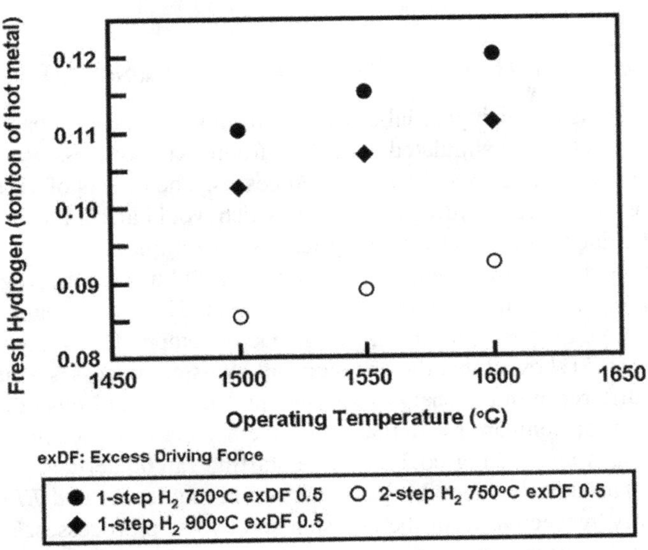

FIGURE 11.5 Fresh hydrogen (tonnes/tonne of iron) required vs. the ironmaking reactor operating temperature at excess driving forces of 0.5 and the hydrogen preheating temperatures of 750°C and 900°C for the one-step process and 750°C for the two-step process.

In the flash ironmaking process, flux calcination was assumed to be performed separately before the flux was fed to the reactor so that the reactor off-gas would not be diluted by carbon oxide gases. When the calcination of $CaCO_3$ and $MgCO_3$ is added to the calculation, the total energy requirements for the one-step and two-step processes would be, respectively, 5.74 and 5.29 GJ/tonne of iron, based on the HHV of hydrogen.

If the ironmaking reactor is operated at 1,600°C, energy requirements increase somewhat to 5.86 and 5.32 GJ/tonne of iron, respectively, for the one-step and two-step processes based on the HHV of hydrogen. The energy requirements at an excess driving force of 1 with the operating temperature of 1,500°C rise to 5.55 and 5.08 GJ/tonne of iron, respectively, for the one-step and two-step processes. Even at an excess driving force of 2 at 1,500°C, the energy requirements would still be only 5.70 and 5.22 GJ/tonne of iron, respectively, for the one-step and two-step processes. This indicates that an increase in excess driving force will not increase the energy requirement greatly because a large amount of the sensible heat in the reactor off-gas will be recovered.

Here, hydrogen and coking coal were considered to be the starting raw materials of fuel/reductant reagent for the energy requirement comparison. However, as for the comparison of energy consumption, it is noted that a more comprehensive evaluation of the energy requirement of the flash ironmaking process and the blast furnace is complicated and depends on the choice of starting raw materials. Depending on the starting material, the energy required for raw material production, such as coal mining, and natural gas pumping, should also be included. Further, the energy required for hydrogen production strongly depends on the production process, such as steam-methane reforming (SMR), coal gasification, or water splitting [Simbeck and Chang, 2002].

11.1.4 Summary on Hydrogen-Based Flash Ironmaking

Process flow sheets for the industrial-scale flash ironmaking process based on hydrogen were constructed and simulated. The flash ironmaking process was modeled in two configurations: one-step and two-step processes. The effects of various operating conditions on the fresh hydrogen amount, which would affect the economic feasibility greatly due to the high hydrogen price, were examined.

The results indicated that the two-step process would use less hydrogen than the one-step process under the same operating condition. The energy balance comparison for the standard case (the ironmaking reactor temperature = 1,500°C, excess driving force = 0.5) showed that the one-step and two-step processes would not have a significant difference in the energy requirement. Energy requirement for the standard case operation compared with that of the average blast furnace process, assuming that hydrogen and coking coal were the starting raw materials, indicated that the hydrogen-based flash ironmaking process would require 7.2 and 7.7 GJ/tonne of iron less energy, respectively, for the one-step and two-step processes. The sensitivity analysis of the energy requirement to operating conditions showed that the flash ironmaking process would have substantially lower energy consumption compared with the blast furnace process within the entire ranges tested. However, it is noted that the energy for hydrogen or coal production must be added for a comprehensive comparison of the energy requirement, and it will depend on which materials are considered as the starting raw materials.

11.2 NATURAL-GAS-BASED FLASH IRONMAKING TECHNOLOGY – REFORMERLESS PROCESS

11.2.1 Introduction

Hydrogen is the simplest and cleanest fuel/reductant for the FIT as described above. However, hydrogen is expensive. Thus, it will be more cost-effective if the flash ironmaking process can directly use natural gas. For this reason, a "reformerless" process in which natural gas is directly oxidized with industrial oxygen in the ironmaking reactor without external reforming was applied to this simulation work.

11.2.2 Flow Sheet Development and Process Simulation

The one-step and two-step processes, described in the previous section for the hydrogen-based flash ironmaking process, were applied to reformerless natural gas-based ironmaking. These processes would have ironmaking steps and off-gas treatment steps to recover and recycle hydrogen. The latter steps were designed based on the hydrogen production process by SMR [Carrara et al., 2010], namely water-gas shift (WGS) reaction and pressure swing adsorption (PSA) were incorporated to recover/generate as much hydrogen as possible from the ironmaking-reactor off-gas that contains CO and H_2O in addition to hydrogen and CO_2. The sensible heat of the off-gas was used for this step.

Block diagrams for METSIM process simulation of the reformerless processes are illustrated in Figures 11.6 and 11.7. In these process models, the same names of unit operations described in Table 11.1 were used. The description of unit operations and streams are shown in Tables 11.9 and 11.10 for the one-step process, and in Tables 11.11 and 11.12 for the two-step process. The annual iron production rate was 1 million tonnes with 300 days of annual operation.

Reactions included in the process simulation of the reformerless one-step and two-step processes are shown in Table 11.13. The details of the simulation work can be found elsewhere [Pinegar et al., 2010]. Here, only the salient results will be presented.

11.2.3 Material and Energy Balances

Summaries of material and energy balances for the industrial-scale reformerless one-step and two-step flash ironmaking processes with an annual production rate of 1 million tonnes, based on the flow diagrams in Figures 11.6 and 11.7, are shown in Tables 11.14 and 11.15, respectively. The results in these tables are based on the simulation with a reactor operating temperature of 1,500°C and excess driving force of 0.5 as the standard case. Material and energy flow diagrams for the one-step and two-step processes for the standard case with all the stream values and conditions are shown in Figures 11.8 and 11.9.

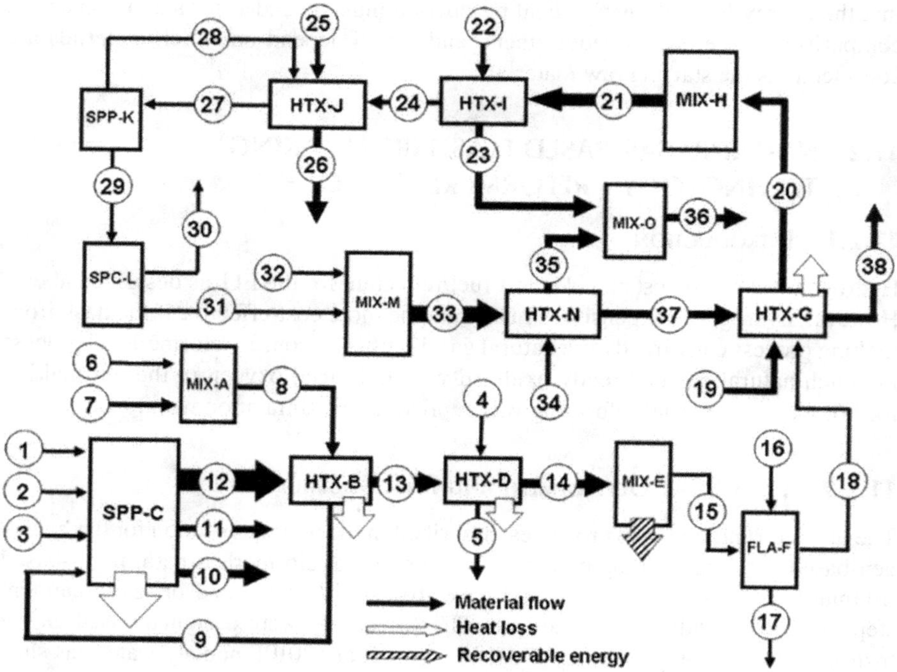

FIGURE 11.6 Block diagram of a METSIM model for the industrial-scale reformerless one-step ironmaking process.

As shown in Table 11.14, in the reformerless two-step process, the required natural gas amount is smaller than that in the one-step process. In the two-step process, wustite is fed to the ironmaking reactor instead of magnetite, which reduces the need for hydrogen used for reduction. Thus, the water vapor generated by reduction is less than in the one-step process. As the same excess driving force was applied in both processes, less hydrogen against the generated water vapor is required to satisfy the excess driving force in the two-step process. This reduces the energy required to heat the products to 1,500°C due to a smaller volume of the off-gas. For these reasons, the two-step process requires less fuel in the ironmaking reactor.

As shown in Table 11.15, although the sum of the heat loss from the ironmaking and prereduction reactors in the two-step process is almost the same as the heat loss from the ironmaking reactor in the one-step process, the total heat loss is smaller in the two-step process. This is because the prereduction reactor in the two-step process operated at 900°C, and thus less energy is required to produce iron. This makes the heat loss from the heat exchanger, through which the off-gas passes to preheat the reactor feed gas, smaller. This consequently reduces the amount of natural gas to provide energy.

As shown in Tables 11.14 and 11.15, the reformerless one-step and two-step processes generate, respectively, 0.98 and 0.78 tonne of CO_2 per tonne of iron, and require 8.68 and 7.17 GJ per tonne of iron of energy. These energy requirements were obtained using the HHV of the natural gas used as the fuel and the raw material for reductants.

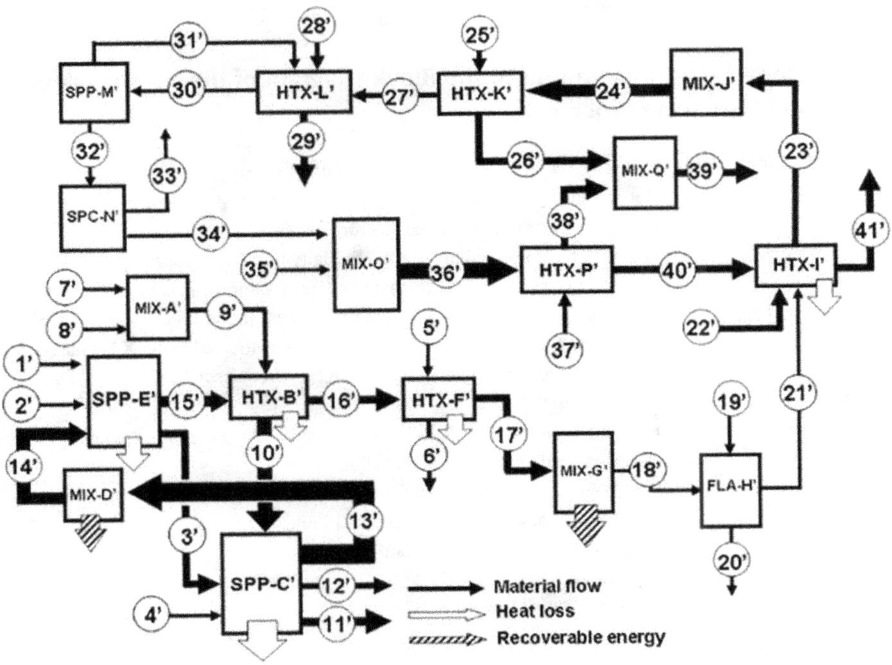

FIGURE 11.7 Block diagram of a METSIM model for the industrial-scale reformerless two-step ironmaking process.

It is noted that the flash ironmaking process in either configuration can save significant amounts of energy even with larger excess driving force and operating temperature than the standard case (the ironmaking reactor temperature = 1,500°C, excess driving force = 0.5). If excess driving force rises to 1, the reformerless one-step and two-step processes will require 10.0 and 8.19 GJ/tonne of iron of energy, respectively, compared with 12.7 GJ/tonne of iron for the average blast furnace. When the operating temperature of the ironmaking reactor is 1,600°C, the energy requirement of the processes will be 9.48 and 7.78 GJ/tonne of iron. If the heat losses in the ironmaking and prereduction reactors are twice the base case assumptions in the standard case operation (108 GJ/h in the one-step process, 82.6 and 25.5 GJ/h in the two-step process), the energy requirements in the one-step and two-step processes will increase to 9.93 and 8.33 GJ/tonne of iron, respectively, still representing substantial energy saving relative to the average blast furnace process, in the case of using the HHV. As for the comparison of energy consumption, it is noted that a comprehensive comparison of energy requirements of the flash ironmaking processes and the blast furnace process is complicated and depends on the choice of starting raw materials, that is, whether or not energy required to produce coking coal, natural gas, and hydrogen is produced from other resources is considered. In this work, the comparison is made by considering coking coal and natural gas as the input materials to the ironmaking plant and does not include electrical power requirements for plant operation.

TABLE 11.9
Descriptions of Unit Operations in the Block Diagram of the Reformerless One-Step Ironmaking Process

Unit Operation Label	Description
MIX-A	Reactor feed gas mixer
HTX-B	Heat exchanger to preheat the reactor feed gas
SPP-C	One-step flash ironmaking reactor (initially termed suspension)
HTX-D	Heat exchanger to preheat fresh natural gas (Natural gas is assumed to be desulfurized, but not simulated in this process model)
MIX-E	Waste heat boiler modeled to recover energy (Modeled by MIX because there is no specified heating/cooling stream HTX requires)
FLA-F	Wet scrubber
HTX-G	Heat exchanger to preheat the water-gas shift (WGS) reactor feed gas
MIX-H	Water-gas shift (WGS) reactor
HTX-I	First boiler to generate steam for the WGS reactor feed gas
HTX-J	Boiler feed water heater using the WGS reactor off-gas sensible heat and energy generated by condensation
SPP-K	Condensed water removal
SPC-L	Pressure swing adsorption (PSA)
MIX-M	PSA tail gas burner
HTX-N	Second boiler to generate steam for the WGS reactor feed gas
MIX-O	Generated steam mixer

11.2.4 Summary on Reformerless Natural-Gas-Based Flash Ironmaking Technology

Process flow sheets for industrial-scale reformerless one-step and two-step ironmaking processes in which natural gas was directly injected into the ironmaking reactor were constructed and simulated. Simulations were performed for a plant with an annual iron production rate of 1 million tonnes with 300 days of operation a year. As the standard case, the ironmaking reactor was assumed to operate at 1,500°C with an excess driving force of 0.5. The prereduction reactor in the two-step process was assumed to operate at 900°C. The feed gas to the ironmaking reactor was preheated to 900°C in the one-step process, and to 750°C in the two-step process. The results showed that the reformerless two-step process required less natural gas than the one-step. This is because the two-step process reduced wustite in the ironmaking reactor in which fuel was burned, and the lower water vapor generation decreased the need for fuel due to less off-gas from the ironmaking reactor. However, it is noted that the added complexity of the two-step process would require greater capital cost, which will be discussed in Chapter 12.

Compared with the average blast furnace process, the reformerless processes are expected to emit 39% and 51% less carbon dioxide than the average blast furnace and consume significantly less energy for all the different cases considered, calculated by using the HHV of the fuel and reductants in the standard one-step and two-step

TABLE 11.10
Descriptions of Streams in the Block Diagram of the Reformerless One-Step Ironmaking Process[a]

No.	Description	No.	Description
1	Iron ore (95.4 Fe_3O_4, 0.4% CaO, 3% SiO_2, 1% Al_2O_3, and 0.2% MgO)	20	WGS reactor feed gas (350°C) (50.8% H_2, 12.5% CO, 34.4% H_2O, 2.0% CO_2, and 0.3% N_2)
2	Flux (90.5% CaO, 2.5% SiO_2, and 7% MgO)	21	WGS reactor outlet gas (454°C) (59.8% H_2, 3.5% CO, 25.3% H_2O, 11.1% CO_2, and 0.3% N_2)
3	Oxygen	22	Boiler feed water (220°C)
4	Fresh natural gas (96% CH_4, 2% C_2H_6, and 2% N_2)	23	Steam (300°C)
5	Preheated natural gas (380°C) (assumed to be desulfurized before mixing with hydrogen)	24	Boiler feed water heater inlet gas (250°C)
6	Recycled hydrogen (30°C)	25	Freshwater (25°C)
7	Same as Stream 5	26	Hot water (220°C) (Boiler feed water is assumed to be provided from this stream)
8	Reactor feed gas (78.5% H_2, 20.7% CH_4, 0.4% C_2H_6, and 0.4% N_2)	27	Boiler feed water heater outlet gas (30°C)
9	Preheated reactor feed gas (900°C)	28	Condensed water (30°C)
10	Molten metal (1,500°C)	29	Pressure swing adsorption (PSA) inlet gas (30°C, 1,800 kPa) (79.9% H_2, 4.7% CO, 14.8% CO_2, 0.4% N_2, and 0.2% H_2O)
11	Slag (1,500°C) (40.9% CaO, 33.6% SiO_2, 10.9% Al_2O_3, 5% MgO, and 9.6% FeO)	30	Same as Stream 6
12	Ironmaking reactor off-gas (1,500°C) (52.3% H_2, 12.9% CO, 32.3% H_2O, 2.1% CO_2, and 0.3% N_2)	31	PSA tail gas (30°C) (30.4% H_2, 16.1% CO, 0.8% H_2O, 51.3% CO_2, and 1.4% N_2)
13	Natural gas preheater inlet gas (896°C)	32	Air
14	Waste heat boiler inlet gas (787°C)	33	Combustion gas used for generating steam (1,323°C) (12.7% H_2O, 27.4% CO_2, 5.0% O_2, and 54.9% N_2)
15	Waste heat boiler outlet gas (90°C)	34	Boiler feed water (220°C)
16	Cooling water (25°C)	35	Steam (300°C)
17	Slurry (50°C)	36	Combined generated steam (300°C)
18	Cleaned gas with saturated water vapor (50°C) (76.1% H_2, 18.8% CO, 1.5% H_2O, 3.1% CO_2, and 0.4% N_2)	37	Combustion gas used for preheating the off-gas from the scrubber (680°C)
19	Steam with half moles of Stream 18 (300°C)	38	Flue gas (300°C)

[a] Shown compositions are for the simulation at an operating temperature of 1,500°C and excess driving force of 0.5 as the standard case.

TABLE 11.11
Descriptions of Unit Operations in the Block Diagram of the Reformerless Two-Step Ironmaking Process

Unit Operation Label	Description
MIX-A’	Reactor feed gas mixer
HTX-B’	Heat exchanger to preheat the reactor feed gas
SPP-C’	Flash ironmaking reactor
MIX-D’	First waste heat boiler modeled to recover energy (Modeled by MIX because there is no specified heating/ cooling stream HTX requires)
SPP-E’	Prereduction reactor
HTX-F’	Heat exchanger to preheat fresh natural gas (Natural gas is assumed to be desulfurized, but not simulated in this process model)
MIX-G’	Second waste heat boiler modeled to recover energy
FLA-H’	Wet scrubber
HTX-I’	Heat exchanger to preheat the water-gas shift (WGS) reactor feed gas
MIX-J’	Water-gas shift (WGS) reactor
HTX-K’	First boiler to generate steam for the WGS reactor feed gas
HTX-L’	Boiler feed water heater using the WGS reactor off-gas sensible heat and energy generated by condensation
SPP-M’	Condensed water removal
SPC-N’	Pressure swing adsorption (PSA)
MIX-O’	PSA tail gas burner
HTX-P’	Second boiler to generate steam for the WGS reactor feed gas
MIX-Q’	Generated steam mixer

configurations, respectively. If the carbon dioxide emissions and energy requirement for $CaCO_3$/$MgCO_3$ calcination to produce flux are considered, their levels would increase accordingly, but not significantly. The simulation also showed that either process would still require much less energy than the blast furnace process at the operating temperature of 1,600°C; or an excess driving force of 1.

The results of these simulations indicate that the reformerless process will significantly reduce the energy consumption and carbon dioxide emissions for iron production *vis-a-vis* the blast furnace technology.

11.3 NATURAL-GAS-BASED FLASH IRONMAKING TECHNOLOGY–IRONMAKING COMBINED WITH STEAM-METHANE REFORMING

Simulations of an industrial-scale ironmaking process combined with a separate SMR process were carried out. The SMR process was simulated to produce (1) hydrogen or (2) syngas to provide fuel/reductant for ironmaking. Ironmaking was simulated in one-step configuration.

TABLE 11.12
Descriptions of Streams in the Block Diagram of the Reformerless Two-Step Ironmaking Process[a]

No.	Description	No.	Description
1'	Iron ore (95.4 Fe_3O_4, 0.4% CaO, 3% SiO_2, 1% Al_2O_3 and 0.2% MgO)	22'	Steam with half moles of Stream 21' (300°C)
2'	Flux (90.5% CaO, 2.5% SiO_2 and 7% MgO)	23'	WGS reactor feed gas (370°C) (48.8% H_2, 10.8% CO, 34.4% H_2O, 5.7% CO_2 and 0.3% N_2)
3'	Reduced iron ore (wustite) and flux (900°C) (91.4% FeO, 3.9% CaO, 3.2% SiO_2, 1.0% Al_2O_3, 0.5% MgO)	24'	WGS reactor outlet gas (451°C) (56.1% H_2, 3.6% CO, 27.1% H_2O, 12.9% CO_2 and 0.3% N_2)
4'	Oxygen	25'	Boiler feed water (220°C)
5'	Fresh natural gas (96% CH_4, 2% C_2H_6 and 2% N_2)	26'	Steam (300°C)
6'	Preheated natural gas (380°C) (Assumed to be desulfurized before mixing with hydrogen)	27'	Boiler feed water heater inlet gas (250°C)
7'	Recycled hydrogen (30°C)	28'	Freshwater (25°C)
8'	Same as Stream 6'	29'	Hot water (220°C) (Boiler feed water is assumed to be provided from this hot water)
9'	Reactor feed gas (75.2% H_2, 23.9% CH_4, 0.5% C_2H_6 and 0.5% N_2)	30'	Boiler feed water heater outlet gas (30°C)
10'	Preheated reactor feed gas (750°C)	31'	Condensed water (30°C)
11'	Molten metal (1,500°C)	32'	Pressure swing adsorption (PSA) feed gas (30°C, 1,800kPa) (76.7% H_2, 4.9% CO, 17.7% CO_2, 0.5% N_2 and 0.2% H_2O)
12'	Slag (1,500°C) (40.9% CaO, 33.6% SiO_2, 10.9% Al_2O_3, 5% MgO and 9.6% FeO)	33'	Same as Stream 7'
13'	Ironmaking reactor off-gas (1,500°C) (51.4% H_2, 14.3% CO, 31.7% H_2O, 2.3% CO_2 and 0.3% N_2)	34'	PSA tail gas (30°C) (26.6% H_2, 15.3% CO, 0.8% H_2O, 55.9% CO_2 and 1.4% N_2)
14'	Prereduction reactor feed gas (1,278°C)	35'	Air
15'	Prereduction reactor off-gas (900°C) (49.1% H_2, 10.9% CO, 33.9% H_2O, 5.8% CO_2 and 0.3% N_2)	36'	Combustion gas used for generating steam (1,248°C) (11.7% H_2O, 30.3% CO_2, 5.0% O_2 and 53.0% N_2)
16'	Natural gas preheater inlet gas (432°C)	37'	Boiler feed water (220°C)
17'	Waste heat boiler inlet gas (333°C)	38'	Steam (300°C)
18'	Waste heat boiler outlet gas (90°C)	39'	Combined generated steam (300°C)
19'	Cooling water (25°C)	40'	Combustion gas used for preheating the off-gas from the scrubber (711°C)
20'	Slurry (50°C)	41'	Flue gas (300°C)
21'	Cleaned gas (50°C) (73.2% H_2, 16.2% CO, 1.5% H_2O, 8.6% CO_2 and 0.5% N_2)		

[a] Shown compositions are for the simulation at an operating temperature of 1,500°C and excess driving force of 0.5 as the standard case.

TABLE 11.13
Reactions Considered in the Reformerless One-Step and Two-Step Ironmaking Processes

Description	Reaction	Reaction Extent
Fuel combustion	(A) $2C_2H_6+7O_2=6H_2O+4CO_2$	1
	(B) $CH_4+2O_2=2H_2O+CO_2$	1
	(C) $2H_2+O_2=2H_2O$	1
	(D) $2CO+O_2=2CO_2$	1
Methane reforming by remaining water vapor	(E) $CH_4+H_2O=3H_2+CO$	Equilibrium at the ironmaking reactor temperature
Reduction of iron oxide by hydrogen	(F) $Fe_3O_4+H_2=3FeO+H_2O$	1
	(G) $FeO+H_2=Fe+H_2O$	0.99
Reverse water-gas shift reaction	(H) $CO_2+H_2=CO+H_2O$	Equilibrium at the ironmaking reactor temperature. 900°C in the prereduction reactor
Slagmaking	(I) $CaO(s)=CaO(l)$	1
	(J) $SiO_2(s)=SiO_2(l)$	1
	(K) $Al_2O_3(s)=Al_2O_3(l)$	1
	(L) $MgO(s)=MgO(l)$	1
Water-gas shift reaction	(M) $CO+H_2O=CO_2+H_2$	Equilibrium at 450°C [Ladebeck and Wagner, 2003]
Steam generation	(N) $H_2O(l)=H_2O(g)$	1
Condensation	(O) $H_2O(g)=H_2O(l)$	Saturation at 30°C, 1,800 kPa

11.3.1 Introduction

SMR of natural gas is currently the most dominant production method of hydrogen which is a promising fuel/reductant gas for FIT. Thus, an industrial-scale plant based on this FIT may operate with a built-in SMR hydrogen production plant.

In the previous section, flow sheets for an industrial-scale flash ironmaking process were developed with reformerless use of natural gas, which would eliminate the need for external SMR to produce hydrogen. In this section, flow sheet development and process simulations of an industrial-scale flash ironmaking process combined with external SMR are discussed to evaluate the advantages and disadvantages of this type of flow sheet from material and energy balance perspective. Process simulation was performed for two cases: (1) hydrogen was produced from SMR and used for ironmaking and (2) syngas produced from reforming was directly injected into the ironmaking reactor. The difference between these processes is in the existence in the former of a WGS reactor and a PSA step [New York State, 2010]. WGS increases the hydrogen content in the reformed gas, and PSA separates hydrogen from other gases. The use of syngas in the ironmaking process was considered as it simplified the flow sheet by the elimination of WGS and PSA steps after reforming. Both cases are discussed with an emphasis on carbon dioxide emissions and energy requirements compared with the blast furnace process.

TABLE 11.14
Material Balances of the Industrial-Scale Reformerless One-Step and Two-Step Ironmaking Processes Producing 1 Million Tonnes of Iron per Year at the Operating Temperature of 1,500°C, the Excess Driving Force of 0.5 and the Reactor Feed Gas Temperature of 900°C in the One-Step Process and 750°C in the Two-Step Process, Compared with the Average Blast Furnace Process

		One-Step Process	Two-Step Process	BF Process[Choi, 2010]
Input (tonnes/tonne of iron)	Natural gas	0.37	0.29	
	Oxygen	0.78	0.58	0.68
	Iron ore	1.46 (magnetite)	1.46 (magnetite)	1.43 (hematite)
	Flux	0.06	0.06	0.29
	Nitrogen			2.21
	Coke			0.46
	New water	2.61	1.92	
	Air	1.61	1.12	
	Total	6.89	5.43	5.07
	Cooling water	20.8	15.8	22.3[b]
Output (tonnes/tonne of iron)	Molten metal	1.00	1.00	1.05 (contains 4.5% C)
	Slag	0.14	0.14	0.21
	Condensed water vapor in slurry	0.83	0.63	
	Hot water not used	1.87	1.53	
	Steam not used	0.51	0.27	
	Total discharged gas	2.54	1.86	3.81
	Carbon dioxide	0.98	0.78	1.60[a]
	Nitrogen	1.25	0.87	2.21
	Total	6.89	5.43	5.07
	Cooling water in slurry	20.8	15.8	22.3[b]
	Carbon dioxide from $CaCO_3$ and $MgCO_3$ calcination	0.04	0.04	

[a] This number does not include the carbon dioxide emissions from ore/coke preparation.

[b] Water requirement for US Steel's Fairfield No.8 Blast Furnace (production rate: 5,500 tonnes/day) [Camlic and Goodman, 1997; United States Steel Corporation, 1985] including water for top gear, stack plates, bosh channels, tuyere jacket, tuyeres and tuyere coolers, hearth and subhearth, and stove valves; as the reference years of water usage is older than that of the production rate, this number may be somewhat larger than the number for a modern plant.

TABLE 11.15

Energy Balances of the Industrial-Scale Reformerless One-Step and Two-Step Ironmaking Processes Producing 1 Million Tonnes of Iron per Year at the Operating Temperature of 1,500°C, the Excess Driving Force of 0.5 and the Reactor Feed Gas Temperature of 900°C in the One-Step Process and 750°C in the Two-Step Process, Compared with the Average Blast Furnace Process

		One-Step Process	Two-Step Process	BF Process[a]
Input (GJ/tonne of iron)	Fuel combustion[b]	13.45	9.62	8.33
	Heat recovery	−4.77	−2.45	−1.32[c]
	Waste heat boiler	−3.39	−1.73	
	Steam not used[d]	−1.38	−0.72	
	Subtotal	8.68	7.17	7.01
	Energy input for ore/coke preparation			5.68
	Total[h]	8.68	7.17	12.7
	Energy input for $CaCO_3$ and $MgCO_3$ calcination	0.26[e]	0.26[e]	
Output (GJ/tonne of iron)	Reduction	0.91	0.91	2.10
	Sensible heat of molten metal	1.27 (1,500°C)[f]	1.27 (1,500°C)[f]	1.35 (1,600°C)
	Sensible heat of slag	0.24 (1,500°C)[f]	0.24 (1,500°C)[f]	0.47 (1,600°C)
	Slurry ($H_2O(l)$)	2.25 (50°C)	1.71 (50°C)	
	Hot water not used	1.57 (220°C)	1.28 (220°C)	
	Flue gas	0.79 (300°C)	0.58 (300°C)	0.26 (90°C)
	$CaCO_3$ decomposition			0.33
	Slagmaking			−0.17
	Heat loss in the reactor(s)	0.78	0.78	2.60
	Heat loss in the heat exchangers	0.73[g]	0.33[g]	0.07[c] (heat loss in heat recovery)
	Reactor feed gas heater	0.40	0.17	
	Natural gas heater	0.21	0.07	
	WGS reactor feed gas heater	0.12	0.09	

TABLE 11.15 (*Continued*)
Energy Balances of the Industrial-Scale Reformerless One-Step and Two-Step Ironmaking Processes Producing 1 Million Tonnes of Iron per Year at the Operating Temperature of 1,500°C, the Excess Driving Force of 0.5 and the Reactor Feed Gas Temperature of 900°C in the One-Step Process and 750°C in the Two-Step Process, Compared with the Average Blast Furnace Process

		One-Step Process	Two-Step Process	BF Process[a]
Output (GJ/tonne of iron)	Steam not used (90°C)[d]	0.14	0.07	
	Subtotal	8.68	7.17	7.01
	Pelletizing			3.01 [Stubbles, 2,000]
	Sintering			0.65 [Stubbles, 2,000]
	Cokemaking			2.02 [Stubbles, 2,000]
	Total[h]	8.68	7.17	12.7
	$CaCO_3$ and $MgCO_3$ decomposition at 1,000°C [Biswas, 1981]	0.26[e]	0.26[e]	

[a] Energy balance was calculated by METSIM based on the material balance.
[b] Fuel combustion energy input in the process was calculated by subtracting the energy used for iron ore reduction from the balance of heat formation of all output components from the entire system and input components into it. The high heating value of natural gas was used for this calculation. If the low heating value is used, the total energy requirements for the one-step and two-step processes will decrease to 4.96 GJ and 4.35 GJ/tonne of iron, respectively.
[c] Heat loss in the heat recovery step was estimated to be 5% of the energy difference between the hot off-gas (calculated to be 426°C) and cooled reactor off-gas (90°C), and the rest was estimated to be the recovered energy.
[d] Energy recovered by steam not used in the process includes the sensible heat of steam from 300°C to 90°C and the energy generated by condensation.
[e] The energy requirement for $CaCO_3$ and $MgCO_3$ calcination to generate CaO and MgO for the flux was calculated by METSIM. Calcination was assumed to be done at 1,000°C separately from the ironmaking process.
[f] If the temperature of iron and slag is replaced by 1,600°C (same as the blast furnace process), the total energy requirements for the one-step and two-step processes increase to 9.48 and 7.78 GJ/tonne of iron, respectively.
[g] Heat losses in a heat exchanger were assumed to be 5% of the total sensible heat of input streams.
[h] Natural gas and coking coal are considered the starting raw materials. The energy requirement for the average blast furnace process includes energy required for ore/coke preparation. This energy requirement comparison does not include electrical power requirements for plant operation.

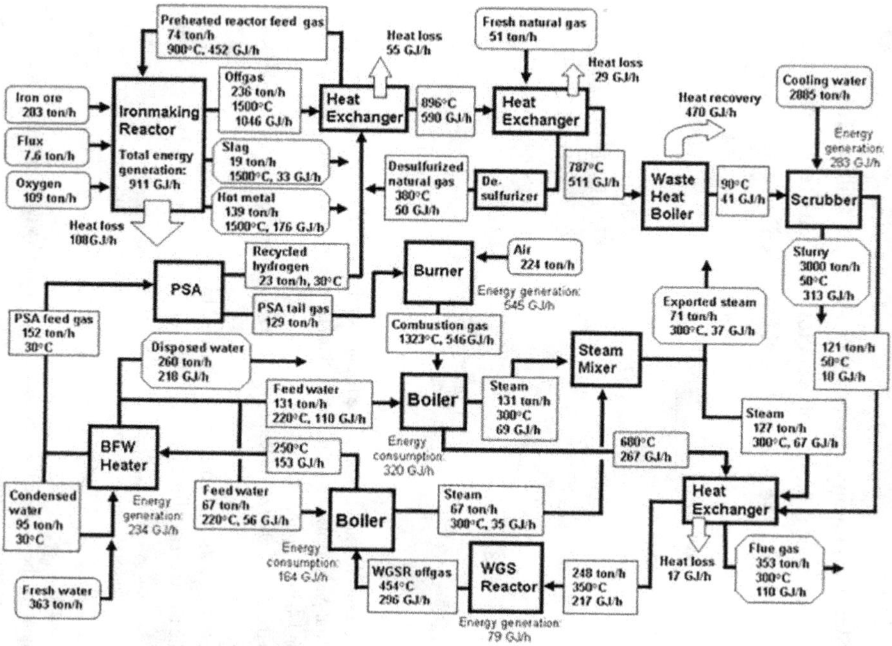

FIGURE 11.8 Material and energy flow diagram for the industrial-scale reformerless one-step ironmaking process producing 1 million tonnes of iron per year at the operating temperature of 1,500°C, the excess driving force of 0.5 and the reactor feed gas temperature of 900°C (300 days of operation in 1 year).

11.3.2 Flow Sheet Development and Process Simulation

The block diagram of the ironmaking section combined with the SMR section used for METSIM process simulation is shown in Figure 11.10. The simulation was performed for only the one-step process. The one-step process is a simpler configuration than the two-step process. The process flow sheet of the ironmaking section used in simulations is the same in either case in which hydrogen or syngas produced from SMR is used, except for some differences in reactions as described below. The block diagram of the SMR section is shown in Figures 11.11 and 11.12; Figure 11.11 was used when hydrogen was produced, and Figure 11.12 was used when syngas was produced. Descriptions of the unit operations and streams in Figure 11.10 are shown in Tables 11.16 and 11.17. The descriptions of the unit operations and streams in Figures 11.11 and 11.12 are given in Tables 11.18 and 11.19. Figures 11.11 and 11.12 use the same labels of unit operations and streams except for some processes eliminated in Figure 11.12. Unit operations in Figures 11.10–11.12 are labeled in the order with which process simulation was performed as METSIM used a sequential modular approach. The temperature of all fresh input streams into the entire system was set to 25°C.

Pressures of most of the streams were not specified in this simulation (the default pressure is 101.3 kPa). The pressure of the stream fed into the PSA step was specified

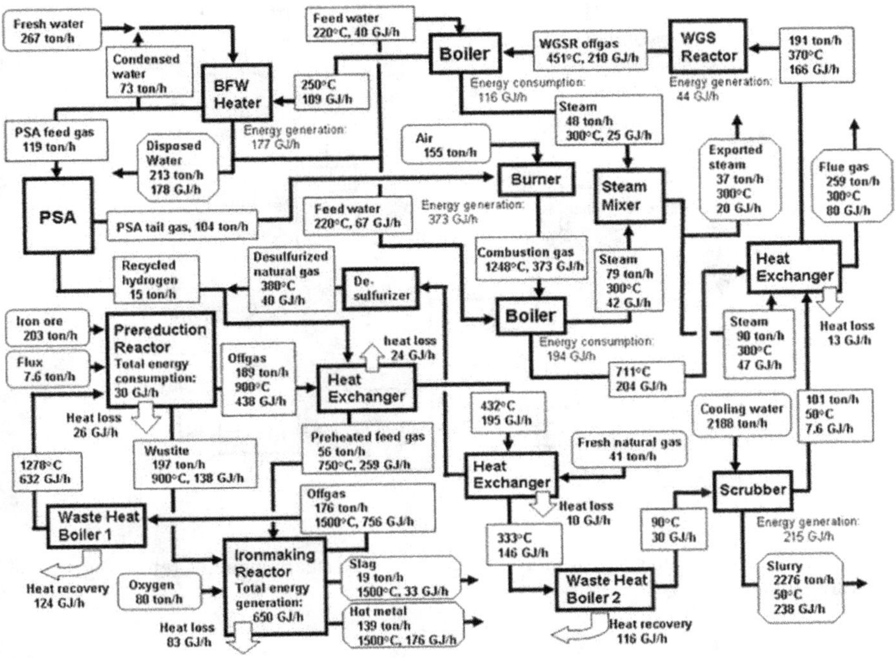

FIGURE 11.9 Material and energy flow diagram for the industrial-scale reformerless two-step ironmaking process producing 1 million tonnes of iron per year at the operating temperature of 1,500°C, the excess driving force of 0.5 and the reactor feed gas temperature of 750°C (300 days of operation in 1 year).

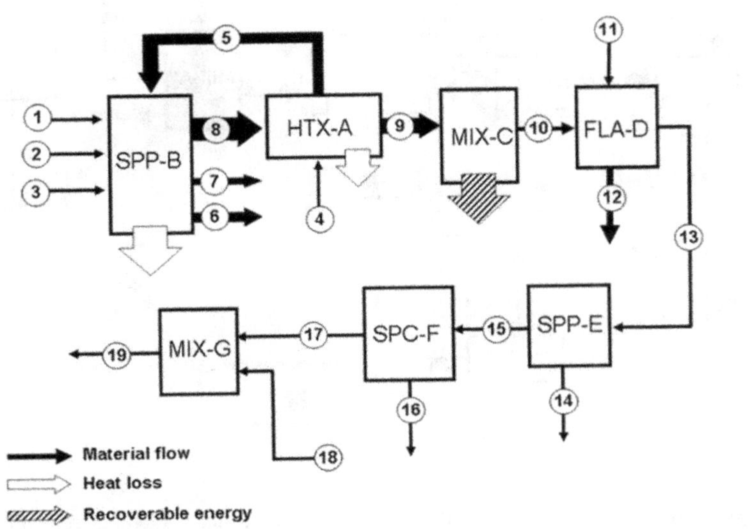

FIGURE 11.10 Block diagram of METSIM model for the ironmaking section in the one-step ironmaking process combined with a steam-methane reforming section.

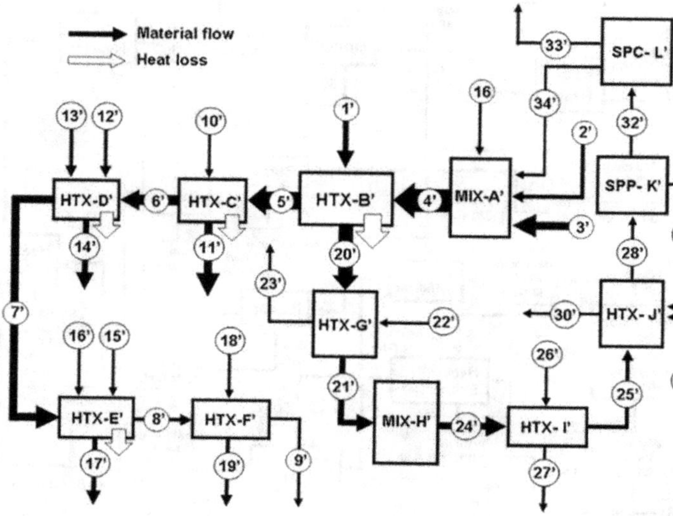

FIGURE 11.11 Block diagram of METSIM model for the steam-methane reforming section for hydrogen production.

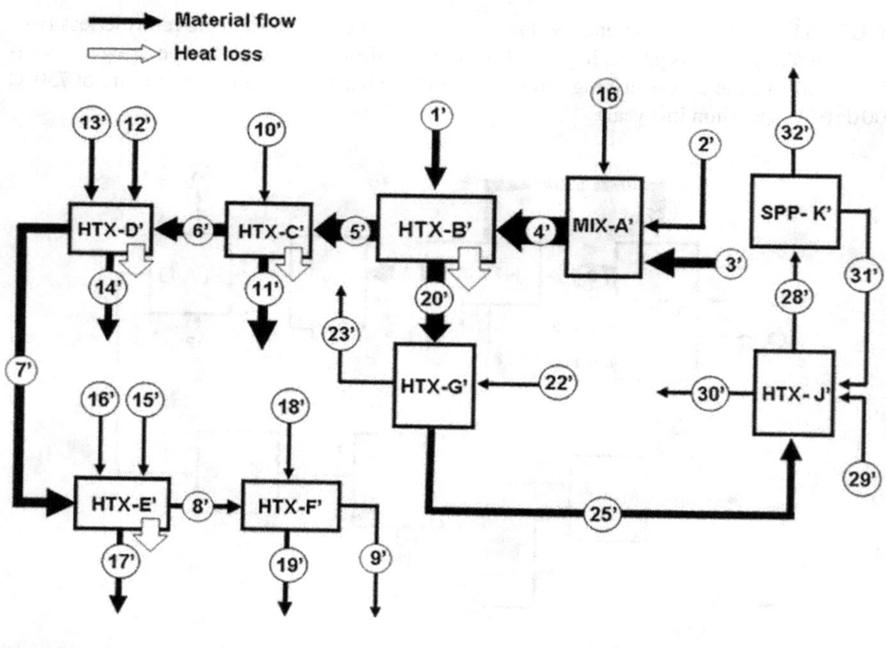

FIGURE 11.12 Block diagram of METSIM model for the steam-methane reforming section for syngas production.

TABLE 11.16
Descriptions of Unit Operations in the Block Diagram of the Ironmaking Section Combined with the Steam-Methane Reforming (SMR) Section

Unit operation label	Description
HTX-A	Heat exchanger to preheat the reactor feed gas
SPP-B	One-step flash ironmaking reactor
MIX-C	Waste heat boiler modeled to recover energy (Modeled by MIX because there is no specified heating/ cooling stream HTX requires)
FLA-D	Wet scrubber
SPP-E	Condensed water removal process
SPC-F	Pressure swing adsorption (PSA) for the cleaned reactor off-gas with H_2 recovery rate of 89%
MIX-G	Reactor feed gas mixer

to calculate the amount of saturated water vapor in it. This specified pressure was assumed to be obtained by compression of the off-gas from the ironmaking reactor or pressure decrease from a higher pressure in the initial process stage in the SMR process.

The iron production rate was set to be 1 million tonnes a year, assuming a year of operation is made up of 300 days, in either case.

11.3.2.1 Ironmaking Section

In the ironmaking section (Figure 11.10), iron ore (Stream 1), flux (Stream 2), oxygen (Stream 3), and preheated reactor feed gas (Stream 5) were fed into the ironmaking reactor (SPP-B). The reactor feed gas was hydrogen when hydrogen was provided from the SMR section, and a mixture of hydrogen and syngas when syngas was provided from the SMR section. The chemical reactions involved in fuel combustion, iron ore reduction, and slagmaking reactions used in the simulation of reactor SPP-B are listed in Table 11.20. The reduction extent was again 0.99. Reactions C', E', F', and H' to K' were used with pure hydrogen, and Reactions B' to K' were used with syngas. The reactor feed gas (Stream 4) was heated to 900°C by a heat exchanger (HTX-A) using the sensible heat of the off-gas (Stream 8) from the reactor. The amount of Stream 18 was adjusted so that the excess driving force in the off-gas out of SPP-B became equal to the specified value.

Other details of the ironmaking section can be found in Pinegar et al. [2013a].

11.3.2.2 Steam-Methane Reforming Section

In the SMR section (Figure 11.11 for hydrogen production and Figure 11.12 for syngas production), the SMR process was modeled with several heat recovery steps so that the sensible heat of the combustion gas or the reformed gas from the reformer was used to provide process heat. Fresh natural gas composition was assumed to be 96% CH_4, 2% C_2H_6, and 2% N_2.

TABLE 11.17
Descriptions of Streams in the Block Diagram of the Ironmaking Section Combined with the Steam-Methane Reforming (SMR) Section[a]

No.	Description	No.	Description
1	Iron ore ($95.4 Fe_3O_4$, 0.4% CaO, 3% SiO_2, 1% Al_2O_3 and 0.2% MgO)	11	Cooling water (25°C)
2	Flux (90.5% CaO, 2.5% SiO_2 and 7% MgO)	12	Slurry (50°C)
3	Oxygen	13	Cleaned gas with saturated water vapor (50°C)
4	Reactor feed fuel/reductant gas (pure H_2 when hydrogen is produced) 86.5% H_2, 6.9% CO, 0.1% H_2O, 4.0% CO_2, 2.2% CH_4 and 0.3% N_2 when syngas is produced)	14	Removed condensed water (30°C)
5	Preheated reactor feed gas (900°C)	15	PSA feed gas (1,800 kPa and 30°C) (99.8% H_2 and 0.2% H_2O when hydrogen is produced/ 80.6% H_2, 16.1% CO, 0.3% H_2O, 2.6% CO_2 and 0.4% N_2 when syngas is produced)
6	Molten metal (1,500°C)	16	PSA tail gas used in the SMR burner (97.9% H_2 and 2.1% H_2O when hydrogen is produced/ 31.4% H_2, 57.2% CO, 0.8% H_2O, 9.3% CO_2, and 1.3% N_2 when syngas is produced)
7	Slag (1,500°C) (40.9% CaO, 33.6% SiO_2, 10.9% Al_2O_3, 5% MgO and 9.6% FeO)	17	Recovered hydrogen (30°C)
8	Ironmaking reactor off-gas (1,500°C) (61.8% H_2 and 38.2% H_2O when hydrogen is produced/ 53.9% H_2, 10.8% CO, 33.3% H_2O, 1.8% CO_2 and 0.2% N_2 when syngas is produced)	18	Produced fuel/reductant gas (pure H_2 or syngas: 73% H_2, 13.8% CO, 0.2% H_2O, 8.0% CO_2, 4.4% CH_4, and 0.5% N_2)
9	Waste heat boiler inlet gas (725°C when hydrogen is produced/ 716°C when syngas is produced)	19	Same as Stream 4 (Disconnected for METSIM simulation for computational convenience)
10	Waste heat boiler outlet gas (90°C)		

[a] Shown compositions are for the simulation at the operating temperature of 1,500°C and an excess driving force of 0.5 as the standard case.

TABLE 11.18
Descriptions of Unit Operations in the Block Diagram of the Steam-Methane Reforming (SMR) Section

MIX-A'	Burner for the reformer
HTX-B'	Steam-methane reformer operating at 850°C
HTX-C'	Air preheating step by the combustion gas
HTX-D'	SMR feed gas preheating step by the combustion gas
HTX-E	Natural gas preheating step by the combustion gas before desulfurization (desulfurization was assumed to take place in a ZnO bed after preheating)
HTX-F'	Boiler using the combustion gas
HTX-G'	First boiler using sensible heat of the reformed gas
MIX-H'	Water-gas shift (WGS) reactor
HTX-I'	Second boiler using sensible heat of the reformed gas
HTX-J'	Boiler feed water heater using the WGS reactor off-gas sensible heat and energy generated by condensation
SPP-K'	Condensed water removal process
SPC-L'	Pressure swing adsorption for hydrogen purification after water vapor removal

The reforming step (Unit operations A' to F') was modeled to use the combustion energy from a burner for (1) reforming and (2) preheating gases used for reforming. This step is the same in both the hydrogen and syngas production processes.

In a burner (MIX-A'), natural gas desulfurized at 380°C (Stream 2'), preheated air at 800°C (Stream 3'), PSA tail gas from the ironmaking section (Stream 16), and tail gas from a PSA step in the SMR section (Stream 34') were added to generate energy by Reactions A' to D'. Using the combustion gas (Stream 4') provided from a burner, a reformer feed gas (Stream 1') fed to the reformer (HTX-B') was reformed by Reactions L' to N' at 850°C [Liu, 2006]. The steam/carbon ratio in the feed gas was set to be 3[Liu, 2006]. The extent of Reaction L' was decided so that the methane content in a reformed gas (Stream 20') would be 3% [Carrara et al., 2010].

Figures 11.11 and 11.12 have different process steps after the reformer. In the SMR section in which hydrogen was produced (Figure 11.11), a WGS reactor (MIX-H'), boiler HTX-I', and a PSA step (SPC-L') were added to purify hydrogen. These steps were removed in the SMR section in which syngas was produced (Figure 11.12). The description provided in the next paragraph is applicable to either case.

Other details of the SMR section are given in Pinegar et al. [2013a].

11.3.3 Material and Energy Balances

Summaries of material and energy balances for the ironmaking section for the industrial-scale one-step flash ironmaking process combined with the SMR section producing 1 million tonnes of iron per year are shown in Tables 11.21 and 11.22, respectively. These balances are compared with a typical average blast furnace operation in each table. The material and energy balances for the SMR section are separately shown in Table 11.23. The results shown in Tables 11.21–11.23 are

TABLE 11.19
Descriptions of Streams in the Block Diagram of the Steam-Methane Reforming (SMR) Section[a]

No.	Description	No.	Description
1'	Preheated SMR feed gas (600°C) (24.4% CH_4, 0.5% C_2H_6, 0.5% N_2, and 74.6% H_2O)	18'	Boiler feed water (220°C)
2'	Desulfurized natural gas used for fuel (380°C) (96% CH_4, 2% C_2H_6, and 2% N_2)	19'	Steam (300°C)
3'	Preheated air (800°C)	20'	Reformed gas (850°C) (49.6% H_2, 9.4% CO, 32.2% H_2O, 5.4% CO_2, 3.0% CH_4, and 0.4% N_2)
4'	Combustion gas used for SMR (2,085°C when hydrogen is produced/ 2,384°C when syngas is produced)	21'	WGS reactor feed gas (385°C) (49.6% H_2, 9.4% CO, 32.2% H_2O, 5.4% CO_2, 3.0% CH_4 and 0.4% N_2)
5'	Combustion gas preheating air to 800°C	22'	Boiler feed water (220°C)
6'	Combustion gas preheating SMR feed gas to 600°C	23'	Steam (300°C)
7'	Combustion gas preheating natural gas to 380°C	24'	WGS reactor off-gas (452°C) (55.7% H_2, 3.3% CO, 26.1% H_2O, 11.6% CO_2, 3.0% CH_4, and 0.3% N_2)
8'	Combustion gas used for producing steam at 300°C	25'	Inlet gas to the boiler feed water heater (250°C) (55.7% H_2, 3.3% CO, 26.1% H_2O, 11.6% CO_2, 3.0% CH_4 and 0.3% N_2 when hydrogen is produced/ 49.6% H_2, 9.4% CO, 32.2% H_2O, 5.5% CO_2, 3.0% CH_4, and 0.3% N_2 when syngas is produced)
9'	Flue gas (300°C)	26'	Boiler feed water (220°C)
10'	Air	27'	Steam (300°C)
11'	Same as Stream 3' (Disconnected for METSIM simulation for computational convenience)	28'	Outlet gas from the boiler feed water heater (30°C)
12'	Desulfurized natural gas used for SMR feed gas (380°C) (96% CH_4, 2% C_2H_6 and 2% N_2)	29'	New water
13'	Steam with three times the molar amount of carbon in Stream 12'	30'	Hot water used for boiler feed water (220°C)
14'	Same as Stream 1'. (Disconnected for METSIM simulation for computational convenience)	31'	Condensed water recycled to boiler feed water (30°C)

(Continued)

TABLE 11.19 (*Continued*)
Descriptions of Streams in the Block Diagram of the Steam-Methane Reforming (SMR) Section[a]

No.	Description	No.	Description
15'	Natural gas used for SMR feed gas ($96\%\ CH_4$, $2\%\ C_2H_6$, and $2\%\ N_2$)	32'	PSA feed gas/syngas provided to the ironmaking step (1,800 kPa and 30°C). ($75.2\%\ H_2$, 4.4% CO, $0.2\%\ H_2O$, $15.6\%\ CO_2$, $4.1\%\ CH_4$ and $0.5\%\ N_2$ when hydrogen is produced/ $73.0\%\ H_2$, 13.8% CO, $0.2\%\ H_2O$, $8.0\%\ CO_2$, $4.4\%\ CH_4$ and $0.5\%\ N_2$ when syngas is produced)
16'	Natural gas used for fuel ($96\%\ CH_4$, $2\%\ C_2H_6$, and $2\%\ N_2$)	33'	Hydrogen (30°C)
17'	Preheated natural gas (380°C)	34'	PSA tail gas used in SMR burner (30°C) ($25.0\%\ H_2$, 13.4% CO, $0.7\%\ H_2O$, $47.2\%CO_2$, $12.2\%\ CH_4$ and $1.5\%\ N_2$)

[a] Shown compositions are for the simulation at the operating temperature of 1,500°C and an excess driving force of 0.5 as the standard case.

TABLE 11.20
Reactions Considered in the Simulation of the One-Step Ironmaking Process Combined with a Steam-Methane Reforming (SMR) Process

Description	Reaction	Reaction Extent
Fuel combustion	(a') $2C_2H_6+7O_2=4CO_2+6H_2O$	1
	(b') $CH_4+2O_2=CO_2+2H_2O$	1
	(c') $2H_2+O_2=2H_2O$	1
	(d') $2CO+O_2=2CO_2$	1
Reduction of iron oxide by hydrogen	(e') $Fe_3O_4+H_2=3FeO+H_2O$	1
	(f') $FeO+H_2=Fe+H_2O$	0.99
Reverse water-gas shift reaction	(g') $CO_2+H_2=CO+H_2O$	Equilibrium at the reactor operating temperature
Slagmaking	(h') $CaO(s)=CaO(l)$	1
	(i') $SiO_2\ (s)=SiO_2\ (l)$	1
	(j') $Al_2O_3(s)=Al_2O_3(l)$	1
	(k') $MgO(s)=MgO(l)$	1
Steam-methane reforming	(l') $CH_4+H_2O=3H_2+CO$	0.825 (to make CH_4 in the reformed gas 3%)
	(m') $C_2H_6+2H_2O=2CO+5H_2$	1
	(n') $CO+H_2O=CO_2+H_2$	Equilibrium at 850°C
Steam generation	(o') $H_2O(l)=H_2O(g)$	1
Water-gas shift reaction	(p') $CO+H_2O=CO_2+H_2$	Equilibrium at 450°C
Condensation	(q') $H_2O(g)=H_2O(l)$	Saturation at 30°C, 1,800 kPa

TABLE 11.21
Material Balances of the Ironmaking Section in the Industrial-Scale One-Step Ironmaking Process Producing 1 Million Tonnes of Iron per Year Combined with the Steam-Methane Reforming (SMR) Section with the Operating Temperature of 1,500°C, the Excess Driving Force of 0.5 and the Reactor Feed Gas Preheating Temperature of 900°C, Compared with the Average Blast Furnace Process

		With H_2 Production from SMR	Direct Use of Syngas in Ironmaking	BF Process [Choi, 2010]
Input (tonnes/tonne of iron)	Hydrogen	0.12		
	Syngas		0.83	
	Oxygen	0.43	0.57	0.68
	Iron ore	1.46 (magnetite)	1.46 (magnetite)	1.43 (hematite)
	Flux	0.06	0.06	0.29
	Nitrogen			2.21
	Coke			0.46
	Total ironmaking	2.07	2.92	5.07
	Cooling water for scrubber	17.9	19.8	
Output (tonnes/tonne of iron)	Molten metal	1.0	1.0	1.05 (contains 4.5% C)
	Slag	0.14	0.14	0.21
	Condensed water vapor	0.91	1.06	
	Total discharged gas	0.02	0.72	3.81
	Carbon dioxide		0.99[a]	1.60[b]
	Nitrogen		0.01	2.21
	Total ironmaking	2.07	2.92	5.07
	Cooling water in slurry	17.9	19.8	22.3[c]
	Carbon dioxide from $CaCO_3$ and $MgCO_3$ calcination	0.04	0.04	

[a] Includes CO_2 converted from CO contained in the PSA tail gas. [b]This number does not include the carbon dioxide emissions from ore/coke preparation. [c]Water requirement for US Steel's Fairfield No.8 Blast Furnace (production: 5,500 tonnes/day) [Camlic and Goodman, 1997; United States Steel Corporation, 1985] including water for top gear, stack plates, bosh channels, tuyere jacket, tuyeres and tuyere coolers, hearth and subhearth, and stove valves; as the reference year of water usage is older than that of the iron production rate, this number may be somewhat larger than the number for a modern plant.

based on simulation of the standard case (the ironmaking reactor operating temperature = 1,500°C, excess driving force = 0.5). Material and energy flow diagrams of the two processes for the standard case with all the stream values and conditions are shown in Figures 11.13 and 11.14. Table 11.22 shows that the ironmaking section using hydrogen from the SMR section would require less energy than that using syngas. When syngas was produced in the SMR section, the ironmaking reactor feed gas included about 86.5% hydrogen, 6.9% carbon monoxide, 0.1% water vapor, 4.0% carbon dioxide, 2.2% methane, and 0.3% nitrogen. This required more feed gas to provide hydrogen against generated water vapor to achieve the specified excess driving force. The reverse WGS reaction decreases the hydrogen content and increases water vapor while consuming energy. For this reason, more fuel was required to satisfy the specified excess driving force and operating temperature when syngas from the SMR section was used than when pure hydrogen was first obtained from SMR. This means that the SMR section producing syngas for ironmaking would require more natural gas than the SMR section providing hydrogen, as shown in Table 11.23, although it would considerably simplify the flow sheet and lower the capital cost.

The energy requirement for the flash ironmaking process using natural gas as the starting raw material is composed of the energy requirements in the ironmaking and SMR sections. (See Tables 11.22 and 11.23). The total energy required to produce iron is 12.7 or 13.5 GJ/tonne of iron, respectively, when hydrogen or syngas is produced from SMR. These requirements were obtained using the HHV of the natural gas used as the fuel and the raw material for reductants. These numbers indicate that the flash ironmaking processes combined with the SMR process would require more energy than the average blast furnace process whether hydrogen or syngas is produced from reforming. If the LHV of natural gas is used, the total energy requirements will be 9.07 GJ/tonne of iron when hydrogen is produced from SMR and 9.15 GJ/tonne of iron when syngas is produced, compared with 12.7 GJ/tonne of iron for the average blast furnace.

The carbon dioxide emissions from the flash ironmaking process combined with the SMR process are shown in Table 11.23. In this process configuration, all the carbon dioxide leaves the process through the flue gas from the SMR section. The SMR section producing hydrogen would emit 1.04 tonnes/tonne of iron of carbon dioxide. The SMR section providing syngas would emit 1.1 tonnes/tonne of iron of carbon dioxide. These amounts represent 35% and 31% carbon dioxide emissions reductions, respectively, compared with the average blast furnace process. This difference in carbon dioxide emissions between the SMR section producing hydrogen and syngas is because of the different amounts of natural gas required in the two cases.

Sensitivity analyses were performed to examine the effects of reactor operating temperature, excess driving force, and the heat loss rate in the ironmaking reactor on the energy requirement of the process. When the reactor operating temperature increased to 1,600°C from 1,500°C, the energy requirement from the ironmaking process combined with the SMR process would increase to 13.7 GJ/tonne of iron when hydrogen was provided from the SMR process, and 15.0 GJ/tonne of iron when syngas was provided. The energy requirement in the two processes was similar to these results when excess driving force rose to 1; 13.7 GJ/tonne of iron when hydrogen was provided, and 15.2 GJ/tonne of iron when syngas was provided, respectively.

TABLE 11.22
Energy Balances of the Ironmaking Section in the Industrial-Scale One-Step Ironmaking Process Producing 1 Million Tonnes of Iron per Year Combined with the Steam-Methane Reforming (SMR) Section with the Operating Temperature of 1,500°C, the Excess Driving Force of 0.5 and the Reactor Feed Gas Preheating Temperature of 900°C, Compared with the Average Blast Furnace Process

		With H_2 Production from SMR	Direct Use of Syngas in Ironmaking	BF Process[a]
Input (GJ/tonne of iron)	Fuel combustion[b]	8.78	10.25	8.33
	Produced hydrogen	0.01 (30°C)		
	Produced syngas		0.01 (30°C)	
	Heat recovery	−2.77	−3.66	−1.32[c]
	Subtotal	6.02	6.60	7.01
	Energy input for ore/coke preparation			5.68
	Total ironmaking	**6.02**	**6.60**	**12.7**
	Energy input for $CaCO_3$ and $MgCO_3$ calcination	0.26[d]	0.26[d]	
Output (GJ/tonne of iron)	Reduction	0.91	0.91	2.09
	Slagmaking			−0.16
	Sensible heat of molten metal	1.27 (1,500°C)[e]	1.27 (1,500°C)[e]	1.35 (1,600°C)
	Sensible heat of slag	0.24 (1,500°C)[e]	0.24 (1,500°C)[e]	0.47 (1,600°C)
	Slurry ($H_2O(l)$)	1.93	2.14	
	Condensed water vapor after scrubber	0.004	0.006	
	Sensible heat of off-gas			0.26 (90°C)
	$CaCO_3$ decomposition			0.33
	PSA tail gas	0.001	0.005	

(*Continued*)

TABLE 11.22 (*Continued*)

Energy Balances of the Ironmaking Section in the Industrial-Scale One-Step Ironmaking Process Producing 1 Million Tonnes of Iron per Year Combined with the Steam-Methane Reforming (SMR) Section with the Operating Temperature of 1,500°C, the Excess Driving Force of 0.5 and the Reactor Feed Gas Preheating Temperature of 900°C, Compared with the Average Blast Furnace Process

Output (GJ/tonne of iron)	With H_2 Production from SMR	Direct Use of Syngas in Ironmaking	BF Processa
Heat loss in the reactor	0.78	0.78	2.6
Heat loss in the heat exchanger	0.34[f]	0.46[f]	0.07[c] (heat loss in heat recovery)
Heat loss in the reactor	0.55	0.79	
Subtotal	6.02	6.60	7.01
Pelletizing			3.01 [Stubbles, 2,000]
Sintering			0.65 [Stubbles, 2,000]
Cokemaking			2.02 [Stubbles, 2,000]
Total ironmaking	6.02	6.60	12.7
$CaCO_3$ and $MgCO_3$ decomposition at 1,000°C [Biswas, 1981]	0.26[d]	0.26[d]	

[a] Energy balance was calculated by METSIM based on the material balance and the heat loss in the reactor.

[b] Fuel combustion energy generated in the process was calculated by subtracting the energy used for iron ore reduction from the balance of heat formation of all output components from the ironmaking section and input components into it. The high heating value of natural gas was used for this calculation. If the low heating value is used, the total energy requirements for the ironmaking section will decrease to 3.79 GJ and 3.99 GJ/tonne of iron, respectively.

[c] Heat loss in the heat recovery step was estimated to be 5% of the energy difference between the hot off-gas (calculated to be 426°C) and cooled reactor off-gas (90°C), and the rest was estimated to be the net heat recovery.

[d] The energy requirement for $CaCO_3$ and $MgCO_3$ calcination to generate CaO and MgO for the flux was calculated by METSIM. Calcination was assumed to be done at 1,000°C separately from the ironmaking process.

[e] If the ironmaking reactor operating temperature is replaced by 1,600°C (same as the blast furnace process), the total energy for the ironmaking section increases to 6.46 (when hydrogen is produced) and 7.21 (when syngas is produced) GJ/tonne of iron, respectively.

[f] Heat losses in a heat exchanger were assumed to be 5% of the total sensible heat of input streams.

TABLE 11.23
Material and Energy Balances of the Steam-Methane Reforming (SMR) Section Combined with the Ironmaking Section

Material Balance

		With H_2 Production from SMR	Direct Use of Syngas in Ironmaking
Input (tonnes/tonne of iron)	Natural gas	0.39	0.41
	Air	4.20	3.81
	PSA tail gas from ironmaking	0.02	0.72
	New water	2.12	2.45
	Total SMR	6.73	7.39
Output (tonnes/tonne of iron)	Hydrogen	0.12	
	Syngas		0.84
	Flue gas	5.07	4.57
	(CO_2 emissions)	(1.04)	(1.10)
	Hot water not used	0.81	1.23
	Steam not used	0.73	0.75
	Total SMR	6.73	7.39

Energy Balance

		With H_2 Production from SMR	Direct Use of Syngas in Ironmaking
Input (GJ/tonne of iron)	Fuel combustion[a]	8.61	8.95
	PSA tail gas from ironmaking	0.001 (30°C)	0.005 (30°C)
	Steam not used[b]	−1.96	−2.03
	Total SMR	6.65[c]	6.92[c]
Output (GJ/tonne of iron)	Steam-methane reforming	3.40	3.53
	Produced hydrogen	0.009 (30°C)	
	Produced syngas		0.01
	Flue gas	1.63 (300°C)	1.43 (300°C)
	Hot water not used	0.68	1.03
	Heat loss in the reformer	0.73	0.71
	Steam not used (90°C)[b]	0.20	0.21
	Total SMR	6.65[c]	6.92[c]

[a] Fuel combustion energy generated in the process was calculated by subtracting the energy used for reforming from the balance of heat formation of all output components from the SMR section and input components into it. The high heating value of natural gas was used for this calculation. If the low heating value is used, the total energy requirements for the SMR section producing hydrogen and that producing syngas will decrease to 5.28 GJ and 5.15 GJ/tonne of iron, respectively.

[b] Energy recovered by steam not used in the process includes the sensible heat of steam from 300°C to 90°C and the energy generated by condensation. [c]If the ironmaking reactor operating temperature is replaced by 1,600°C (same as the blast furnace process), the total energy for the SMR process increases to 7.24 (when hydrogen is produced) and 7.76 (when syngas is produced) GJ/tonne of iron, respectively.

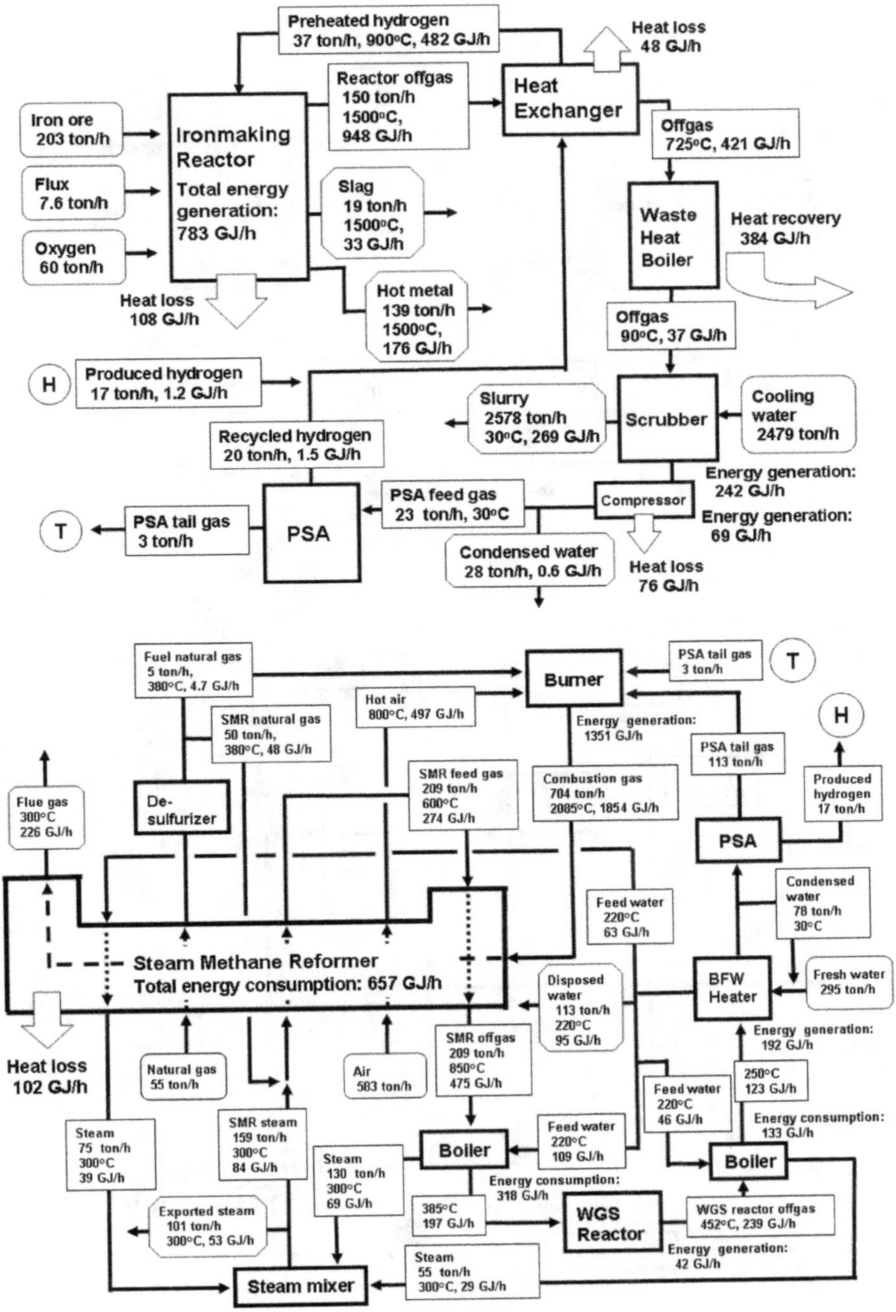

FIGURE 11.13 Material and energy flow diagram for the industrial-scale one-step ironmaking process producing 1 million tonnes of iron per year at the operating temperature of 1,500°C, the excess driving force of 0.5 and the reactor feed gas temperature of 900°C combined with the steam-methane reforming process producing hydrogen (300 days of operation in 1 year).

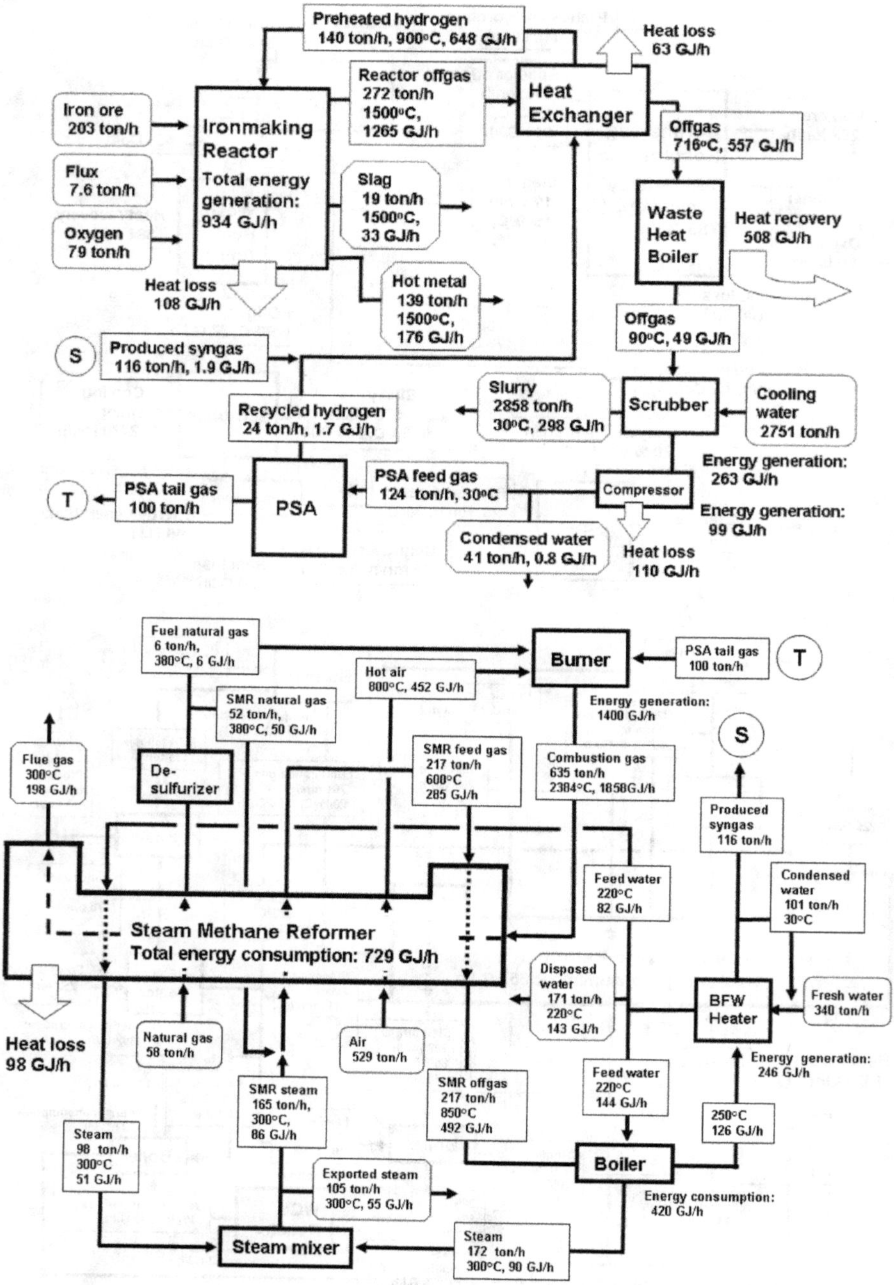

FIGURE 11.14 Material and energy flow diagram for the industrial-scale one-step ironmaking process producing 1 million tonnes of iron per year at the operating temperature of 1,500°C, the excess driving force of 0.5 and the reactor feed gas temperature of 900°C combined with the steam-methane reforming process producing syngas (300 days of operation in 1 year).

Water usage in the scrubber was 17.9 tonnes or 19.8 tonnes/tonne of iron, respectively, for hydrogen or syngas usage, which is marginally less than the reformerless one-step process (20.8 tonnes of water/tonne of iron, shown in Section 11.2) or the average blast furnace process [Camlic and Goodman, 1997; United States Steel Corporation, 1985]. However, it can be expected that both processes will require water other than for a scrubber, such as coolant for a reactor.

Table 11.24 shows a summarized comparison of the natural gas and energy requirements along with carbon dioxide emissions for the one-step flash ironmaking processes based on natural gas supply (reformerless, hydrogen from SMR and syngas from SMR) and the average blast furnace process. It indicates that the flash ironmaking process combined with the SMR process would require more natural gas than the reformerless process. This is because of the energy required for endothermic reforming and heat losses from the additional equipment, which increases the need for fuel. Compared with the average blast furnace process, the flash ironmaking processes combined with the SMR process would consume 0%–6% more energy, based on the HHV of natural gas (28%–29% less energy if the LHV is used), although

TABLE 11.24
Comparisons of Carbon Dioxide Emissions and Energy Requirement between the One-Step Flash Ironmaking Processes Based on Natural Gas with Different Process Configurations and the Average Blast Furnace Process

	Reformerless Use of Natural Gas	With H_2 Production from SMR	Direct Use of Syngas from SMR	BF Process
Natural gas requirement (tonnes/tonne of iron)	0.37	0.39	0.41	
Carbon dioxide emissions (tonnes/tonne of iron)	0.98	1.04[a]	1.10[a]	1.60[b]
Energy requirement				
Total[c] (GJ/tonne of iron)	8.94	12.96	13.76	12.7
Ironmaking section (GJ/tonne of iron)		6.28	6.86	
SMR section (GJ/tonne of iron)		6.65	6.92	

[a] Carbon dioxide emissions from the SMR section to which the PSA tail gas from the ironmaking section was fed. Carbon dioxide emissions from flux calcination are not included.

[b] Carbon dioxide emissions from ore/coke preparation are not included.

[c] Natural gas and coking coal are considered as the starting raw materials. The energy requirement for the blast furnace process includes energy required for ore/coke preparation. This energy requirement comparison does not include electrical power required for the plant operations.

carbon dioxide emissions would be smaller. Thus, the flash ironmaking process with reformerless use of natural gas, described in the previous section, would be more environment-friendly and would achieve the reduction of energy consumption in the steel industry.

11.3.4 Summary on Natural-Gas-Based Flash Ironmaking Technology – Combined with Steam-Methane Reforming

Flow sheets for the industrial-scale one-step flash ironmaking process producing 1 million tonnes of iron per year combined with a built-in SMR process were constructed and simulated. Using the same steam-methane reformer, the one-step ironmaking process was constructed and simulated in the cases in which (1) hydrogen was first produced and (2) syngas from the SMR process was directly injected into the ironmaking reactor. The ironmaking reactor operating at 1,500°C with an excess driving force of 0.5 was used as the standard case.

The results showed that the process using hydrogen would require less energy than the process using syngas. This was caused by more required energy to satisfy the excess driving force and operating temperature when syngas was fed to the ironmaking section. As the process using syngas required more fuel than using hydrogen, more natural gas was required in the SMR section when syngas was produced.

When natural gas was designated as the starting raw material, the energy requirement in the flash ironmaking process in which hydrogen from the SMR section was used was almost the same as (considerably lower than, if the LHV of natural gas is used) the average blast furnace process starting with coal. However, the energy requirement was 0.8 GJ/tonne of iron higher than the average blast furnace process when syngas was produced from the SMR section. Although energy saving was not achieved, the carbon dioxide emissions from the process were estimated to be 35% less when hydrogen was produced, and 31% less when syngas was produced from the SMR section, respectively, compared with the average blast furnace process. The energy requirement and carbon dioxide emissions would slightly increase when those from the separated flux calcination process were added.

The total energy requirement for the process was estimated to be similar when the reactor operating temperature was increased to 1,600°C or when excess driving force was increased to 1 without changing other conditions.

Compared with the reformerless one-step process, the ironmaking process combined with the SMR process required more natural gas and energy due to energy required for endothermic reforming and heat losses from additional equipment. Among the flash ironmaking process based on natural gas, the reformerless process would be the most environment-friendly, which would reduce energy requirements and carbon dioxide emissions significantly compared with the blast furnace process. A more comprehensive comparison of the energy requirement and carbon dioxide emissions must be performed considering the energy required for mining coal and pumping natural gas, and electrical power required for plant operations.

12 Economic Analysis of Flash Ironmaking Technology

12.1 INTRODUCTION

Economic feasibility analysis is essential to decide whether a proposed process is worth investment. Economic feasibility can be analyzed by the probable capital and operating costs determined by capital costs of similar processes and current material prices. The mass balance of the simulated process is used for estimating capital and operating costs.

The economic feasibility analysis was performed for the standard operating case described in Chapter 11 (the ironmaking reactor temperature = 1,500°C and excess driving force = 0.5). The economic feasibility of the natural-gas-based process with external reforming was performed only for the case when hydrogen is produced from steam-methane reforming (SMR) since the case of using syngas produced from SMR consumes more natural gas, although the capital cost would be reduced due to the elimination of water-gas shift (WGS) reaction and pressure swing adsorption (PSA).

The following description is aimed to describe the general idea and method of "discounted cash flow analysis." The term "discounted" is equivalent to "present worth," and the term "cash flow" is used to refer to the net inflow or outflow of money that occurs during a specified operation period such as a month or year in economic evaluation work. The future value, which is the sum of the accrued interest and the initial investment, is more than the present value if compound interest is positive. Thus, the present value has more influence on the current economic situation than the future value has, and investors typically evaluate a project's economic potential using present value calculations.

In the following sections, the method used to estimate capital and operating costs for NPV estimation is described. Then, net present value (NPV) estimation and sensitivity analysis with respect to item costs are discussed.

12.2 METHODS OF ESTIMATION FOR ECONOMIC FEASIBILITY ANALYSIS

12.2.1 Capital Cost Estimation

For simple estimation, the capital cost of a plant or a certain process step is determined by the following standard scaling-up equation [Peters and Timmerhaus, 1991]:

DOI: 10.1201/9781003342199-12

$$(\mathit{Capital\ cost\ of\ scaled-up\ plant})$$

$$= (\mathit{Base\ capital\ cost}) \times \left(\frac{\mathit{Scaled\ up\ production\ rate}}{\mathit{Base\ production\ rate}} \right)^{0.6} \tag{12.1}$$

The base capital cost and production rate used for this equation were found from available open references. Since the base cost was taken from reference years, the capital cost obtained by Eq. (12.1) was further modified to 2,010 dollars considering the change in the Chemical Engineering Plant Cost Index (CEPCI), as follows [Peters and Timmerhaus, 1991]:

$$(\mathit{Capital\ cost\ of\ the\ plant\ at\ present\ time}) = (\mathit{capital\ cost\ at\ the\ time\ of\ the\ reference})$$

$$\times \left(\frac{\mathit{CEPCI\ at\ present\ time}}{\mathit{CEPCI}\ \text{at the time of thereference}} \right) \tag{12.2}$$

In this calculation, 2,010 was set to be the present time.

For the estimation of the capital cost of the ironmaking section in the 1 million tonne scale one-step process, the capital cost of the 500,000 tonnes scale one-step flash ironmaking process ($149 million) was used as the base cost in Eq. (12.1). This base cost was estimated by Kim referring to the "Oxycup process" producing 175,000 tonnes of iron a year 25 [Varnbuler and Lemperle, 2010], which has a somewhat similar furnace structure, if not the process concept, to the flash ironmaking process. An industrial plant with a capacity of 500,000 tonnes a year was chosen because this scale was thought to be reasonable compared with other new ironmaking processes, such as COREX (300,000 tonnes/year) and Hismelt (800,000 tonnes/year).

The capital cost of the ironmaking section in the two-step process was estimated using the capital cost of the ironmaking section in the one-step process. First, the cost of the ironmaking reactor was assumed to be 75% of the capital cost of the ironmaking section in the one-step process. Then, the cost of the prereduction reactor was assumed to be 75% of the cost of the ironmaking reactor in the one-step process. The sum of capital costs of the ironmaking section in the one-step process and prereduction reactor was estimated as the capital cost of the two-step process. The capital costs of the ironmaking processes estimated by Eqs. (12.1) and (12.2) were $298 and $466 million, respectively, for the one-step and two-step processes, in 2,010 dollars.

The capital cost of the energy recovery power generation section was calculated using the capital cost of a heat recovery steam boiler and steam turbine in an industrial combined cycle power plant from 1998 ($54.4 million, 108.88 MW) [U.S. Department of Energy, 1998]. The power generation rate was determined from the recovered energy. The energy equivalent to the generated power was estimated to be 35% [U.S. Department of Energy, 1998] of the recovered energy in the waste heat boiler. The energy recovered from the second waste heat boiler in the two-step process was not considered in power generation due to low temperature. The capital cost of the waste heat boiler/steam turbine section was calculated by Eqs. (12.1) and (12.2) using power generation rate in each process.

12.2.2 Operating Cost Estimation

Material costs were estimated based on material balances with most of the material prices taken from 2,010 values. Water used in the process was assumed to be recycled internally and was not counted in material costs. Non-material operating costs (labor, maintenance, and electricity) in the one-step ironmaking section were referred to other processes such as a blast furnace process and MIDREX [U.S. Department of Energy. 2000]. Operating costs for steam turbine power generation were estimated from the operating costs of an industrial boiler and steam turbine ($20/MWh) [U.S. Environmental Protection Agency, 2011].

12.2.2.1 Estimation Procedure of Net Present Value

NPV is the cumulative present worth of positive or negative investment cash flow at a specified discount rate [Stermole and Stermole, 2000] and is calculated by the following equation [Stermole and Stermole, 2000]:

$$(NPV)=\sum_{i=0}^{n}\frac{C_i}{(1+r)^i} \tag{12.3}$$

where n is the project period, C_i is the annual cash flow at the end of the year i, and r is the discount rate. The discount rate used in this calculation is an acceptable investment rate of return for other existing processes, which is referred to as "internal rate of return," the rate of return where NPV in a certain project period becomes zero.

The NPV estimation was performed for the industrial-scale, hydrogen-based processes producing 1 million tonnes of iron per year under the standard operation case (the ironmaking reactor temperature = 1,500°C, excess driving force = 0.5). For the estimation, the plant operation period was set to be 15 years, the discount rate was set to be 10%, all cash flows occurred at the end of each yearly period, and capitalization was spent equally at the beginning of the project and the end of year 1. The plant operation was assumed to start at the beginning of year 2 at full capacity. The price of iron was set to be $512/tonne [Metalprices.com, 2010a] and that of electricity was $66.9/MWh to calculate annual inflow.

12.2.2.2 Carbon Dioxide Emission Credit

Assuming that the CO_2 emissions from the current average blast furnace are the maximum emissions allowed, the difference in emissions between the current average blast furnace and the flash ironmaking technology was assumed to be sold at a certain unit price of carbon dioxide, ranging from $0 to $100/tonne of CO_2. CO_2 emissions from the average blast furnace process including the CO_2 emissions from the preparation of coke and iron ore pellets (7%) were calculated to be 1,724 kg CO_2/tonne of iron [Sohn, 2008; Choi, 2010]. CO_2 emissions from the reformerless one-step and two-step processes were estimated to be 978 kg and 780 kg/tonne of iron [Pinegar et al., 2012], respectively, and that from the ironmaking process combined with the SMR process was 1,045 kg/tonne of iron [Pinegar et al., 2013a]. CO_2 emissions from flux ($CaCO_3$/$MgCO_3$) calcination are not included in these numbers but are small [Pinegar et al., 2011; 2012].

Calculated CO_2 emission credits obtained from the CO_2 emissions balance with each CO_2 unit price were counted as annual cash inflow. Although the exact CO_2 emission credit is currently unknown, the traded CO_2 unit price was determined referring to the current CO_2 capture and storage cost, typically \$30 to \$90/tonne of CO_2 from a power plant in 2006, and projected to decrease to \$25/tonne of CO_2 by 2030 [*Organization for Economic Co-Operation and Development, and International Energy Agency, 2006*].

12.3 NET PRESENT VALUE RESULTS AND SENSITIVITY ANALYSIS – HYDROGEN-BASED FLASH IRONMAKING

A summary of capital costs and production rates for the hydrogen-based, one-step and two-step ironmaking processes is shown in Table 12.1 with references to capital costs and production rates. A summary of operating cost estimation for the ironmaking process is shown with material prices in Table 12.2.

The NPV estimations for the industrial-scale one-step and two-step processes based on hydrogen are shown in Table 12.3. Both processes show negative NPV value due to expensive operating costs. This is in contrast to operating the flash ironmaking process by using natural gas, either in a reformerless mode or by first reforming it to produce hydrogen as shown in the next section.

12.4 NET PRESENT VALUE RESULTS AND SENSITIVITY ANALYSIS – NATURAL-GAS-BASED FLASH IRONMAKING

The methods of estimation of capital and operating costs and NPV as well as carbon dioxide emission credit described in Section 12.2 were applied here.

12.4.1 Capital Cost Estimation

The capital costs of the heat exchanger, WGS reactor, and PSA in the hydrogen recycling section in the reformerless ironmaking process were estimated using the capital

TABLE 12.1
Summary of the Capital Cost in 2010 for the Industrial-Scale Ironmaking Processes Based on Hydrogen with Different Configurations

	Referenced Base Cost (\$ Million) [Capacity, Year]	One-Step Process (\$ Million) [Capacity]	Two-Step Process (\$ Million) [Capacity]
Ironmaking	149 [500,000 tonnes of iron/year, 2004]	298 [1 MM tonnes/year]	466 [1 MM tonnes/year]
Power generation	54.4 [108.88 MW, 1998]	43 [37.9 MW]	16 [7.2 MW]
Total capital cost (\$ million)		**341**	**482**

TABLE 12.2
Material Prices Used and a Summary of Operating Cost for the Industrial-Scale Ironmaking Processes Based on Hydrogen with Different Configurations

	Reference Price ($/unit)	One-Step Process ($/tonne of iron)	Two-Step Process ($/tonne of iron)
Hydrogen	$2.5/kg [Mytelka, and Boyle, 2008]	257	214
Industrial O_2	$0.08/Nm^3	24	16
Iron ore concentrate	$167/tonne [Index mundi, 2010]	244	244
Flux	$80.9/tonne [USGS, 2010]	5	5
Ironmaking operating cost		26[a]	52[a]
Labor		2	4
Maintenance		15	30
Electricity		9	18
Power generation operating cost	$20/MWh [US EPA, 2010a]	6	1
Total operating cost ($/tonne of iron)		**562**	**532**

[a] Operating cost for ironmaking was estimated based on other ironmaking processes in the reference [U.S. Department of Energy, 2000). For labor and electricity, Labor Cost Index (110.0) [Bureau of Labor Statistics, 2010] for 2,010 dollars and industrial electricity price ($66.9/MWh) [U. S. Energy Information Administration, 2010] was further applied to calculate cost increase relative to 2,000 dollars.

TABLE 12.3
The Result of 2,010 NPV Estimation for the Industrial-Scale Ironmaking Process Based on Hydrogen at the Iron Price of $512/tonne Shown with the Effect of Carbon Dioxide Emission Credit

	NPV ($ Million)	
CO_2 emission credit	One-step process	Two-step process
No CO_2 emission credit	–$546	–$575
$13/tonne of CO_2	–$394	–$423
$25/tonne of CO_2	–$249	–$278
$50/tonne of CO_2	$48	$19
$75/tonne of CO_2	$346	$317
$100/tonne of CO_2	$643	$614

costs of these process steps in a Pumped Carbon Mining Substitute Natural Gas production plant from 2004 (heat exchanger = $1 million, WGS reactor = $3 million and PSA = $8 million) [HCE, 2004]. Eq. (12.1) was used to estimate each step's capital cost by substituting the amount of hydrogen produced from PSA and that produced in the reference PSA (3,669 kg of H_2/h) as a base. The capital costs of boilers in the hydrogen recycling section in the reformerless process were estimated by Eq. (12.1) based on the capital cost of the heat recovery steam generator ($17.2 million, 515,097 kg/h) [U.S. Department of Energy, 1998]. These capital costs were further modified by Eq. (12.2) to 2,010 dollars.

The capital cost for the SMR section was estimated using the capital cost estimation for an industrial hydrogen production plant from 2002 [Simbeck and Chang, 2002]. As the built-in SMR section is connected to the ironmaking section, the capital cost for a compressor was not considered in the capital cost of the SMR section. For the estimation, the capital cost of $62.8 million in 2,002 dollars with 135,000 kg of hydrogen production was used as the base cost for Eq. (12.1), and Eq. (12.2) was further used to modify it to 2,010 dollars.

A summary of capital costs for the flash ironmaking processes based on natural gas supply is shown in Table 12.4 with reference capital costs and production rates.

12.4.2 Operating Cost Estimation

In this economic feasibility analysis, "material cost" is defined as the cost used for raw materials (iron ore concentrate, flux), fuel, and oxygen. Most material prices were taken from available prices in 2010. Water used in the process was assumed to be recycled internally and was not counted in material costs.

The electricity costs were recalculated using a current value ($66.9/MWh in August 2010) [U. S. Energy Information Administration, 2010]. Non-material operating costs in the two-step ironmaking section were estimated to be twice the non-material operating costs in the one-step ironmaking section. The operating cost for steam turbine power generation was calculated using information from an industrial boiler and steam turbine ($20/MWh) [U.S. Environmental Protection Agency, 2010a].

To estimate the non-material operating cost for hydrogen production in the SMR section, it was assumed that the non-material operating cost would be one-sixth of natural gas cost [Eccleston, 2010] in the SMR hydrogen production without using the PSA tail gas from the ironmaking section. For the standard case, the natural gas required in the SMR section would be 0.44 tonnes/tonne of iron when the PSA tail gas from the ironmaking section was not used for the SMR burner.

A summary of estimated operating costs for the flash ironmaking process based on natural gas is shown with material prices in Table 12.5.

12.4.3 Summary of Net Present Value Estimation

A summary of the NPV estimation with the effect of CO_2 emission credit is shown in Table 12.6. The result shows that the flash ironmaking process based on natural gas would be economically feasible at the natural gas price of $5/million Btu [Metalprices.com, 2010b] and the iron price of $512/tonne.

Among these processes, the reformerless one-step process is most economical because it would require the least capital and operating costs due to the elimination of the reforming process and less equipment than the two-step process. The capital

TABLE 12.4
Summary of the Capital Cost in 2010 for the Industrial-Scale Ironmaking Processes Based on Natural Gas with Different Configurations [1 million tonnes iron/year]

	Reference	One-Step Reformerless Process	Two-Step Reformerless Process	Ironmaking with Built-in SMR Process
		Base Cost ($ Million) [capacity, year]		Ironmaking Section Capital Cost ($ Million) [capacity]
Ironmaking	149 [500,000 tonnes of iron/year, 2004] [Varnbuler and Lemperle, 2010]	298 [1,000,000 tonnes/year]	466 [1,000,000 tonnes/year]	298 [1,000,000 tonnes/year]
Power generation	54.4 [108.88 MW, 1998] [US DOE, 1998]	49 [45.8 MW]	22 [12.1 MW]	43 [37.4 MW]
Hydrogen recycle		66[b]	51[b]	30[b]
Heat exchanger	1 [3,669 kg of H_2/h, 2004] [HSE, 2004]	4 [22,711 kg/h]	3 [15,014 kg/h]	
WGS reactor	3 [3,669 kg of H_2/h, 2004] [HSE, 2004]	12 [22,711 kg/h]	9 [15,014 kg/h]	
PSA	8 [3,669 kg of H_2/h, 2004] [HSE, 2004]	31 [22,711 kg/h]	25 [15,014 kg/h]	30 [20,574 kg/h]
Boiler 1	17.2[515,097 kg of steam/h, 1998] [US DOE, 1998]	8 [67,339 kg/h]	6 [47,543 kg/h]	
Boiler 2		11 [130,988 kg/h]	8 [79,331 kg/h]	
		Base Cost ($ Million) [capacity, year]		Steam-Methane Reforming Section Capital Cost ($ Million) [capacity]
Hydrogen production	62.8 [135,000 kg of H_2/day, 2002] [Simbeck and Chang, 2002]			180 [403,816 kg/day]
Total capital cost ($ million)		**413**	**539**	**551**

[a] Chemical Engineering Plant Cost Index (CEPCI): 2010: 585.9, 2004:444.2, 2002: 395.6, 1998: 389.5[5].

[b] From the simulation, the amount of hydrogen from PSA was calculated to be 22,711 kg/h in the one-step process, 15,014 kg/h in the two-step process, and 20,574 kg/h in the hydrogen recycling section of the ironmaking process.

TABLE 12.5

Material Prices Used and a Summary of Operating Cost for the Industrial-Scale Ironmaking Processes Based on Natural Gas with Different Configurations [1 million tonnes iron/year]

	Reference Price [$/unit]	One-Step Reformerless Process	Two-Step Reformerless Process	Ironmaking with Built-in SMR Process
		Ironmaking Section Operating Cost ($/tonne of iron)		
Natural gas	$5/Million Btu [Metalprices.com, 2010b] ($254/tonne)a	93	75	
Industrial O_2 [blast furnace route steelmaking costs, 2,010]	$0.08/Nm3	44	32	24
Iron ore concentrate	$167/tonne [Index mundi, 2010]	244	244	244
Flux	$80.9/tonne [USGS, 2010]	5	5	5
Ironmaking operating cost		26[b]	52[b]	26[b]
Labor		2	4	2
Maintenance		15	30	15
Electricity		9	18	9
Power generation operating cost	$20/MWh [US EPA, 2011]	7	2	5
	Reference price [$/unit]	Steam-methane reforming section operating cost ($/tonne of iron)		
Natural gas	$5/Million Btu ($254/tonne)			100
H_2 production operating cost				19[c]
Total operating cost ($/tonne of iron)		419	410	423

[a] Natural gas cost ($/tonne) was calculated for the combustion energy (HHV) of the natural gas used in the process simulation (96% CH_4, 2% C_2H_6, and 2% N_2).

[b] Operating cost for ironmaking was estimated deriving from other ironmaking processes in the reference (U.S. Department of Energy. *Ironmaking Process Alternatives Screening Study Vol.* 2, 2000, October). For labor and electricity, Labor Cost Index (110.0) [Bureau of Labor Statistics, 2010] for 2010 dollars and industrial electricity price ($66.9/MWh) [U. S. Energy Information Administration, 2010] was further applied to calculate cost increases relative to 2,000 dollars.

[c] Operating cost for H_2 production is one-sixth of natural gas costs for the SMR section when PSA tail gas from the ironmaking section is not sent to the SMR burner [Eccleston, 2010]. In this case, required natural gas in the SMR section was 0.44 tonne/tonne of iron, which was larger than 0.39 tonne/tonne of iron required when PSA tail gas from the ironmaking section was used as f

TABLE 12.6
The Result of 2010 Net Present Value (NPV) Estimation for the Industrial-Scale Ironmaking Process Based on Natural Gas Shown with the Effect of Carbon Dioxide Emission Credit (Iron Price = $512/tonne; 1 million tonnes iron/year)

	NPV ($ Million)		
Carbon Dioxide Emission Credit	**Reformerless One-Step Process**	**Reformerless Two-Step Process**	**Ironmaking with Built-in SMR Process**
No CO_2 emission credit	$401	$232	$214
$13/tonne of CO_2	$471	$315	$277
$25/tonne of CO_2	$533	$398	$332
$50/tonne of CO_2	$657	$557	$449
$75/tonne of CO_2	$789	$723	$567
$100/tonne of CO_2	$920	$882	$685

[a] CO_2 emissions were 978 kg, 780 kg, and 1,045 kg/tonne of iron, respectively, from the reformerless one-step process, reformerless two-step process, and the ironmaking process combined with the SMR hydrogen production process.

[b] In this calculation, the amount of CO_2 emissions from the average blast furnace process was assumed to be the maximum allowance of emissions [estimated to be 1,724 kg CO_2/tonne of iron, including the CO_2 emissions from ore/coke preparation (7% of the total emissions)] [Sohn, 2008; Choi, 2010]. CO_2 emissions without considering ore/coke preparation were estimated to be 1,603 kg/tonne of iron (from Chapter 11).

cost of the reformerless one-step process was more than $100 million less than the reformerless two-step process and the ironmaking process combined with the SMR hydrogen production process.

The ironmaking process combined with the SMR hydrogen production process was expected to be less economical than the reformerless processes. As described above, it was because the capital cost increased by the built-in SMR process ($180 million) had a great effect on its NPV estimation. Furthermore, the operating cost was higher than the reformerless processes due to additional natural gas injection for fuel to the SMR burner and the cost for operation and maintenance of the SMR process.

For the reformerless one-step process, which is expected to be the most economical among the ironmaking processes based on natural gas, a sensitivity analysis using item costs was performed to examine their effects on the estimated NPV. The results are shown in Figures 12.1 and 12.2. Figure 12.1 shows the sensitivity to natural gas, capital, and non-material operating costs on the estimated NPV. Figure 12.2 shows the sensitivity to iron price.

12.4.4 Conclusions on Natural-Gas-Based Flash Ironmaking

The reformerless one-step process would be the most economical process due to the smallest capital and operating costs. Although the two-step process would use

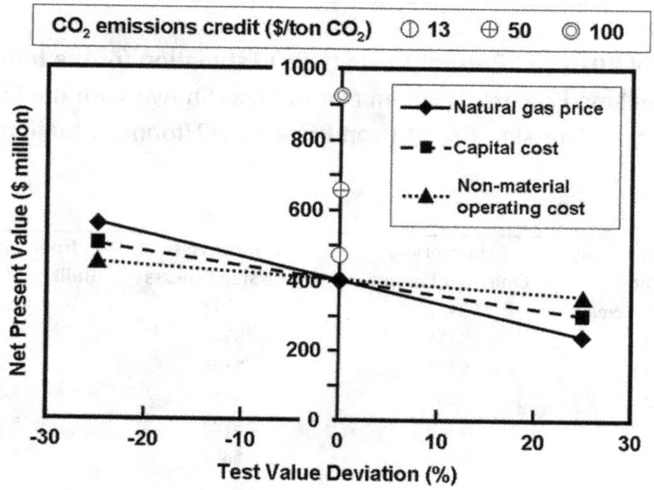

FIGURE 12.1 The result of 2,010 net present value estimation for the industrial-scale reformerless one-step ironmaking process with a variation by the difference of CO_2 emission credit and sensitivity analysis using natural gas price, capital, and non-material operating costs.

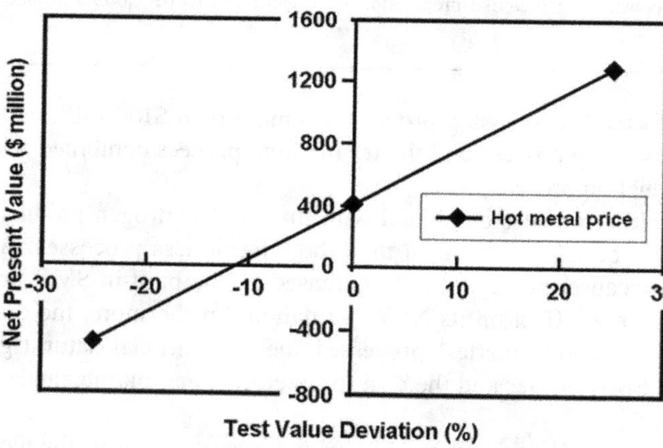

FIGURE 12.2 The result of sensitivity analysis using iron price on the 2,010 NPV estimation for the industrial-scale reformerless one-step ironmaking process.

less natural gas than the one-step process, it was less economical because of the additional capital costs for the prereduction reactor and doubled non-material operating cost than the one-step process. The ironmaking process combined with the SMR process was less economical because the effect of the capital cost for the SMR process on the NPV was significant, and it required more operating costs than the reformerless process.

The sensitivity analysis performed on the estimated NPV for the reformerless one-step ironmaking process showed that the natural gas price would affect the NPV more strongly than capital and non-material operating costs.

From these economic analysis results, the flash ironmaking process based on natural gas supply would be economically feasible, and the capital investment would be significantly lower than for an ironmaking plant based on the blast furnace and coke oven. However, it must be noted that the capital cost of each process would change significantly depending on the difficulty of construction of the component steps, and the capital costs applied in this economic analysis might have significant uncertainties.

References

Abdelghany A., Fan D.-Q., Elzohiery M., and Sohn H. Y., "Experimental investigation and computational fluid dynamics simulation of a novel flash ironmaking process based on partial combustion of natural gas in a reactor," *Steel Res Int.,* 2019, *90*, 1900126. https://doi.org/10.1002/srin.201900126.

Abdelghany A., Fan D.-Q., and Sohn H. Y., "Novel flash ironmaking technology based on iron ore concentrate and partial combustion of natural gas: A CFD study," *Metall. Mater. Trans. B* 2020, *51*, 2046–2056.

Abd Rashid R. Z., Salleh H. M., Ani M. H., Yunus N. A., Akiyama T., and Purwanto, H., "Reduction of low grade iron ore pellet using palm kernel shell," *Renew. Energy*. 2014, *63*, 617–623, https://doi.org/10.1016/j.renene.2013.09.046.

Alexander D. C. D., Brown L., Nimlos M. R., and Daily J. W., "Design and characterization of an entrained flow reactor for the study of biomass pyrolysis chemistry at high heating rates," *Energy Fuel.*, 2001, *15* (5), 1276–1285.

Altúzar P., and Valenzuela R., "Avrami and kissinger theories for crystallization of metallic amorphous alloys," *Mater. Lett.* 1991, *11* (3), 101–104.

ANSYS Inc., *Fluent Theory Guide*, Canonsburg, USA, 2019.

Atsushi M., Uemura H., and Sakaguchi T., *Kobelco Technol. Rev.* December 2010, 29, 50–57, http://www.kobelco.co.jp/english/ktr/pdfktr_29/050-057.pdf.

Avila M. R., Honkanen M., Raiko R., and Oksanen A., "Coal char combustion in O_2/N_2 and O_2/CO_2 conditions in a drop tube reactor: an optical study, Industrial Combustion Journal of the International Flame Research Foundation," *Article Number 201201,* May 2012, pp. 1–13.

Avrami M., "Kinetics of phase change I. General theory," *J. Chem. Phys.* 1939, *7* (12), 1103–1112.

Avrami M., "Kinetics of phase change II. Transformation-time relations for random distribution of nuclei," *J. Chem. Phys.* 1940, *8* (2), 212–224.

Avrami M., "Granulation, phase change, and microstructure kinetics of phase change III," *J. Chem. Phys.* 1941, *9* (2), 177–184.

Babich A., Senk D., Gudenau H.W., and Mavrommatis K., Th., *Ironmaking*,RWTH Aachen, Aachen, 2009.

Bains P., Psarras P., and Wilcox J., "CO_2 capture from the industry sector,". *Prog. Energy Combust. Sci.* 2017, *63*, 146–172.

Barlow R. S., and Carter C. D., "Raman/Rayleigh/LIF measurements of nitric oxide formation in turbulent hydrogen jet flames," *Combust Flame* 1994, *97* (3), 261–280.

Barlow R. S., and Carter C. D., "Relationships among nitric oxide, temperature, and mixture fraction in hydrogen flames,"*Combust Flame* 1996, *104* (3), 288–299.

Bartlett J., Holtzapple A., and Rempel C., "A Brief Overview of the Process Modeling/Simulation and Design Capabilities of METSIM," *COM 2014-Conference of Metallurgists Proceedings* 2014, ISBN: 978-1-926872-24-7.

Battle T., Srivastava U., Kopfle J., Hunter R., and McClelland J., The direct reduction of iron, *Treatise on Process Metallurgy, Industrial Processes Part A,* Chapter 1.2. vol. 3, pp. 89–176, Elsevier, Oxford, UK and Waltham, MA, USA, 2014.

Bergman T. L., Lavine A. S., Incropera F. P., and DeWitt D. P., *Fundamentals of Heat and Mass Transfer*, 7th Ed., Willey, USA, pp. 795–796, 2011,

Biswas A. K., *Principles of Blast Furnace lronmaking: Theory and Practice*, Cootha Publishing House, Brisbane., 1981.

Blast Furnace Route Steelmaking Costs, 2010 Conversion Cost for BOF Steelmaking, http://www.steelonthenet.com/steel_cost_bof.html, accessed September 2010.

Boldyrev V. V., "Topochemistry of thermal decompositions of solids," *Thermochim. Acta* 1986, *100* (1), 315–338.

Brent A. D., Mayfield P. L. J., Honeyands T.A., "The Port Hedland FINMET (R) project - Fluid bed production of high quality virgin iron for the 21st century," In *Proceedings of the International Conference on Alternative Routes of Iron and Steelmaking (ICARISM '99),* Misra V. N., Holmes R. J., Eds., The Australasian Institute of Mining and Metallurgy, Victoria, Australia, 1999, p. 111.

Bureau of Labor Statistics 2010, Manufacturing, Table 9 Employment Cost Index for wages and salaries for private industry workers by occupational group and industry, Employment Cost Index – June 2010, Office of Compensation Levels and Trends, Washington DC.

Burgo J. A., "The Manufacture of Pig Iron in the Blast Furnace," In *The Making, Shaping and Treating of Steel, 11th Ed.*, Wakelin D. H., Ed., The AISE Steel Foundation, Pittsburgh, PA, USA,1999, pp. 712–713.

Cai J. M., Liu R. H., and Shen F. J., "Kinetic analysis of solid state reactions without neglecting temperature integral at starting temperature: relationship between conversion degree at maximum reaction rate and reaction mechanism,"*J. Energy Inst.*, 2009, *82* (2), 109–113.

Camlic R. L., and Goodman N. J., 1997, *Upgrade of Process Control System at U.S. Steel Fairfield No.8 Blast Furnace, Iron and Steel Engineer*, http://steellibrary.com/BookContents/PR-PM0197-1%201%20View%20Abstract.pdf, accessed October 2010.

Carrara A., Perdichizzi A., and Barigozzi G., "Simulation of a hydrogen production steam reforming industrial plant for energetic performance prediction," Int. *J. Hydrog. Energy*, 2010, *35* (8), 3499–3508.

Chase M. W., Jr., Curnutt J. L., Downey J. R., Jr., McDonald R. A., Syverud A. N., and Valenzuela E. A., JANAF thermochemical tables, 1982 supplement, *J. Phys. Chem. Ref. Data*, 1982, *11*, 695, https://doi.org/10.1063/1.555666.

Chatterjee A., *Beyond the Blast Furnace,* CRC Press, Boca Raton, FL, pp. 1, 1993.

Chatterjee A., *Sponge Iron Production by Direct Reduction of Iron Oxide,* 2nd Ed., PHI Learning Private Limited, New Delhi, pp. 65–66, 2012.

Chaubal P. C., and Sohn H. Y., "Intrinsic kinetics of the oxidation of chalcopyrite particles under isothermal and non–isothermal conditions," *Metall. Trans. B 1986, 17B*, 51–60 .

Chen F., Mohassab Y., Tao J., and Sohn H. Y., "Hydrogen reduction kinetics of hematite concentrate particles relevant to a novel flash ironmaking process," *Metall. Mater. Trans. B* 2015a, *46B* (3), 1133–1145, http://link.springer.com/article/10.1007/s11663-015-0332-z.

Chen F., Mohassab Y., Zhang S., and Sohn H. Y., "Kinetics of the reduction of hematite concentrate particles by carbon monoxide relevant to a novel flash ironmaking process," *Metall. Mater. Trans. B, 2015b, 46B* (4), 1716–1728, http://link.springer.com/article/10.1007/s11663-015-0345-7

Choi M. E., *Suspension Hydrogen Reduction of Iron Ore Concentrate, Ph.D. Dissertation*, University of Utah, Salt Lake City, Utah, 2010.

Choi M. E., and Sohn H. Y., "Development of green suspension ironmaking technology based on hydrogen reduction of iron oxide concentrate: rate measurements," *Ironmaking Steelmaking* 2010, *37*, 81–88.

Christian J. W., *The Theory of Transformations in Metals and Alloys,* Vol. 1. Pregamon Press, Oxford, 1975.

Chui E. H., and Raithby G. D., "Computation of radiant heat transfer on a nonorthogonal mesh using the finite," *Numer. Heat Tr. B-Fund.*, 1993, 23 (3), 269–288.

Clayton C. K., Sohn H. Y., and Whitty K. J., "Oxidation kinetics of cuprous oxide in oxygen carriers for chemical looping with oxygen uncoupling," *Ind. Eng. Chem. Res 2014, 53* (8), 2976–2986. http://dx.doi.org/10.1021/ie402495a.

Clift R., Grace J. R., Weber M. E., *Bubbles, Drops, and Particles*, Academic Press, New York, 1978.

Coats A. W., and Redfern J. P., "Kinetics parameters from thermogravimetric data," *Nature,* 1964, *201*, 68–69.

Coats A. W., and Redfern J. P., J. "Structural relaxation of stacked ultrathin polystyrene films," *Polym. Sci., Part B: Polym. Lett* 1965, *3*, 917–920.

Contreras W., Hardy C., Tovar K., Piwetz A. M., Harris C. R., Tullos E. E., Bymaster A., McMichael J., and Laurenzi I. J., "Life cycle greenhouse gas emissions of crude oil and natural gas from the Delaware Basin," *J. Clean. Prod.* 2021, *328*, 129530, https://doi.org/10.1016/j.jclepro.2021.129530

Coppalle A., and Vervisch P., "The total emissivities of high temperature flames," *Combust. Flame* 1983, *49* (1–3), 101–108.

Cumbrera F. L., Sánchez-Bajo F., "The use of the JMAYK kinetic equation for the analysis of solid-state reactions: critical considerations and recent interpretations," *Thermochim. Acta* 1995, *266*, 315–330.

Davenport W. G., Jones D. M., King M. J., and Partelpoeg E. H., *Flash Smelting: Analysis, Control and Optimization*, TMS, Warrendale, PA, 2001.

Dinan T., *Trade-Offs in Allocating Allowances for CO_2 emissions, Economic and Budget Issue Brief, Congressional Budget Office*, Washington, DC, 2007.

Duarte P. E., Becerra J., Lizcano C., and Martinis A., "Energiron direct reduction ironmaking – Economical, flexible, environmentally friendly" *Steel Times International* 2010, 34 (3), pp. 25–26, 28, 30.

Durst F., et al., "The development lengths of laminar pipe and channel flows," Transactions of the ASME, *J. Fluids Eng.* 2005, *127* (6), 1154–1160.

Eccleston E., Personal Communication Technip, USA, Los Angeles, CA, April 2010.

Eklund D. R., Drummond J. P., and Hassan H. A. "Calculation of supersonic turbulent reacting coaxial jets," *AIAA Journal* 1990, *28*, 1633–1641.

Ellingham H. J. T., "Reducibility of oxides and sulphides in metallurgical processes," *J. Soc. Chem. Ind. Trans* 1944, *63*, 125–160.

Elzohiery M., *Flash Reduction of Magnetite Concentrate Related to a Novel Flash Ironmaking Process*, Ph.D. Dissertation, University of Utah, Salt Lake City, Utah, 2018.

Elzohiery M,, Fan D.-Q., Mohassab Y., and Sohn H. Y., "Experimental investigation and computational fluid dynamics simulation of the magnetite concentrate reduction using methane-oxygen flame in a laboratory flash reactor," *Metall. Mater. Trans. B.* 2020, *51* (3), 1003–1015. https://doi.org/10.1007/s11663-020-01809-9

Elzohiery M., Fan D.-Q., Mohassab Y., and Sohn H. Y., "Kinetics of hydrogen reduction of magnetite concentrate particles at 1623 – 1873 K relevant to flash ironmaking," *Ironmaking Steelmaking 2021*, *48* (5), 485–42 https://doi.org/10.1080/03019233.2020.1819942

Elzohiery M., Sohn H. Y., and Mohassab Y., "Kinetics of hydrogen reduction of magnetite concentrate particles in solid state relevant to flash ironmaking," *Steel Res. Int.* 2017, *88* (2), 14, https://doi.org/10.1002/srin.201600133.

e-metallicus, http://e-metallicus.com/en/news/world/posco-ignites-the-largest-blast-furnace-in-the-world.html, accessed July 27, 2017.

Erofeyev B. V., "A generalized equation of chemical kinetics and its application in reactions involving solids," *Dokl. Akad. Nauk SSSR* 1946, *52*, 511.

Evans J. W., and Koo C.-H., "The Reduction of Metal Oxides," in *Rate Processes of Extractive 3Metallurgy*, Sohn H. Y., and Wadsworth M. E., Eds., Plenum Press, NY, 1979, pp. 285–321.

Fan D.-Q., *Computational Fluid Dynamics analysis and Design of Flash Ironmaking Reactors,* Ph.D. Dissertation, University of Utah, Salt Lake City, Utah, 2019.

Fan D.-Q., Elzohiery M., Mohassab Y., and Sohn H. Y., "Rate-Enhancement effect of CO in magnetite concentrate particle reduction by H_2+CO mixtures," *Ironmaking Steelmaking 2021a*, *48* (9), 1064–1075, https://doi.org/10.1080/03019233.2021.1915645.

Fan D.-Q., Elzohiery M., Mohassab Y., and Sohn H. Y., "The kinetics of carbon monoxide reduction of magnetite concentrate particles through CFD Modeling," *Ironmaking Steelmaking 2021b, 48* (7), 769–778, https://doi.org/10.1080/03019233.2020.1861857.

Fan D.-Q., Mohassab Y., Elzohiery M., and Sohn H. Y., "Analysis of the hydrogen reduction rate of magnetite concentrate particles in a drop tube reactor through CFD modeling," *Metall. Mater. Trans. B* 2016a, *47B* (3), 1669–1680, https://doi.org/10.1007/s11663-016-0603-3.

Fan D.-Q., Sohn H. Y., Mohassab Y., and Elzohiery M., "Computational fluid dynamics simulation of the hydrogen reduction of magnetite concentrate in a laboratory flash reactor," *Metall. Mater. Trans. B* 2016b, *47B* (6), 3489–3500, http://link.springer.com/article/10.1007/s11663-016-0797-4.

Fan D.-Q., Sohn H. Y., and Elzohiery M., "Analysis of the reduction rate of hematite concentrate particles by H_2 or CO in a drop-tube reactor through CFD Modeling," *Metall. Mater. Trans. B 2017, 48B*, 2677–2684, http://rdcu.be/ybnM, https://link.springer.com/content/pdf/10.1007%2Fs11663-017-1053-2.pdf.

Ferziger J. H., and Peric M., *Computational Methods for Fluid Dynamics*, Springer-Verlag, Berlin, 2002.

Friedman H. L., "Kinetics of thermal degradation of char-forming plastics from thermogravimetry - application to a phenolic resin," *J. Polym. Sci., Part C: Polym. Lett.* 1964, *6* 183–195.

Fruehan R. J., Fortini O., Paxton H.W., and Brindle R., 2000, *Theoretical Minimum Energies to Product Steel for Selected Conditions*, Department of Energy, Washington, DC, USA, http://www1.eere.energy.gov/industry/steel/pdfs/theoretical_minimum_energies.pdf, accessed on July 27, 2013.

Galwey A. K., and Brown M. E., Chapter 3 "Kinetic models for solid-state reactions", In: *Thermal Decomposition of Ionic Solids: Chemical Properties and Reactivities of Ionic Crystalline Phases*, Elsevier, Amsterdam, 1999, pp. 75–138.

Geerdes M., Chaigneau R., and Lingiardi O., *Modern Blast Furnace Ironmaking: An Introduction*, 4th Ed., IOS Press, Amsterdam, 2020.

Geerdes M., Toxopeus H., and Van Der Vliet C., *Modern Blast Furnace Ironmaking—An Introduction*, IOS Press, Amsterdam, 2009.

General Electric Company Research and Development Center, 1970, *Heat Transfer and Fluid Flow Data Book*, Vol. 1, Schenectady, NY.

Gran I. R., and Magnussen B. F., "A numerical study of bluff- body stabilized diffusion flame. Part 2. Influence of combustion modeling and finite-rate chemistry," *Combust. Sci. Technol.* 1996, *119* (1–6), 191–217.

Guo D. B., Zhu L. D., Guo S., Cui B. H., Luo S. P., Laghari M., Chen Z. H., Ma C. F., Zhou Y., Chen J., Xiao B., Hu M., Luo S. Y., "Direct reduction of oxidized iron ore pellets using biomass syngas as the reducer," *Fuel Process. Technol. 2016, 148,* 276–281, https://doi.org/10.1016/j.fuproc.2016.03.009.

Guo X. J., Process Modeling and Intelligent System in Copper Smelter — The Concept of Future Smart Smelter, *Proceedings of the Copper 2003-Cobre 2003 the 5th International Conference Vol.4. Pyrometallurgy of Copper: Hermann Schwarze Symposium* (book 2), Montreal, Canada, Canadian Institute of Mining, Metallurgy and Petroleum, 2003, pp. 165–177.

Hahn Y. B., and Sohn H. Y., "Mathematical modeling of sulfide flash smelting process: Part I. Model development and verification with laboratory and pilot plant measurements for chalcopyrite concentrate smelting," *Metall. Trans. B* 1990a, *21* (6), 945–958.

Hahn Y. B., and Sohn H. Y., "Mathematical modeling of sulfide flash smelting process: Part II. Quantitative analysis of radiative heat transfer," Metall. *Mater. Trans. B* 1990b, *21* (6), 959–966.

Hasanbeigi A., Arens M., Price L., "Alternative emerging ironmaking technologies for energy-efficiency and carbon dioxide emissions reduction: a technical review," *Renew. Sustain. Energ Rev.* 2014, *33*, 645–658.

Hasanbeigi A., Morrow W., Sathaye J., et al., Assessment of energy efficiency improvement and CO_2 emission reduction potentials in the iron and steel industry in China. Report, 2012, Berkeley, CA, Lawrence Berkeley National Laboratory Report LBNL-LBNL-5535E.

HCE, LLC, Pumped Carbon Mining (PCM) Substitute Natural Gas (SNG) Production Cost Estimate Underground Coal Gasification with Above Ground Processing, HCEI-11-04-2, Proprietary Information, November, 2004, accessed September 2010 http://www.hceco.com/HCEI1104002.pdf.

Husain R., Sneyd S., and Weber P., Circored and Circofer—Two New Fine Ore Reduction Processes *In Proceedings of the International Conference on Alternative Routes of Iron and Steelmaking (ICARISM '99)*, Misra V. N., Holmes R. J., Eds., The Australasian Institute of Mining and Metallurgy, Victoria, Australia, 1999, pp. 123–129.

Index Mundi. 2010. Iron Ore Monthly Price, accessed September 2010, http://www.indexmundi.com/commodities/?commodity=iron-ore.

Institute for Industrial Productivity. 2019, accessed on 30 May 2019, http://www.ietd.iipnetwork.org/content/direct-reduced-iron.

International Energy Agency (IEA). 2007, Tracking industrial energy efficiency and CO_2 emissions, Paris: OECD/IEA.

IUPAC: Compendium of Chemical Terminology, 2nd ed., McNaught A. D., and Wilkinson A., Eds., Blackwell Scientific Publications, Oxford, 1997, p. 85.

Jander W. and Anorg Z., "Kinetic model for solid-state reactions," *Allg. Chem.* 1927, 163, 1–30.

Jones W. P., and Lindstedt R. P., "Global reaction schemes for hydrocarbon combustion," Combust. *Flame* 1988, *73* (3), 233–249.

Kee R. J., Rupley F. M., Meeks E., and Miller J. A., Report No. 96-8216, 1996, Sandia National Laboratories, Albuquerque, NM.

Kee R. J., Rupley F. M., Wang C., and Adigun O., *CHEMKIN Collection, Release 3.6*, Reaction Design, Inc., 2000, San Diego, CA.

Kemori N., Ojima Y., and Kondo, Variation of the composition and size of copper concentrate particles in the reaction shaft In: *Flash Reaction Processes*, Robertson D. G. C., Sohn H. Y., and Themelis N. J., Eds., Proceedings of Center for Pyrometallurgy Conference, University of Utah, Published by the center for pyrometallurgy, Rolla, MO, 1988, pp. 47–69.

Khatami R., Stivers C., Joshi K., Levendis Y.A., and Sarofim A.F., "Combustion behavior of single particles from three different coal ranks and from sugar cane bagasse in O_2/N_2 and O_2/CO_2 atmospheres," *Combust. Flame* 2012, *159* (3), 1253–1271.

Khawam A., Flanagan D. R., "Basics and applications of solid-state kinetics: A pharmaceutical perspective," *J. Pharm. Sci.* 2006a, *95*, 472–498.

Khawam A., and Flanagan D. R., "Solid-state kinectic models: Basics and mathematical fundamentals," *J. Phys. Chem. B* 2006b, *110* (35), 17315–17328, https:doi.org/10.1021/jp062746a

Kikuchi S., Ito S., Kobayashi I., Tsuge O., Tokuda K., "ITmk3 Process," *Kobelco Technology Review No. 29*, 2010, pp. 77–84. http://www.kobelco.co.jp/english/ktr/pdf/ktr_29/077-084.pdf

Kim B.-S., and Sohn H. Y., "A novel cyclic reaction system involving CaS and $CaSO_4$ for converting sulfur dioxide to elemental sulfur without generating secondary pollutants: 3. Kinetics of the hydrogen reduction of calcium sulfate powder to calcium sulfide," *Ind. Eng. Chem. Res.* 2002a, *41*, 3092–3096.

Kim B.-S., and Sohn H. Y., "A novel cyclic process using $CaSO_4$/CaS pellets for converting sulfur dioxide to elemental sulfur without generating secondary pollutants: Part II. Hydrogen reduction of calcium sulfate pellets to calcium sulfide," *Metall. Mater. Trans. B* 2002b, *33B*, 717–721, http://www.springerlink.com/content/g3h8777k4657x4w2/fulltext.pdf

Kimura H., Liu S., Moats M. S., and Sohn H. Y., "Process Simulation for a Novel Green Ironmaking Technology with Greatly Reduced CO_2 Emission and Energy Consumption," Preprint, 2010 SME Annual Meeting, Phoenix, AZ, February 28 – March 3 2010.

Kimura T., Ojima Y., Mori Y., and Ishii Y., "Reaction mechanism in a flash smelting reaction shaft." In: *The Reinhardt Schuhmann International Symposium on Innovative Technology and Reactor Design in Extraction Metallurgy*, 9 November 1986. TMS/AIME, Colorado Springs, Colorado, pp. 403–418.

Kirsch B. L., Richman E. K., Riley A. E., Tolbert S. H. J., "In-situ x-ray diffraction study of the crystallization kinetics of Mesoporous Titania films," *Phys. Chem. B* 2004, 108, 12698–12706.

Knacke O., Kubaschewski O., and Hasselman K., *Thermochemical Properties of Inorganic Substances,* Springer-Verlag, Berlin, 1991

Kubaschewski O., and Alcock C. B., *Metallurgical Thermochemistry*, 5th Ed., Pergamon, Oxford, 1979.

Ladebeck J. R., and Wagner J. P., Chapter 16 "Catalyst development for water-gas shift", In: *Catalyst Development for Water-Gas Shift, Handbook of Fuel Cells – Fundamentals, Technology, and Applications*, W. Vielstich, A. Lamm, and H. A. Gasteiger, Eds., John-Wiley and Sons Inc, Chichester, England, 2003, pp. 190–201, ISBN: 0-471-49926-9.

Lehto J., "Determination of kinetic parameters for finnish milled peat using drop tube reactor and optical measurements techniques" *Fuel* 2007, *86* (12–13), 1656–1663.

Lewis G. N., and Randall M., revised by Pitzer K. S., and Brewer L., *Thermodynamics,* 2nd Ed. McGraw-Hill, New York, 1961.

Zhang L., Binner E., Qiao Y., and Li C.-Z., "High-speed camera observation of coal combustion in zir and O_2/CO_2 mixtures and measurement of burning coal particle velocity," *Energy Fuels* 2010, *24* (1), 29–37.

Lin H. K., and Sohn H. Y., "Mixed–Control Kinetics of Oxygen Leaching of Chalcopyrite and Pyrite from Porous Primary Ore Fragments," *Metall. Trans. B* 1987, *18B*, 497–504 .

Liu J. A., Kinetics, *Catalysis, and Mechanism of Methane Steam Reforming,* M.S. thesis, Worcester Polytechnic Institute, Department of Chemical Engineering, 2006.

Macauley, D. "Options increase for non-BF ironmaking," *Steel Times Int.* 1997, 21, 20.

Manning, C. P. and Fruehan, R. J. "Emerging technologies for iron and steelmaking," *JOM* 2001, 53 (10), 36–43.

Matejak E. -M., 2018, S. POSCO Ignites the Largest Blast Furnace in the World, accessed 03/30/2018, http://e-metallicus.com/en/news/world/posco-ignites-the-largest-blast-furnace-in-the-world.html.

Metalprices.com, Average pig iron price, Basic Pig Iron Foundry Midwest (USD/MT) in September 2010, Chicago Iron Tables, accessed September 2010a; http://www.metalprices.com/freesite/metals/fe/fe_chicago.asp#Tables

Metalprices.com, Natural gas price, Natural Gas Prices 90 day 29 Jun, -29 Sep, 2010b, accessed September 2010; http://www.metalprices.com/FreeSite/metals/ng/ng.asp

Metius, G. E., McClelland, J.M., Hornby-Anderson S., "Comparing CO_2 emissions and energy demands for alternative ironmaking routes," *Steel Times Int.* 2006, *30*, 32–36.

Midrex, *World Direct Reduction Statistics*. 2016, accessed 03/30/2018, https://www.midrex.com/assets/user/news/MidrexStatsBook2016.pdf

Mohassab Y., Elzohiery M., Chen F., and Sohn H. Y., "Determination of Total Iron Content in Iron Ore and DRI: Titrimetric Method versus ICP-OES Analysis," *EPD Congress, 2016, Proceedings of TMS 2016*, Nashville, TN, Allanore A., Bartlett L., Wang C., Zhang L., and Lee J., Eds., Springer, Cham,2016, pp 125–133.

Moon I. J., Rhee C.-H., Min D.-J., "Reduction of Hematite Compacts by H_2-CO Gas Mixtures," *Steel Res* 1998, *69* (8), 302–306.

Moore J. J., *Chemical Metallurgy*, 2nd Ed., 1990, Elsevier, Cambridge, MA.

Mytelka L. K., and Boyle G., *Making Choices about Hydrogen Transport issues for Developing Countries*, United Nations University Press, Tokyo, Japan, 2008, p. 68.

New York State Energy Research and Development Authority, *Hydrogen Fact Sheet Hydrogen production- Steam Methane Reforming (SMR)*, Albany, NY, accessed October 2010, http://www.getenergysmart.org/files/hydrogeneducation/6hydrogenproductionsteammethanereforming.pdf.

Olivas-Martinez M., *Computational Fluid Dynamic Modeling of Chemically Reacting Gas-Particle Flows,* Ph.D. Dissertation, 2013, University of Utah, Salt Lake City, Utah,

Organization for Economic Co-Operation and Development, and International Energy Agency, 2006, CO_2 *Capture & Storage, IEA Energy Technology Essentials*, International Energy Agency, Paris, France.

Orth A., Anastasijevic N., Eichberger H., "Low CO_2 Emission Technologies for Iron and Steelmaking as well as Titania Slag Production," *Miner. Eng.* 2007, *20*, 854–861.

Pankratz L. B., Stuve J. M., and Gokcen N. A., *Thermodynamic Data for Mineral Technology,* U.S. Department of the Interior, Bureau of Mines Bulletin, 1984, p. 677.

Pérez-Maqueda L. A., Sánchez-Jiménez P. E., and Criado J. M., "Kinetic analysis of solid-state reactions: Precision of the activation energy calculated by integral methods," *Int. J. Chem. Kinetics* 2005, *37* (11), 658–666.

Peters M. S., Timmerhaus K. D., *Plant Design and Economics for Chemical Engineers*, 4th Ed., McGraw-Hill Companies, July 1991; pp. 163–169.

Pinegar H. K., Moats M. S., and Sohn H. Y. "Flowsheet development, process simulation and economic feasibility analysis for novel suspension ironmaking technology based on natural gas: Part 3 – Economic feasibility analysis." *Ironmak Steelmak.* 2013b, *40*, 44–49.

Pittsburgh Post-Gazette, 2011, Penn State forecasts boom for Marcellus Shale, http://www.post-gazette.com/pg/11202/1161933-503.stm (accessed May 2011); see also http://www.eia.gov/analysis/studies/usshalegas/.

Prout E. G., Tompkins, F. C. "The thermal decomposition of potassium permanganate," *Trans. Faraday Soc.,* 1944, *40*, 488.

Ranz W. E. and W. R. "Modeling the interaction between a thermal flow and a liquid," *Marshall: Chem. Eng. Prog,* 1952, *vol. 48* (4), pp. 173–180.

Reid, R. C., Prausnitz, J. M., and Sherwood, T. K. *The Properties of Gases and Liquids.* McGraw-Hill, New York, 1988.

Remus R., Aguado Monsonet M.A., Roudier S., and Delgado Sancho L. "Best Available Techniques (BAT) In Reference Document for Iron and Steel Production. Industrial Emissions Directive 2010/75/EU (Integrated Pollution Prevention and Control);" Joint Research Centre (JRC) of the European Union, Seville, Spain, 2013, pp. 304–305.

Rhee K. I., and Sohn H. Y. "The selective chlorination of iron from ilmenite ore by CO–C_{12} mixtures: Part I. Intrinsic kinetics," *Metall. Trans. B* 1990a, 21B, 321–330.

Rhee K. I., and Sohn H. Y. "The selective carbochlorination of iron from titaniferous magnetite ore in a fluidized bed," *Metall. Trans. B* 1990b, 21B, 341–347.

Roine A. 2002, HSC Chemistry for Windows, *Version 5.1*, Outokumpu Research Oy, Pori.

Roller P. W. "The theoretical volume decrease on reduction of hematite to magnetite," *Trans Iron Steel Inst. Jpn* 1986, *26* (9), 834–835.

Roy S. *Hydrogen Reduction of Iron Ore Concentrate in Loose Layers and Compacts,* Ph.D. Dissertation, University of Utah, Salt Lake City, Utah, 2022.

Sarkar S., Bhattacharya R., Roy G. G., and Sen P. K. "Modeling MIDREX based process configurations for energy and emission analysis," *Steel Res. Int.* 2018, *89*, 1700248. https://doi.org/10.1002/srin.201700248.

Sestak J. Thermophysical properties of solids, vol. XII *Part D of Comprehensive Analytical Chemistry,* Svehla G. Ed., Elsevier, Amsterdam, 1984, pp. 190–192.

Shamsuddin M., and Sohn H. Y. "Constitutive topics in physical chemistry of high-temperature nonferrous metallurgy – a review: Part 2. Reduction and refining," *JOM*, 2019, *71* (9), 3266–3276.

Shih S.-M., and Sohn H. Y. "Nonisothermal determination of the intrinsic kinetics of oil generation from oil shale," *Ind. Eng. Chem. Process Des. Dev*. 1980, *19* (3), 420–426.

Shih T.-H., Liou W. W., Shabbir A., Yang Z., and Zhu J. "A new k-ε eddy viscosity model for high reynolds number turbulent flows," *Comput. Fluids* 1995, *24* (3), 227–238.

Simbeck D., and Chang E. Hydrogen Supply: Cost Estimate for Hydrogen Pathways-Scoping Analysis, NREL/SR-540-32525, National Renewable Energy Laboratory, Golden, Colorado, November 2002.

Simon P., "An international forum for thermal studies," *J. Therm. Anal. Cal.* 2004, *76*, 123–132.

Simone M., Biagini E., Galletti C., and Tognotti L., "Evaluation of global biomass devolatilization kinetics in a drop tube reactor with CFD aided experiments," *Fuel* 2009, *88* (10), 1818–1827.

Smith J. M., and Van Ness H. C. *Introduction to Chemical Engineering Thermodynamics,* 2nd Ed., McGraw-Hill, New York, 1959.

Smith T. F., Shen Z. F., and Friedman J. N., "Evaluation of coefficients for the weighted sum of gray gases model," *J. Heat Transfer* 1982, *104* (4), 602–608.

Sohn H. Y. *Class Notes, University of Utah, Salt Lake City*, Utah, unpublished work. 1980.

Sohn H. Y. "Suspension ironmaking technology with greatly reduced energy requirement and CO_2 emissions." *Steel Times Int.* 2007, 31, 68–72.

Sohn H. Y. *Suspension Hydrogen Reduction of Iron Oxide Concentrate,* American Iron and Steel Institute Technology Roadmap Program Final Project Report for Phase 1, DE-FC36-97ID13554, American Iron and Steel Institute, March 2008.

Sohn H.Y. "On the rate expressions for "reversible" gas-solid reactions." *Metall. Mater. Trans. B* 2014, *45* (5), 1600–1602. http://dx.doi.org/10.1007/s11663-014-0113-0

Sohn H. Y. "A non-linear temperature-time program for non-isothermal kinetic measurements," *Metall. Mater. Trans. B* 2016, *47B* (2), 1203–1208. http://link.springer.com/article/10.1007/s11663-015-0551-3.

Sohn H. Y. *Fluid–Solid Reactions*, Elsevier, Cambridge, MA 02139, 536 pp., 2020a. https://www.elsevier.com/books/fluid-solid-reactions/sohn/978-0-12-816466-2

Sohn H. Y. "Energy consumption and CO_2 emissions in ironmaking and development of a novel flash technology," *Metals*, 2020b, *10* (1), 54. https://doi.org/10.3390/met10010054.

Sohn H.Y., and Choi M.E. "Development of a novel gas suspension ironmaking technology with greatly reduced energy consumption and CO_2 emissions," collected proceedings: supplemental proceedings, *vol. 1* Materials Processing and Properties, *International Symposium on High-Temperature Metallurgical Processing*, Drelich J., Hwang J. Y., Jiang T., Downey J., Eds., TMS: Warrendale, PA, 2010, pp. 347–354.

Sohn H.Y., Choi M.E. Steel industry and carbon dioxide emissions - A novel ironmaking process with greatly reduced carbon footprint. In *Carbon Dioxide Emissions: New Research Carpenter*, M., Shelton, E.J., Eds., Nova Science Publishers, Hauppauge, NY, USA, 2012.

Sohn H. Y. and Emami S., "Kinetics of dehydrogenation of the Mg-Ti-H hydrogen storage system," *Int. J. Hydrog. Energy* 2011, 36, 8344–8350.

Sohn H. Y. and Kim D., "The law of additive reaction times applied to the pellets," *Metall. Trans. B* 1984, *15B*, 403–406.

Sohn H. Y. and Kim D., "Intrinsic kinetics of the reaction between zinc sulfide and water vapor," *Metall. Trans. B* 1987, *18B*, 451–457.

Sohn H. Y., Olivas-Martinez M. Methods for calculating energy requirements for processes in which a reactant is also a fuel: need for standardization. *JOM* 2014, *66*, 1557–1564. http://dx.doi.org/10.1007/s11837-014-1120-y

Sohn H. Y., and Perez-Fontes S. E. "Computational fluid dynamics modeling of hydrogen-oxygen flame," *Int. J. Hydrogen Energy* 2016, *41* (4), 3284–3290. doi:10.1016/j.ijhydene.2015.12.013

Sohn H. Y., and Sohn H.–J. "The effect of bulk flow due to volume change in the gas phase on gas-solid reactions: Initially nonporous solids," *Ind. Eng. Chem. Process Des. Dev.* 1980, *19* (2), 237.

Sohn H. Y., and Won S. "Intrinsic kinetics of the hydrogen reduction of Cu_2S," *Metall. Trans. B* 1985, *16B*, 831–839.

Sohn H. Y., and Yang H. S. "The effect of reduced pressure on oil shale retorting: I. kinetics of oil generation," *Ind. Eng. Chem. Process Des. Dev.* 1985, *24*, 265–270.

Sohn H. Y., Zhou L., and Cho K. "Intrinsic kinetics and mechanism of rutile chlorination by $CO+Cl_2$ mixtures," *Ind. Eng. Chem. Research* 1998, *37*, 3800–3805.

Sohn H. Y., Choi M. E., Zhang Y., and Ramos J. E. "Suspension reduction technology for iron-making with low CO_2 emission and energy requirement," *Iron & Steel Technol. (AIST Trans).* 2009a, *6*, 158–165.

Sohn H. Y., Choi M. E., Zhang Y., and Ramos J. E. "Suspension Ironmaking Technology with Greatly Reduced CO_2 Emission and Energy Requirement," In: *Energy Technology Perspectives: Carbon Dioxide Reduction and Production from Alternative Sources*, Neelameggham N. R., Reddy R. G., Belt C. K., Vidal E. E., Eds., (also available in Collected Proceedings CD, TMS 2009b Annual Meeting), TMS: Warrendale, PA, 2009b, pp. 93–101.

Sohn H. Y., Elzohiery M., and Fan D. Q., "Development of the flash ironmaking technology (FIT) for green ironmaking with low energy consumption," *J. Energy Power Technol.* 2021, *3* (3), 042. https://doi.org/10.21926/jept.2103042. https://www.lidsen.com/journals/jept/jept-03-03-042#.

Spreitzer D., Schenk J. "Reduction of iron oxides with hydrogen—A review," *Steel Res. Int.* 2019, *90* (10), 1900108.

Starink M. J. "Kinetic equations for diffusion-controlled precipitation reactions," *J. Mater. Sci.* 1997, *32* (15), 4061–4070.

Stephenson R. L. and Smailer R. M., *Direct Reduced Iron – Technology and Economics of Production and Use,* The Iron & Steel Soc. of AIME, Warrendale, PA, 1980.

Stermole F. J., and Stermole J. M. *Economic Evaluation and Investment Decision Methods,* 10th Ed., Investment Evaluations Corporation, Lakewood, CO, 2000, pp. 6–11.

Stubbles J. *Energy Use in the U.S. In Steel Industry: An Historical Perspective and Future Opportunities,* Department of Energy, Washington, DC, 2000. http://www1.eere.energy.gov/manufacturing/resources/steel/pdfs/steel_energy_use.pdf. accessed on July 27, 2013.

Stull D. R. and Prophet H. 1971. *JANAF Thermochemical Tables.* 2nd Ed., Defense Technical Information Center, Fort Belvoir, Virginia, and subsequent supplements.

Tacke K.-H., Steffen R. "Hydrogen for the reduction of iron ores-state of the art and future aspects," *Stahl Eisen.* 2004, *124* (4), 45–52.

Themelis N. J., Wu L., and Jiao Q. "Some aspects of mathematical modeling of flash smelting phenomena," In: *Flash Reaction Processes*, Robertson D. G. C., Sohn H. Y., and Themelis N. J. Eds., Proceedings of Center for Pyrometallurgy Conference, University of Utah, Published by the center for pyrometallurgy, Rolla, MO, 1988, pp. 263–285.

Tolvanen H., and Raiko R. "An experimental study and numerical modeling of combusting two coal chars in a drop-tube reactor: A comparison between N_2/O_2, CO_2/O_2 and $N_2/CO_2/O_2$ atmospheres," *Fuel* 2014, *124*, 190–201.

Tsay Q. T., Ray W. H., and Szekely J. "The modeling of hematite reduction with hydrogen plus carbon monoxide mixtures: part I. The behavior of single pellets," *AIChE J.* 1976, *22* (6), 1064–1072.

Turkdogan E. T., Olsson R. G., and Vinters J. V. "Gaseous reduction of iron oxides: part II. Pore characteristics of iron reduced from hematite in hydrogen," *Metall. Mater. Trans. B* 1971, *2* (11), 3189–3196.

UN Climate Technology Centre & Network 2022 Smelt reduction for iron and steel sector, Connecting Countries to Climate Technology Solutions, accessed May 31, 2022. https://www.ctc-n.org/technologies/smelt-reduction-iron-and-steel-sector.

United States Steel Corporation, *The Making, Shaping and Treating of Steal*, 10th Ed., Association of Iron and Steel Engineers, March 1985.

U.S. Department of Energy, Chapter 6- Natural Gas Combined Cycle, Market-Based Advanced Coal Power Systems, Washington DC, December 1998, p. 32, accessed September 2010, http://fossil.energy.gov/programs/powersystems/publications/MarketBasedPowerSystems/sec6.pdf

U.S. Department of Energy, 2006, Hydrogen posture plan - an integrated research, development and demonstration plan [Internet]. https://www.hydrogen.energy.gov/pdfs/hydrogen_posture_plan_dec06.pdf

U.S. Department of Energy, 2000, Ironmaking Process Alternatives Screening Study vol. 2, LG job No. 010529.01, Washington DC.

U. S. Energy Information Administration Independent Statistics and Analysis, Table 5.6.A 2010 and 2009 Average Retail Price of Electricity to Ultimate Customers by End-Use Sector, by State, Washington DC, accessed August 2010, http://www.eia.doe.gov/cneaf/electricity/epm/table5_6_a.html.

U.S. Energy Information Administration 2011 Annual Energy Outlook [Internet]. Washington, DC: http://www.eia.gov/forecasts/aeo/index.cfm

U.S. Environmental Protection Agency, Combined Heat and Power (CHP) Level 1 Feasibility Analysis, Washington, DC, p 11, accessed September 2010a, www.epa.gov/chp/documents/sample_fa_industrial.pdf.

U.S. Environmental Protection Agency, 5.4 Wet Scrubbers, accessed October 2010b, http://www.epa.gov/ttn/caaa/t1/reports/sect5-4.pdf.

U.S. Environmental Protection Agency Combined Heat and Power (CHP) Level 1 Feasibility Analysis, Washington, DC, p. 11, accessed January. 2011, www.epa.gov/chp/documents/sample_fa_industrial.pdf.

U.S. Geological Survey 2007, *Minerals Yearbook 2005—Iron Ore,* (Report, U.S. Department of the Interior) U.S. Geological Survey), Reston, VA, USA.

U.S. Geological Survey, *Lime, Minerals Yearbook 2008*, Advanced release, U.S Department of Interior, February 2010, p. 10.

Varaksin A. Y. "Particle-Laden Channel Flows," In: Varaksin, A.Y. (Ed.), *Turbulent Particle-Laden Gas Flows. Atomic, optical, and plasma physics, vol 41*, Springer, New York, 2007, pp. 22–23.

Varnbuler C. B., Lemperle M. Zero waste – Zero Cost Concept for Integrated Steel Mills Oxycup Process for Steel Mill Waste Oxides, ThyssenKrupp Stahl AG Küttner, accessed September 2010, http://www.sternasia.com/file/Zero_Waste.pdf.

Vyazovkin S. "Computational Aspects of Kinetic Analysis. 2000. Part C." *The ICTAC Kinetics* "Project—the light at the end of the tunnel?" *Thermochim. Acta 355*:155–163.

Vyazovkin S., Dranca "Ten lectures on theoretical rheology. By markus reiner," *I. J. Phys. Chem. B* 2005, *109*, 18637.

Wang H., and Sohn H. Y. "Effects of reducing gas on swelling and iron whisker formation during the reduction of iron oxide compact," *Steel Res. Int.* 2012, *83*, 903.

Wang H., and Sohn H. Y. "Hydrogen reduction kinetics of magnetite concentrate particles relevant to a novel flash ironmaking process," Metall. *Mater. Trans. B*, 2013, *44B*, 133–145. http://dx.doi.org/10.1007/s11663-012-9754-z

Weiland F., Nordwaeger M., Olofsson I., Wiinikka H., and Nordin A. "Entrained flow gasification of torrefied wood residues," *Fuel Process. Technol* 2014, *125*, 51-58. https://doi.org/10.1016/j.fuproc.2014.03.026

Welch A. J. E. Solid–solid reactions. In *Chemistry of the Solid State,* Chapter 12, Garner W. E. Ed., 1955, Academic Press, New York.

World Steel Association. 2018a. *Monthly Iron Production 2018–2017*, accessed March 30, 2018, https://www.worldsteel.org/en/dam/jcr:e7064b8b-d1ef-4ef7-861e-cf3c3768e6ef/Iron+February+2018.pdf.

World Steel Association. 2018b. *Steel's Contribution to a Low Carbon Future and Climate Resilient Societies*, accessed 03/30/2018, https://www.worldsteel.org/en/dam/jcr:66fed386-fd0b-485e-aa23-b8a5e7533435/Position_paper_climate_2018.pdf.

World Steel Association. 2018c. *World Steel in Figures. 2018*, accessed on 16 December 2019, https://www.worldsteel.org/en/dam/jcr:f9359dff-9546-4d6b-bed0-996201185b12/World+Steel+in+Figures+2018.

Worldsteel Association. 2019. *Steel Statistical Yearbook 2019*, accessed June 2, 2022, https://worldsteel.org/wp-content/uploads/Steel-Statistical-Yearbook-2019-concise-version.pdf.

Worrell E., Price L., Neelis M., Galitsky C., and Nan Z. 2008, *World Best Practice Energy Intensity Values for Selected Industrial Sectors; LBNL-62806*, Lawrence Berkeley National Laboratory, Berkeley, CA, accessed on 16 Dec. 2019, http://eaei.lbl.gov/sites/all/files/industrial_best_practice_en.pdf.

Wunderlich, B. *Macromolecular Physics: Crystal Nucleation, Growth, Annealing,* Academic Press: New York, 1976.

Xiao X., Sichen D., Sohn H. Y., and Seetharaman S. "Determination of kinetics parameters using differential thermal analysis–application to the decomposition of $CaCO_3$," *Metall. Mater. Trans. B* 1997, *28B*, 1157–1164. http://www.springerlink.com/content/7615654j00750228/fulltext.pdf

Yang Y., Raipala K. and Holappa L. "Ironmaking," *Treatise on process metallurgy, vol. 3 industrial processes part A*, Chapter 1.1. 2014, pp. 2–88, Elsevier, Oxford, UK and Waltham, MA.

Yi S.-H., Choi M.-E., Kim D.-H., Ko C.-K., Park W.-I., and Kim S.-Y. 2019, "FINEX® as an environmentally sustainable ironmaking process," *Ironmaking Steelmaking 46*:7, 625–631. https://doi.org/10.1080/03019233.2019.1641682

Yilmaz C., Turek T. "Modeling and simulation of the use of direct reduced iron in a blast furnace to reduce carbon dioxide emissions," *J. Clean Prod.* 2017, *164*, 1519–1530. https://doi.org/10.1016/j.jclepro.2017.07.043

Yuan Z., Sohn H. Y., and Olivas-Martinez M. "Re-oxidation kinetics of flash reduced iron particles in H_2-H_2O(g) atmosphere relevant to a novel flash ironmaking process," *Metall. Mater. Trans. B* 2013, *44B*, 1520–1530. http://www.springerlink.com/openurl.asp?genre=article&id=doi:10.1007/s11663-013-9910-0

Yuan Z. and Sohn H. Y., "Re-oxidation kinetics of flash reduced iron particles in O_2-N_2 gas mixtures relevant to a novel flash ironmaking process," *ISIJ Int.*2014, *54*, 1235–1243. https://doi.org/10.2355/isijinternational.54.1235

Zsako J., "An international forum for thermal studies," *J. Therm. Anal.*, 1996, -*46*, 1845–1864.

Züttel A., Borgschulte A., and Schlapbach L. (eds.) *Hydrogen as a Future Energy Carrier.* https://doi.org/10.1002/9783527622894, 2008.

Index

For Product Safety Concerns and Information please contact our EU
representative GPSR@taylorandfrancis.com
Taylor & Francis Verlag GmbH, Kaufingerstraße 24, 80331 München, Germany

www.ingramcontent.com/pod-product-compliance
Lightning Source LLC
Chambersburg PA
CBHW052011230326
41598CB00078B/2773
9781032378381